Studies in Classification, Data Analysis, and Knowledge Organization

For further volumes:
http://www.springer.com/series/1564

Myra Spiliopoulou • Lars Schmidt-Thieme
Ruth Janning
Editors

Data Analysis, Machine Learning and Knowledge Discovery

Editors
Myra Spiliopoulou
Faculty of Computer Science
Otto-von-Guericke-Universität Magdeburg
Magdeburg
Germany

Lars Schmidt-Thieme
Ruth Janning
Institute of Computer Science
University of Hildesheim
Hildesheim
Germany

ISSN 1431-8814
ISBN 978-3-319-01594-1 ISBN 978-3-319-01595-8 (eBook)
DOI 10.1007/978-3-319-01595-8
Springer Cham Heidelberg New York Dordrecht London

Library of Congress Control Number: 2013954389

Printed on acid-free paper

Springer is part of Springer Science+Business Media (www.springer.com)

Preface

This volume contains the revised versions of selected papers presented during the 36th Annual Conference (GfKl 2012) of the German Classification Society (Gesellschaft für Classification-GfKl). The conference was hosted by the University of Hildesheim, Germany, together with the Otto-von-Guericke-University of Magdeburg, Germany, in August 2012. The GfKl 2012 focused on advances in data analysis, statistics, knowledge discovery and machine learning. Problems from data analysis in marketing, finance, biostatistics, bioinformatics and interdisciplinary domains were considered. Besides 9 plenaryand semi-plenary talks, more than 130 talks took place in 3 days. With participants from 20 countries this GfKl conference provided once again an international forum for discussions and mutual exchange of knowledge with colleagues from different fields of interest.

In recent years, we saw the core topics of the conference crystallize themselves into thematic areas. In 2012, for the first time, these areas were made explicit and their coordination was undertaken by dedicated Area Chairs:

- Statistics and Data Analysis (SDA), organized by Hans-Hermann Bock, Christian Hennig and Claus Weihs
- Machine Learning and Knowledge Discovery (MLKD), organized by Lars Schmidt-Thieme and Myra Spiliopoulou
- Data Analysis and Classification in Marketing (DACMar), organized by Daniel Baier and Reinhold Decker
- Data Analysis in Finance (DAFin), organized by Michael Hanke and Krzysztof Jajuga
- Biostatistics and Bio-informatics, organized by Anne-Laure Boulesteix and Hans Kestler
- Interdisciplinary Domains (InterDom), organized by Andreas Hadjar, Sabine Krolak-Schwerdt and Claus Weihs
- Workshop Library and Information Science (LIS2012), organized by Frank Scholze

As expected, the lion's share among the contributions to these areas came from Germany, followed by Poland, but we had contributions from all over the world, stretching from Portugal to Ukraine, from Canada and the USA to Japan and Thailand.

For every GfKl conference a best paper award is granted to two excellent papers of the postconference proceedings. We are proud to announce the best paper awards of GfKl 2012:

- Nguyen Hoang Huy, Stefan Frenzel and Christoph Bandt (University of Greifswald) with *Two-Step Linear Discriminant Analysis for Classification of EEG Data* (page 51); and
- Tobias Voigt, Roland Fried, Michael Backes and Wolfgang Rhode (University of Dortmund) with *Gamma-Hadron-Separation in the MAGIC Experiment* (page 115).

We would like to congratulate the prize winners and would like to thank the best paper awards jury members for their excellent work.

Organizing such a conference with its parallel, interleaved events is not an easy task. It requires coordination of many individuals and on many issues, and lives from the tremendous effort of engaged scientists and of the dedicated teams in Hildesheim and Magdeburg. We would like to thank the Area Chairs for their hard work in conference advertisement, author recruitment and submissions evaluation. We are particularly indebted to the Polish Classification Society for its involvement and presence in the GfKl 2012. We would like to thank the EasyChair GfKl 2012 administrator, Miriam Tödten, for her assistance during submissions, evaluation and camera-ready preparation and contribution to the abstracts volume, and Silke Reifgerste, the financial administrator of the KMD research lab at the Otto-von-Guericke University Magdeburg for her fast and competent treatment of all financial matters concerning the Magdeburg team. Further we would like to thank Kerstin Hinze–Melching (University of Hildesheim) for her help with the local organization, Jörg Striewski and Uwe Oppermann, our technicians in Hildesheim, for technical assistance, Selma Batur for help with the abstract proceedings and preparation of the conference and our assistants in Hildesheim for the conference Fabian Brandes, Christian Brauch, Lenja Busch, Sarina Flemnitz, Stephan Reller, Nicole Reuss and Kai Wedig.

This proceedings volume of the GfKl conference in 2012 with 49 contributions is the result of postconference paper submission and a two-round reviewing phase. We would like to thank all reviewers for their rigorous and timely work. We would like to thank all Area Chairs and the LIS Workshop Chair for the organization of the areas during the conference, for the coordination of the reviewing phase and for the paper shepherding towards final acceptance. Furthermore, we would like to thank Martina Bihn and Ruth Milewski of Springer-Verlag, Heidelberg, for their support and dedication to the production of this volume. We also would like to thank Patrick Jähne for assistance with editing tasks to create this volume.

Last but not least, we would like to thank all participants of the GfKl 2012 conference for their interest and activities which made the 36th Annual Conference and this volume an interdisciplinary possibility for scientific discussion.

Magdeburg, Germany — Myra Spiliopoulou
Hildesheim, Germany — Lars Schmidt-Thieme and Ruth Janning

Contents

Part VII LIS WORKSHOP: Workshop on Classification and Subject Indexing in Library and Information Science

Conference Organization

Program Committee Chairs
Myra Spiliopoulou, Otto-von-Guericke-University of Magdeburg, Germany
Lars Schmidt-Thieme, University of Hildesheim, Germany

Local Organizers
Lars Schmidt-Thieme, University of Hildesheim, Germany
Ruth Janning, University of Hildesheim, Germany

Scientific Program Committee

AREA Statistics and Data Analysis
Hans-Hermann Bock, RWTH Aachen, Germany (Area Chair)
Christian Hennig, University of College London, UK (Area Chair)
Claus Weihs, TU Dortmund, Germany (Area Chair)
Patrick Groenen, Erasmus University of Rotterdam, Netherlands
Bettina Gruen, Johannes Kepler University of Linz, Austria

AREA Machine Learning and Knowledge Discovery
Lars Schmidt-Thieme, University of Hildesheim, Germany (Area Chair)
Myra Spiliopoulou, Otto-von-Guericke-University of Magdeburg, Germany (Area Chair)
Martin Atzmueller, University of Kassel, Germany
Joao Gama, University of Porto, Portugal
Andreas Hotho, University of Wuerzburg, Germany
Eyke Hüllermeier, University of Marburg, Germany
Georg Krempl, Otto-von-Guericke-University of Magdeburg, Germany
Eirini Ntoutsi, Ludwig-Maximilians-University of Munich, Germany
Thomas Seidl, RWTH Aachen, Germany

AREA Data Analysis and Classification in Marketing
Daniel Baier, BTU Cottbus, Germany (Area Chair)
Reinhold Decker, University of Bielefeld, Germany (Area Chair)

AREA Data Analysis in Finance
Michael Hanke, University of Liechtenstein, Liechtenstein (Area Chair)
Krzysztof Jajuga, Wroclaw University of Economics, Poland (Area Chair)

AREA Data Analysis in Biostatistics and Bioinformatics
Anne-Laure Boulesteix, Ludwig Maximilian University of Munich, Germany (Area Chair)
Hans Kestler, University of Ulm, Germany (Area Chair)
Harald Binder, University of Mainz, Germany
Matthias Schmid, University of Erlangen, Germany
Friedhelm Schwenker, University of Ulm, Germany

AREA Data Analysis in Interdisciplinary Domains
Andreas Hadjar, University of Luxembourg, Luxembourg (Area Chair)
Sabine Krolak-Schwerdt, University of Luxembourg, Luxembourg (Area Chair)
Claus Weihs, TU Dortmund, Germany (Area Chair)
Irmela Herzog, LVR, Bonn, Germany
Florian Klapproth, University of Luxembourg, Luxembourg
Hans-Joachim Mucha, WIAS, Berlin, Germany
Frank Scholze, KIT Karlsruhe, Germany

LIS'2012
Frank Scholze, KIT Karlsruhe, Germany (Area Chair)
Ewald Brahms, University of Hildesheim, Germany
Andreas Geyer-Schulz, KIT Karlsruhe, Germany
Stefan Gradmann, HU Berlin, Germany
Hans-Joachim Hermes, TUniv. Chemnitz, Germany
Bernd Lorenz, FHÖV Munich, Germany
Michael Mönnich, KIT Karlsruhe, Germany
Heidrun Wiesenmüller, HDM Stuttgart, Germany

Contributors

Klaus Ambrosi Institut für Betriebswirtschaft und Wirtschaftsinformatik, Universität Hildesheim, Hildesheim, Germany

Michael Backes Physics Faculty, TU Dortmund University, Dortmund, Germany,

Daniel Baier Chair of Marketing and Innovation Management, Institute of Business Administration and Economics, Brandenburg University of Technology Cottbus, Cottbus, Germany

Maheen Bakhtyar Asian Institute of Technology Bangkok, Bangkok, Thailand

Michel Ballings Department of Marketing, Ghent University, Ghent, Belgium

Christoph Bandt Department of Mathematics and Computer Science, University of Greifswald, Greifswald, Germany

Udo Bankhofer Department of Quantitative Methods for Business Sciences, Ilmenau University of Technology, Ilmenau, Germany

Hans-Georg Bartel Department of Chemistry, Humboldt University Berlin, Berlin, Germany

Nadja Bauer Chair of Computational Statistics, Faculty of Statistics, TU Dortmund, Dortmund, Germany

Tim Beige Chair of Computational Statistics, TU Dortmund, Germany

Bernd Bischl Chair of Computational Statistics, Department of Statistics, TU Dortmund, Germany

Matthew Bolaños Southern Methodist University, Dallas, TX, USA

Justyna Brzezińska Department of Statistics, University of Economics in Katowice, Katowice, Poland

Andre Busche ISMLL—Information Systems and Machine Learning Lab, University of Hildesheim, Hildesheim, Germany

Krisztian Buza Department of Computer Science and Information Theory, Budapest University of Technology and Economics, Budapest, Hungary

Darrell Conklin Department of Computer Science and AI, Universidad del País Vasco UPV/EHU, San Sebastián, Spain

Nam Dang Tokyo Institute of Technology, Tokyo, Japan

Juan José del Coz University of Oviedo, Gijón, Spain

Jens Dolata Head Office for Cultural Heritage Rhineland-Palatinate (GDKE), Mainz, Germany

Kai Eckert University of Mannheim, Mannheim, Germany

Markus Eichhoff Chair of Computational Statistics, Faculty of Statistics, TU Dortmund, Dortmund, Germany

John Forrest Microsoft, Redmond, WA, USA

Stefan Frenzel Department of Mathematics and Computer Science, University of Greifswald, Greifswald, Germany

Roland Fried Faculty of Statistics, TU Dortmund University, Dortmund, Germany

Klaus Friedrichs Chair of Computational Statistics, Faculty of Statistics, TU Dortmund, Dortmund, Germany

Bernhard Gschrey Chair for Methods in Empirical Educational Research, TUM School of Education, and Centre for International Student Assessment (ZIB), Technische Universität München, Munich, Germany

Sonja Hahn Friedrich-Schiller-Universität Jena, Institut für Psychologie, Jena, Germany

Michael Hahsler Southern Methodist University, Dallas, TX, USA

Christian Hennig Department of Statistical Science, University College London, London, UK

Irmela Herzog The Rhineland Commission for Archaeological Monuments and Sites, The Rhineland Regional Council

Kay F. Hildebrand European Research Center for Information System (ERCIS), Münster, Germany

Ruben Hillewaere Artificial Intelligence Lab, Department of Computing, Vrije Universiteit Brussel, Brussels, Belgium

Thomas Hörstermann University of Luxembourg, Walferdange, Luxembourg

Tomáš Horváth ISMLL—Information Systems and Machine Learning Lab, University of Hildesheim, Hildesheim, Germany

Eyke Hüllermeier Philipps-Universität Marburg, Marburg, Germany

Nguyen Hoang Huy Department of Mathematics and Computer Science, University of Greifswald, Greifswald, Germany

Katsumi Inoue National Institute of Informatics, Tokyo, Japan

Ruth Janning ISMLL—Information Systems and Machine Learning Lab, University of Hildesheim, Hildesheim, Germany

Dieter William Joenssen Department of Quantitative Methods for Business Sciences, Ilmenau University of Technology, Ilmenau, Germany

Daniel Kasper Chair for Methods in Empirical Educational Research, TUM School of Education, and Centre for International Student Assessment (ZIB), Technische Universität München, Munich, Germany

Hans A. Kestler Research Group Bioinformatics and Systems Biology, Institute of Neural Information Processing, Ulm University, Ulm, Germany

Dominik Kirchhoff Chair of Computational Statistics, Faculty of Statistics, TU Dortmund, Dortmund, Germany

Florian Klapproth University of Luxembourg, Walferdange, Luxembourg

Sabine Krolak-Schwerdt University of Luxembourg, Walferdange, Luxembourg

Tatjana Lange Hochschule Merseburg, Geusaer Straße, Merseburg, Germany

Christina Lichtenthäler Institute for Advanced Study, Technische Universität München, Garching, Germany

Hanna Lukashevich Fraunhofer IDMT, Ehrenbergstr. 31, Ilmenau, Germany

Bernard Manderick Artificial Intelligence Lab, Department of Computing, Vrije Universiteit Brussel, Brussels, Belgium

Romain Martin University of Luxembourg, Walferdange, Luxembourg

Oliver Meyer Chair of Computational Statistics, Department of Statistics, TU Dortmund, Germany

Anneke Minke Institut für Betriebswirtschaft und Wirtschaftsinformatik, Universität Hildesheim, Hildesheim, Germany

Karl Mosler Universität zu Köln, Köln, Germany

Sandrine Mouysset University of Toulouse, IRIT-UPS, Toulouse, France

Pavlo Mozharovskyi Universität zu Köln, Köln, Germany

Hans-Joachim Mucha Weierstrass Institute for Applied Analysis and Stochastics (WIAS), Berlin, Germany

Joseph Noailles University of Toulouse, IRIT-ENSEEIHT, Toulouse, France

Marcin Pełka Department of Econometrics and Computer Science, Wrocław University of Economics, Wroclaw, Poland

Magnus Pfeffer Stuttgart Media University, Stuttgart, Germany

Silke Rehme FIZ Karlsruhe, Eggenstein-Leopoldshafen, Germany

Wolfgang Rhode Physics Faculty, TU Dortmund University, Dortmund, Germany,

Dominique Ritze Mannheim University Library, Mannheim, Germany

Günther Rötter Institute for Music and Music Science, TU Dortmund, Dortmund, Germany

Günter Rudolph Chair of Algorithm Engineering, TU Dortmund, Dortmund, Germany

Daniel Ruiz University of Toulouse, IRIT-ENSEEIHT, Toulouse, France

Susanne Rumstadt Chair of Marketing and Innovation Management, Institute of Business Administration and Economics, Brandenburg University of Technology Cottbus, Cottbus, Germany

Julia Schiffner Chair of Computational Statistics, Department of Statistics, TU Dortmund, Germany

Lars Schmidt-Thieme ISMLL—Information Systems and Machine Learning Lab, University of Hildesheim, Hildesheim, Germany

Michael Schurig Chair for Methods in Empirical Educational Research, TUM School of Education, and Centre for International Student Assessment (ZIB), Technische Universität München, Munich, Germany

Michael Schwantner FIZ Karlsruhe, Eggenstein-Leopoldshafen, Germany

Sebastian Selka Chair of Marketing and Innovation Management, Institute of Business Administration and Economics, Brandenburg University of Technology Cottbus, Cottbus, Germany

Robin Senge Philipps-Universität Marburg, Marburg, Germany

Sina Stubben Chair for Methods in Empirical Educational Research, TUM School of Education, and Centre for International Student Assessment (ZIB), Technische Universität München, Munich, Germany

Clovis Tauber University of Tours, Hopital Bretonneau, Tours, France

Thomas Terhorst Institute of Journalism, TU Dortmund, Germany

Matthias Trendtel Chair for Methods in Empirical Educational Research, TUM School of Education, and Centre for International Student Assessment (ZIB), Technische Universität München, Munich, Germany

Ali Ünlü Chair for Methods in Empirical Educational Research, TUM School of Education, and Centre for International Student Assessment (ZIB), Technische Universität München, Munich, Germany

Dirk Van den Poel Department of Marketing, Ghent University, Ghent, Belgium

Igor Vatolkin Chair of Algorithm Engineering, TU Dortmund, Dortmund, Germany

Sascha Voekler Institute of Business Administration and Economics, Brandenburg University of Technology Cottbus, Cottbus, Germany

Tobias Voigt Faculty of Statistics, TU Dortmund University, Dortmund, Germany

Claus Weihs Chair of Computational Statistics, Department of Statistics, TU Dortmund, Germany

Lena Wiese University of Göttingen, Göttingen, Germany

Holger Wormer Institute of Journalism, TU Dortmund, Germany

Part I
AREA Statistics and Data Analysis: Classification, Cluster Analysis, Factor Analysis and Model Selection

On Limiting Donor Usage for Imputation of Missing Data via Hot Deck Methods

Udo Bankhofer and Dieter William Joenssen

Abstract Hot deck methods impute missing values within a data matrix by using available values from the same matrix. The object from which these available values are taken for imputation is called the donor. Selection of a suitable donor for the receiving object can be done within imputation classes. The risk inherent to this strategy is that any donor might be selected for multiple value recipients. In extreme cases one donor can be selected for too many or even all values. To mitigate this donor over usage risk, some hot deck procedures limit the amount of times one donor may be selected for value donation. This study answers if limiting donor usage is a superior strategy when considering imputation variance and bias in parameter estimates.

1 Introduction

Missing data is a problem prevalent in many real empirical investigations. With observations missing, conventional statistical methods cannot simply be applied to the data without proxy. Explicit provisions must be made within the analysis.

Three principle strategies exist to deal with missing data: elimination, imputation, and direct analysis of the observed data. While direct analysis utilizes special methods to, for example, estimate parameters, elimination and imputation methods create a complete data matrix that may be analyzed using standard statistical procedures. Elimination methods remove objects or attributes missing data from analysis. Imputation methods replace the missing values with estimates (Allison 2001).

U. Bankhofer · D.W. Joenssen (✉)
Department of Quantitative Methods for Business Sciences, Ilmenau University of Technology, Helmholtzplatz 3, 98693 Ilmenau, Germany
e-mail: Udo.Bankhofer@TU-Ilmenau.de; Dieter-William.Joenssen@TU-Ilmenau.de

M. Spiliopoulou et al. (eds.), *Data Analysis, Machine Learning and Knowledge Discovery*, Studies in Classification, Data Analysis, and Knowledge Organization,
DOI 10.1007/978-3-319-01595-8_1,

This paper deals with imputation. These methods use available information to estimate imputation values. The simplest techniques replace missing values with eligible location parameters. Beyond that, multivariate methods, such as regression, may be used to identify imputation values. Another category of imputation methods use imputation classes. An advantage of imputing missing values from a pool of similar objects, is that less restrictive assumptions about the missingness mechanism can be made. While imputation methods usually require either the MCAR (missing completely at random) mechanism or under special circumstances allow for the presence of the MAR (missing at random) mechanism, imputation methods utilizing imputation classes can lead to valid results under the NMAR (not missing at random) mechanism (Andridge and Little 2010; Bankhofer 1995).

Methods using imputation classes are categorized as either cold deck or hot deck methods. While cold deck methods impute values by obtaining these from external sources (e.g. similar studies, expert opinions), hot deck methods replace missing data with values that are available within the current matrix (Sande 1983). The object from which these available values are taken is called the donor. Selection of a suitable donor for the receiving object, called the recipient, may be done within the prior constructed classes. The replication of values leads to the problem that a single donor might be selected to accommodate multiple recipients. This poses the inherent risk that too many or even all missing values are imputed utilizing only a single donor. Due to this, some hot deck variants limit the amount of times any donor may be selected to donate its values (donor limit). This inevitably leads to the question under which conditions a donor limit is sensible and whether or not an appropriate value for a limit exists. This study aims to answer these questions.

2 Review of Literature

The theoretical effects of a donor limit were first investigated by Kalton and Kish (1981). Based on combinatorics, they come to the conclusion that selecting a donor from the donor pool without replacement leads to a reduction in the imputation variance, the precision with which any parameter is estimated from the post-imputation data matrix. A possible effect on an imputation introduced bias was not discussed. Two more arguments in favor of a donor limit are made. First, the risk of exclusively using one donor for all imputations is removed (Sande 1983). Second, the probability of using one donor with an extreme value or values too often is reduced (Bankhofer 1995; Strike et al. 2001). Based on these arguments and sources, recommendations are made by Kalton and Kasprzyk (1986), Nordholt (1998), Strike et al. (2001), and Durrant (2009).

In contrast, Andridge and Little (2010) reason that imposing a donor limit inherently reduces the ability to choose the most similar, and therefore most appropriate, donor for imputation. Not limiting the times a donor can be chosen may thus increase data quality. Generally speaking, a donor limit makes results dependent on the order of object imputation. Usually, the imputation order will correspond to the sequence of the objects in the data set. This property is undesirable, especially in deterministic hot decks, as object sorting or random reordering may lead to different results. Thus, from a theoretical point of view, it is not clear whether or not a donor limit has a positive or negative impact on the data quality.

Literature on this subject provides only studies that compare hot decks with other imputation methods. These studies include either only drawing the donor from the donor pool with replacement (Barzi and Woodward 2004; Roth and Switzer III 1995; Yenduri and Iyengar 2007) or without replacement (Kaiser 1983). Based on this review of literature it becomes apparent, that the consequences of imposing a donor limit have not been sufficiently examined.

3 Study Design

Hence, considering possible advantages that a donor limit has, and possible effects that have not been investigated to date, the following questions will be answered by this study. First, can post-imputation parameter estimates be influenced by a donor limit? Second, what factors influence if a hot deck with donor limit leads to better results?

By considering papers where authors chose similar approaches (Roth and Switzer III 1995; Strike et al. 2001) and further deliberations, a series of factors are identified that might influence whether or not a donor limit affects parameter estimates. Factors varied are:

- *Imputation class count:* Imputation classes are assumed to be given prior to imputation and data is generated as determined by the class structure. Factor levels are two and seven imputation classes.
- *Object count per imputation class:* The amount of objects characterizing each imputation class is varied between 50 and 250 objects per class.
- *Class structure:* To differentiate between well- and ill-chosen imputation classes, data is generated with a relatively strong and relatively weak class structure. Strong class structure is achieved by having classes overlap by 5 % and inner-class correlation of 0.5. Weak class structure is achieved by an intra-class overlap of 30 % and no inner-class correlation.
- *Data matrices:* Data matrices of nine multivariate normal variables are generated dependent on the given class structure. Three of these variables are then transformed to a discrete uniform distribution with either five or seven possible values, simulating an ordinal scale. The next three variables are converted to a nominal scale so that 60 % of all objects are expected to take the value one, with the remaining values being set to zero. General details on the NORTA type transformation are described by Cario and Nelson (1997).
- *Portion of missing data:* Factor levels include 5, 10, and 20 % missing data points and every object is assured to have at least one data point available (no subject non-response).

- *Missingness mechanism:* Missingness mechanisms of the type MCAR, MAR, and NMAR are generated as follows: Under MCAR a set amount of values are chosen as missing. Under MAR missing data is generated as under MCAR but using two different rates based on the value of one binary variable, which is not subject to missingness. The different rates of missingness are either 10 % higher or lower than the rates under MCAR. NMAR modifies the MAR mechanism to also allow missingness of the binary variable. To forgo possible problems with the simultaneous imputation methods and the donor limitation of once, at least 50 % of all objects within one class are guaranteed to be complete in all attributes.
- *Hot deck methods:* The six hot deck methods considered are named "SeqR," "SeqDW," "SeqDM," "SimR," "SimDW," and "SimDM" according to the three attributes that define each procedure. The prefixes denote whether attributes are imputed sequentially or simultaneously. Sequential imputation of attributes selects a donor for every attribute exhibiting missing values anew, while simultaneous imputation selects one donor that will accommodate all of the recipients missing values. The postfixes indicate a random (R) or minimum distance function (D) donor selection. The type of adjustment made to compensate for missingness when computing distances is indicated by the last part of the suffix. "W" indicates that each pairwise distance is reweighted by the amount of variables relative to the amount of variables available in both donor and recipient. "M" denotes that location parameter imputation is performed before calculation of pairwise distances. The former method assumes that the missing part of the pairwise distance would contribute averagely to the total distance, while the latter method assumes that missing value is close to the average for this attribute. To account for variability and importance, prior to aggregating the Manhattan distances, variables are weighted with the inverse of their range.

Next to the previously mentioned factors, two static and two dynamic donor limits are evaluated. The two static donor limits allow either a donor to be chosen once or an unlimited number of times. For the dynamic cases, the limit is set to either 25 or 50 % of the recipient count.

To evaluate imputation quality, a set of location and/or variability measures is considered (c.p. Nordholt 1998). For the quantitative variables mean and variance, for the ordinal variables median and quartile distance, and for the binary variables the relative frequency of the value one are computed.

Using a framework implemented in R version 2.13.1, 100 data matrices are simulated for every factor level combination of "imputation class count", "object count per imputation class", "class structure", and "ordinal variables scale." For every complete data matrix, the set of true parameters are computed. Each of these 1,600 data matrices is then subjugated to each of the three missingness mechanisms. All of the matrices with missing data are then imputed by all six hot decks using all four donor limits. Repeating this process ten times creates 3.456 million imputed data matrices, for which each parameter set is recalculated.

Considering every parameter in the set, the relative deviation Δp between the true parameter value p_T and the estimated parameter value p_I, based on the imputed data matrix, is calculated as follows:

$$\Delta p = \frac{p_I - p_T}{p_T} \tag{1}$$

To analyze the impact of different donor limits on the quality of imputation, the differences in the absolute values of Δp, that can be attributed to the change in donor limitation, are considered. Due to the large data amounts that are generated in this simulation, statistical significance tests on these absolute relative deviations are not appropriate. Cohen's d measure of effect (Cohen 1992; Bortz and Döring 2009) is chosen as an alternative. The calculation of d, for this case, is as follows:

$$d = \frac{|\Delta \bar{p}_1| - |\Delta \bar{p}_2|}{\sqrt{\frac{s_1^2 + s_2^2}{2}}} \tag{2}$$

$\Delta \bar{p}_1$ and $\Delta \bar{p}_2$ are the means of all relative deviations calculated via (1) for two different donor limits. s_1^2 and s_2^2 are the corresponding variances in the relative deviations. Using absolute values for $\Delta \bar{p}_1$ and $\Delta \bar{p}_2$ allows interpreting the sign of d. A positive sign means that the second case of donor limitation performed better than the first, while a negative sign means the converse. Cohen (1992) does not offer a threshold value above which an effect is nontrivial. He does, however, consider effects around 0.2 as small. Fröhlich and Pieter (2009) consider 0.1, the smallest value for which Cohen presents tables, as threshold for meaningful differences.

4 Results

Based on the simulations results, the formulated research questions are now answered. Section 4.1 answers the question of whether post-imputation parameter estimates are affected by a donor limit. The next Sect. 4.2 answers what factors influence whether a donor limit is advantageous.

4.1 Analysis of Donor Limitation Advantages

If a donor limit does not influence the hot deck method's effectiveness, any of the four donor limits investigated would be equally capable of delivering optimal imputation results. To this end, the relative frequency, with which a particular donor limit leads to the smallest imputation bias, is tabulated. The results are given in Table 1.

Table 1 Frequency distribution of minimum imputation bias

Evaluated parameter		Donor limit			
		Once	25 %	50 %	Unlimited
Quantitative variables	Mean	42.71 %	20.22 %	18.48 %	18.60 %
	Variance	54.05 %	17.79 %	13.04 %	15.12 %
Ordinal variables	Median	46.41 %	21.53 %	14.47 %	17.59 %
	Quartile distance	56.83 %	16.24 %	12.94 %	13.99 %
Binary variables	Relative frequency	49.42 %	18.94 %	15.07 %	16.57 %

The table shows that in most cases donor selection without replacement leads to the best parameter estimation. Measures of variability are more strongly affected than location parameters. The table also indicates that for all parameters the frequency first decreases with a more lenient donor limit and then increases again with unlimited donor usage. This reveals the situation dependent nature of advantages offered by donor limitation. Apparently, given a set of situation defining factors, an optimal donor limit exists.

4.2 Analysis of Main Effects

Table 2 shows the cross classification of effects, between the two static donor limits, for all factor levels against the set of evaluated parameters. Effect sizes with an absolute value larger than 0.1 are in bold, with negative values indicating an advantage for the most stringent donor limit.

Upon analysis of the results, the first conclusion that can be reached is that, independent of any chosen factors, there are no meaningful differences between using a donor limit and using no donor limit in mean and median estimation. In contrast to this, parameters measuring variability are more heavily influenced through the variation of the chosen factors. Especially data matrices with a high proportion of missing data, as well as those imputed with SimDM will profit from a donor limitation. Also a high amount of imputation classes speaks for a limit on donor usage.

The effects that dimensions of the data matrix and the object amount per imputation class have are ambiguous. Class structure and usage any of the random hot deck or SeqDW have no influence on whether a donor limit is advantageous. Fairly conspicuous is the fact that SimDW leads to partially positive effect sizes meaning that leaving donor usage unlimited is advantageous. This might lead to interesting higher order effects, the investigation of which are beyond the scope of this paper.

Table 2 Effect sizes for each factor

		Quantitative variables		Ordinal variables		Binary variables
		Mean	Variance	Median	Quartile distance	Relative frequency
Imputation	2	*0.000*	*−0.068*	*−0.001*	*−0.029*	*−0.072*
class count	7	*0.000*	**−0.147**	*−0.003*	**−0.115**	*−0.090*
Object count per	50	*0.000*	**−0.112**	*−0.001*	*−0.073*	*−0.028*
imputation class	250	*0.000*	*−0.090*	*−0.005*	*−0.041*	**−0.141**
Class	Strong	*0.000*	*−0.092*	*−0.001*	*−0.072*	*−0.072*
structure	Weak	*0.000*	*−0.094*	*−0.001*	*−0.045*	*−0.080*
	(100 × 9)	*0.000*	*−0.082*	*−0.001*	*−0.030*	*−0.034*
Data matrix	(350 × 9)	*0.000*	**−0.177**	*−0.005*	**−0.152**	*−0.022*
dimension	(500 × 9)	*0.000*	*−0.064*	*−0.004*	*−0.030*	**−0.130**
	(1750 × 9)	*0.001*	**−0.146**	*−0.006*	*−0.065*	**−0.162**
Portion of	5 %	*0.000*	*−0.025*	*0.000*	*−0.013*	*−0.011*
missing data	10 %	*0.000*	*−0.071*	*0.000*	*−0.037*	*−0.051*
	20 %	*0.000*	**−0.148**	*0.000*	**−0.100**	**−0.129**
Missingness	MCAR	*0.001*	*−0.088*	*−0.001*	*−0.053*	*−0.065*
mechanism	MAR	*0.000*	**−0.100**	*0.000*	*−0.066*	*−0.086*
	NMAR	*0.001*	*−0.091*	*0.000*	*−0.058*	*−0.077*
	SimDW	*−0.001*	**0.153**	*−0.002*	*0.025*	*0.075*
	SimDM	*−0.004*	**−0.339**	*0.005*	**−0.214**	**−0.338**
Hot Deck	SeqDW	*0.001*	*−0.007*	*−0.003*	*0.000*	*−0.005*
method	SeqDM	*0.000*	*−0.088*	*0.010*	**−0.133**	*−0.041*
	SimR	*0.000*	*−0.001*	*−0.001*	*−0.004*	*0.000*
	SeqR	*0.000*	*−0.001*	*0.000*	*−0.001*	*−0.003*

5 Conclusions

The simulation conducted shows distinct differences between hot deck imputation procedures that make use of donor limits. Limiting donor usage is not advantages under all circumstances, as allowing for unlimited donor usage leads to best parameter estimates under some circumstances.

In some situations, a stringent donor limit leads to better parameter estimates. Splitting the data into a higher amount of imputation classes leads to a better estimation of variance and quartile distance for quantitative and ordinal variables, respectively. For few objects per imputation class, the variance of quantitative variables is better estimated with a donor limit, while binary variables with more objects per imputation class profit from a donor limit. This is also the case for data matrices with high amounts of missingness. With MAR present, donor limited imputation offers a slight advantage for the estimation of quantitative variables' variance. Estimation of location, such as mean and median, are not influenced by a donor limit.

Next to the data's properties, the hot deck variant used to impute missing data plays an important role. Dependent on the method chosen, a donor limit is either advantageous, disadvantageous or without effect on parameter estimation. SimDM and SeqDM both perform better with, while SimDW performs better without a donor limit. Both random hot decks and SeqDW are unaffected by limiting donor usage.

The results are expected to generalize well for a wide range of situations, with an especially valuable conclusion being that different donor limits can lead to very different results. Nonetheless, the study does have some important limitations. First, while the simulated missingness mechanisms, where two groups have different response propensities, cover a broad range of situations, many more MAR and NMAR mechanism types are conceivable. Second, the amount of imputation classes do not represent the extreme cases of what is to be expected. In practice, especially when data matrices are larger, the amount of imputation classes can be larger. Third, good imputation results are undoubtedly dependent on the definition of similarity. This covers not only the distance function, but also the definition of imputation classes. The effect of either being ill defined can be expected to outweigh the effect of choosing the proper donor limit. However, the investigation of these factors is well beyond the scope afforded by this paper.

Careful selection of a donor limit will, in certain situations, improve parameter estimation. Since for some cases neither donor selection with or without replacement will lead to best imputation results, there must be an optimal, situation dependent donor limit. Thus, the development of further recommendations, based on a more detailed investigation of the underlying interactions between the factors, which may culminate in a data driven donor limit selection procedure, is an interesting point for further research.

References

Allison, P. D. (2001). *Missing data*. Sage University Papers Series on Quantitative Applications in the Social Sciences. London: Thousand Oaks.

Andridge, R. R., & Little, R. J. A. (2010). A review of hot deck imputation for survey non-response. *International Statistical Review*, *78*, 40–64.

Bankhofer, U. (1995). Unvollständige Daten- und Distanzmatrizen in der Multivariaten Datenanalyse. Bergisch Gladbach: Eul.

Barzi, F., & Woodward, M. (2004). Imputations of missing values in practice: Results from imputations of serum cholesterol in 28 cohort studies. *American Journal of Epidemiology*, *160*, 34–45.

Bortz, J., & Döring, N. (2009). Forschungsmethoden und Evaluation für Human- und Sozialwissenschaftler. Berlin: Springer.

Cario, M. C., & Nelson, B. L. (1997). *Modeling and generating random vectors with arbitrary marginal distributions and correlation matrix* (pp. 100–150). Northwestern University, IEMS Technical Report, 50.

Cohen, J. (1992). A power primer. *Quantitative Methods in Psychology*, *112*, 155–159.

Durrant, G. B. (2009). Imputation methods for handling item-nonresponse in practice: Methodological issues and recent debates. *International Journal of Social Research Methodology, 12*, 293–304.

Fröhlich, M., & Pieter, A. (2009). Cohen's Effektstärken als Mass der Bewertung von praktischer Relevanz – Implikationen für die Praxis. *Schweiz. Z. Sportmed. Sporttraumatologie, 57*, 139–142.

Kaiser, J. (1983). The effectiveness of hot-deck procedures in small samples. In *Proceedings of the Section on Survey Research Methods* (pp. 523–528). American Statistical Association. Toronto.

Kalton, G., & Kasprzyk, D. (1986). The treatment of missing survey data. *Survey Methodology, 12*, 1–16.

Kalton, G., & Kish, L. (1981). Two efficient random imputation procedures. In *Proceedings of the Survey Research Methods Section* (pp. 146–151). American Statistical Association.

Nordholt, E. S. (1998). Imputation: Methods, simulation experiments and practical examples. *International Statistical Review, 66*, 157–180.

Roth, P. L., & Switzer III, F. S. (1995). A Monte Carlo analysis of missing data techniques in a HRM setting. *Journal of Management, 21*, 1003–1023.

Sande, I. (1983). Hot-deck imputation procedures. In W. Madow, H. Nisselson, I. Olkin (Eds.), *Incomplete data in sample surveys* (Vol. 3). *Theory and bibliographies* (pp. 339–349). New York: Academic.

Strike, K., Emam, K. E., Madhavji, N. (2001). Software cost estimation with incomplete data. *IEEE Transactions on Software Engineering, 27*, 890–908.

Yenduri, S., & Iyengar, S. S. (2007). Performance evaluation of imputation methods for incomplete datasets. *International Journal of Software Engineering and Knowledge Engineering, 17*, 127–152.

The Most Dangerous Districts of Dortmund

Tim Beige, Thomas Terhorst, Claus Weihs, and Holger Wormer

Abstract In this paper the districts of Dortmund, a big German city, are ranked concerning their level of risk to be involved in an offence. In order to measure this risk the offences reported by police press reports in the year 2011 (Presseportal, http://www.presseportal.de/polizeipresse/pm/4971/polizei-dortmund?start=0, 2011) were analyzed and weighted by their maximum penalty corresponding to the German criminal code. The resulting danger index was used to rank the districts. Moreover, the socio-demographic influences on the different offences are studied. The most probable influences appear to be traffic density (Sierau, Dortmunderinnen und Dortmunder unterwegs—Ergebnisse einer Befragung von Dortmunder Haushalten zu Mobilität und Mobilitätsverhalten, Ergebnisbericht, Dortmund-Agentur/Graphischer Betrieb Dortmund 09/2006, 2006) and the share of older people. Also, the inner city parts appear to be much more dangerous than the outskirts of the city of Dortmund. However, can these results be trusted? Following the press office of Dortmund's police, offences might not be uniformly reported by the districts to the office and small offences like pick-pocketing are never reported in police press reports. Therefore, this case could also be an example how an unsystematic press policy may cause an unintended bias in the public perception and media awareness.

T. Beige (✉) · C. Weihs
Chair of Computational Statistics, TU Dortmund, Germany
e-mail: tim.beige@gmx.de; claus.weihs@T-Online.de

T. Terhorst · H. Wormer
Institute of Journalism, TU Dortmund, Germany
e-mail: ThomasTerhorst@gmx.de; holger.wormer@udo.edu

M. Spiliopoulou et al. (eds.), *Data Analysis, Machine Learning and Knowledge Discovery*, Studies in Classification, Data Analysis, and Knowledge Organization,
DOI 10.1007/978-3-319-01595-8_2,

1 Introduction

The paper is the result of a project together with data journalists, who were particularly interested in the prejudice of some citizens of Dortmund that the district Innenstadt Nord is especially dangerous. To find an answer to this question some methodological problems have to be overcome such as how danger can actually be measured. Section 2 gives information about the used response variables. In Sect. 3 different influential factors on danger in a district are discussed as well as outcomes of regression models. Finally, the development and analysis of a danger index is presented in Sect. 4.

2 Responses

How can danger be measured? This project deals with different types of offences. It is based on a data set which is created by the Dortmund police press releases of the year 2011 (see www.presseportal.de). In our case, incidents or non-crime reports such as demonstrations, announcements, public relations and similar press reports are not considered. Overall, 1,053 press releases are cataloged. Each data row or rather each offence contains several variables. The variable *offence* specifies the type of offence. This nominally scaled variable includes the considered characteristic attributes of this project, which are recorded in Table 3 among other aspects. Figure 1 shows a descriptive analysis of the frequency of offences which are used as response variables y later in Sect. 3.3, see also Table 3. Furthermore, there is a nominally scaled variable specifying *the district* in which the offence took place. These are the districts of Dortmund as defined by official statistics. In addition to these variables, *the street* is given for a more accurate description of the crime scene. If a street is located in several districts it has been allocated to that district in which the longer part of the street is situated.

There are deficits in the response data set. Some of the variables have missing values, others are inaccurate. Only the complete 825 press releases are used for counting the offences in the 12 districts of the city of Dortmund.

3 Modeling the Offences

This section describes the statistical modeling of offences.

3.1 Factors of Influence

Let us first discuss different factors x_i of influence. First, socio-demographic factors are considered such as *population density* given in inhabitants per hectare,

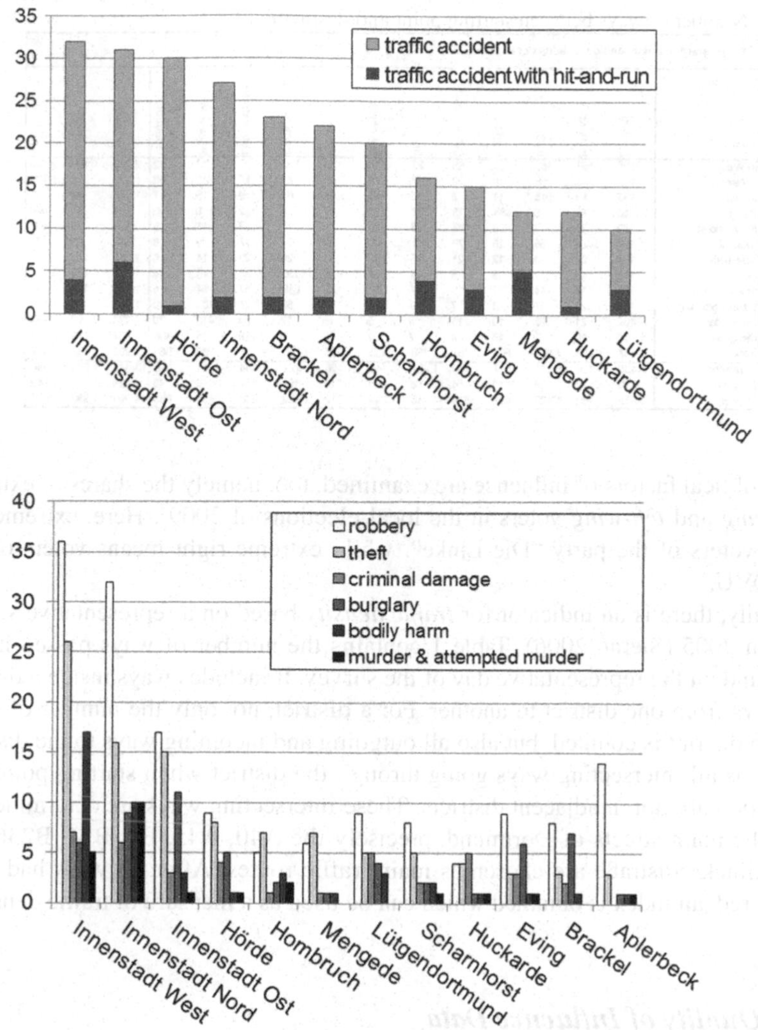

Fig. 1 Types of offences used in the study and their occurrence in the districts

the registered *number of inhabitants* in the district, the number of recorded *unemployed people* in the district, the *unemployment rate*, i.e. the unemployed people as a share of labor force (15–65 years old people), as well as the number of *social security recipients* (persons getting SGB-II). Furthermore, the *share of migrants* in the population is considered as a characteristic of the social structure of a district.

In addition, as a geographic factor the *area* size given in hectare is available in the data set. The demographic factors examined in the project are all ratios to labor force, including the *youth rate* for under 15 year old people, the *old rate* for over 65 year old people, as well as the *very old rate* for over 80 year old people. Moreover,

Table 1 Number of ways between starting point and destination

Table: starting points and destinations of all ways															
to ... / from ...	In-West	In-Nord	In-Ost	Eving	Scharnhorst	Brackel	Aplerbeck	Hörde	Hombruch	Lütgendortmund	Huckarde	Mengede	Dortmund	Outside	Overall
In-West	1,014	168	163	53	65	123	115	126	170	126	92	62	2,277	149	2,426
In-Nord	162	646	68	39	22	27	19	40	21	17	21	18	1,100	73	1,173
In-Ost	190	73	548	11	8	68	43	72	63	19	15	7	1,117	81	1,198
Eving	63	49	9	398	28	17	13	7	8	2	8	10	612	56	668
Scharnhorst	80	25	13	33	492	88	26	9	14	7	11	5	803	75	878
Brackel	146	23	66	16	73	761	60	39	24	17	9	8	1,242	85	1,327
Aplerbeck	128	22	40	12	21	65	707	91	26	29	14	6	1,161	122	1,283
Hörde	143	48	71	9	8	44	81	618	130	38	18	8	1,216	126	1,342
Hombruch	191	18	44	7	15	27	33	122	689	57	14	6	1,223	122	1,345
Lütgendortmund	142	20	17	3	7	20	21	35	53	717	103	20	1,158	169	1,327
Huckarde	102	21	23	10	15	9	5	18	12	89	440	30	774	60	·834
Mengede	76	16	9	13	4	7	7	10	5	19	35	406	607	69	676
Total	2,437	1,129	1,071	604	758	1,256	1,130	1,187	1,215	1,137	780	586	13,290	1,187	14,477
Dortmund	2,438	1,136	1,071	607	760	1,257	1,140	1,196	1,221	1,144	781	590	13,341	1,205	14,546
Outside	97	47	58	50	54	58	104	116	108	144	49	43	928	75	1,003
Overall	2,535	1,183	1,129	657	814	1,315	1,244	1,312	1,329	1,288	830	633	14,269	1,280	15,549

some political factors of influence are examined, too, namely the shares of extreme *right-wing* and *left-wing* voters in the local elections of 2009 . Here, extreme left means voters of the party "Die Linke", while extreme right means voters of the party DVU.

Finally, there is an indicator for *traffic density* based on a representative survey study in 2005 (Sierau 2006). Table 1 contains the number of ways passed inside Dortmund on the representative day of the survey. It includes ways inside a district and ways from one district to another. For a district, not only the number of ways inside a district is counted, but also all outgoing and incoming ways to the district, as well as all intersecting ways going through the district when starting point and destination are not in adjacent districts. These intersecting ways are determined by using the main streets of Dortmund, precisely the A40, A45, B1, B54, B236 and the Malinckrodtstraße are chosen as main traffic routes. After all ways had been considered, an index is obtained which can be used as a measure of traffic density.

3.2 *Quality of Influence Data*

The factors of influence in the data set are not of identical quality. All variables except of the traffic density and the election results are taken from the annual statistical report of Dortmund from 2011. Similarly accurate are the election results which have been determined by the city of Dortmund. They are based solely on elections in 2009. The traffic density relies on a study in 2005 (see Sierau 2006, p. 32) and is estimated using the method described in Sect. 3.1. This variable may not be that accurate, particularly since the estimates are based solely on the motion profile of Dortmund citizens. Any commuters or tourists are not considered.

Table 2 Correlations between possible factors of influence

	Ways	UR	Pop d	Migr.	Yr	Or	Right	Left
Traffic density	1.00							
Unemployment rate	−0.07	1.00						
Population density	0.77	0.31	1.00					
Share of migrants	−0.05	*0.95*	0.33	1.00				
Youth rate	−0.80	0.52	−0.48	0.54	1.00			
Old rate	−0.23	*−0.85*	−0.50	*−0.82*	−0.20	1.00		
Extreme right-wing voters	−0.47	0.28	−0.32	0.19	0.43	−0.35	1.00	
Extreme left-wing voters	0.10	*0.95*	0.44	*0.95*	0.40	*−0.86*	0.14	1.00

3.3 Regression Results for the Factors of Influence

For statistical modeling we use a linear model of the form $y = \beta_0 + \beta_1 x_1 + \cdots + \beta_k x_k + \varepsilon$, where the response y and the influential factors x_i are observed for the 12 districts of the city of Dortmund, and ε is normally distributed with zero mean and variance σ^2. Model fitting is based on the AIC criterion with backward selection. For ease of interpretation, it is helpful that the variables included in the model are (at least nearly) uncorrelated. Therefore, the Bravais–Pearson correlations of the influential factors are examined first. The variable *social security recipients* is particularly highly correlated with *unemployed people* and *unemployment rate*, and also *population density* with *area* as well as the *very old rate* with the *old rate*. It is obvious that one of these variables may (at least nearly) sufficiently explain the corresponding other variables. Therefore it is merely sufficient to put just one of these variables in the model. Also the variable *population* is not included in the model, but only the corresponding rate "population density" in order to avoid size effects in model building. Based on this, we decided to only consider eight rates in the model, namely *traffic density*, *unemployment rate*, *population density*, *share of migrants*, *youth rate*, *old rate*, and the shares of the right-wing and left-wing voters. Please note, however, that due to an outlier there are still some high correlations between some influential factors (see Table 2).

Using these eight factors in the linear model let us first check, whether the normality assumption of the linear model is fulfilled. Figure 2 shows a QQ-plot illustrating that the model residuals are quite normal, e.g., for the response variable criminal damage. Also for the other offences, the assumption is reasonably fulfilled.

As examples, let us roughly discuss the greatest influences on traffic accidents and robberies in the models. A more detailed model discussion will be given for the danger index in Sect. 4. Not surprisingly, for traffic accidents this is traffic density (p-value in t-test equal to 0.0068). Higher traffic density leads to more accidents. But also a higher old rate increases the risk of accidents. Even on robberies the indicator for traffic appears to be significantly positive. That can be explained by the fact that when there are more people the probability of an attack is higher. On the contrary, a high old rate reduces risk of robberies. In general, the traffic density is usually

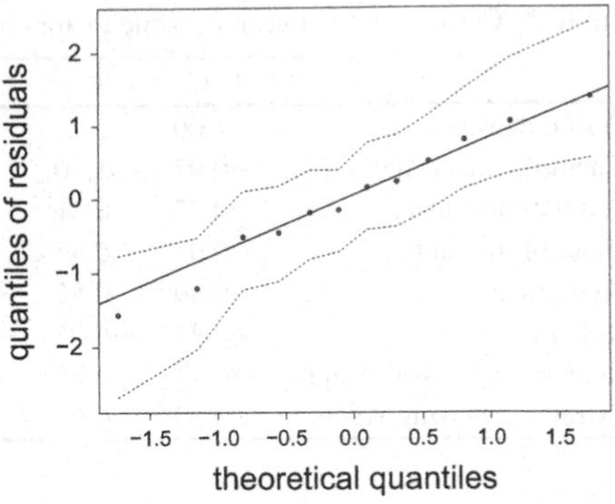

Fig. 2 QQ-plot for the model with the response variable criminal damage with an estimated confidence interval at the 95 % level for each data point

positively significant because it is an indicator for the number of people which spend time in the district. The goodness of fit of the models was always high. Having realized one outlier in the influential factors, L_1-methods could be an alternative for model building. Fortunately, elimination of the outlier did not change the regression coefficients too much. For model building the software R was used (R Development Core Team 2011).

4 Danger Index

Finally, a danger index is created whose value indicates the overall danger for each district. In order to weight the different types of crime, the German criminal code is used. Each offence obtains the maximal penalty as a weight (see Table 3). After that, the weighted offences are summed up as an index for each district. Table 4 shows the values of the calculated danger index for the districts of Dortmund while Fig. 3 illustrates the results using a grey-shaded city map. According to this index, Innenstadt Nord has the highest danger potential while Huckarde would be the safest district in Dortmund. Let us examine which influential factors affect the overall danger. The variables selected by backward selection with AIC are shown in Table 5 together with their estimated coefficients $\hat{\beta}$ and p-values. Moreover, we report the corresponding values $\hat{\beta}_s$ for the model without the insignificant factors. This elimination only leads to a small decrease of R^2 from 0.952 to 0.944. Our modeling shows that a high share of extreme right-wing voters as well as a high old rate is associated with lower danger in a district. Contrary to this, the traffic density significantly increases the risk of becoming a victim of crime or accidents.

Table 3 Offences with their maximum penalty in years

Offence	Max penalty	Offence	Max penalty
Traffic accident with hit-and-run	3	Bodily harm	5
Dangerous disruption of air traffic	10	Attempted bodily harm	5
Traffic accident	0 (0.1)[a]	Criminal damage	2
Wrong way driver	5	Suicide	0
Robbery	5	Attempted suicide	0
Aggravated robbery	10	Murder	∞ (25)[b]
Attempted robbery	5	Attempted murder	∞ (25)[b]
Burglary	10 (5)[c]	Missing	0
Theft	5	Breach of public peace	3
Attempted theft	5	Drug possession	5
Hit-and-run	3	Animal murder	3
Receiving stolen goods	5		

[a] In the German criminal code (StGB) no imprisonment for traffic accidents is provided. To add this delict nevertheless, it is evaluated with the weak factor of 0.1

[b] In accordance to sec. 211 §(1) murder is in Germany liable by life imprisonment. This penalty means to be liable for an indefinite period, but at least 15 years. In addition to this, there are 5 years probation because of sec. 56a §(1), plus 2 years suspension period of eligibility of the convict in according to sec. 57a §(4). Therefore, the total penalty for murder is 22 years. Afterwards the convict can put into an application for releasing from custody. Currently, these applications are usually accepted after the second or third application, therefore after 2 or 4 years. For calculating the danger index, murder is thus indicated by 22+3=25

[c] According to StGB sec. 243 §(1) No. 1 burglary can be penalized with up to 10 years in prison. This occurs only in exceptional cases. In general, burglary is a case of section 243 §(1) No. 2 and is listed with 5 years imprisonment as maximum penalty

Table 4 Danger index for 2011

District	Danger index	District	Danger index
Innenstadt Nord	666.50	Lütgendortmund	134.70
Innenstadt West	635.80	Hörde	133.90
Innenstadt Ost	318.50	Scharnhorst	126.80
Hombruch	160.20	Brackel	123.10
Eving	156.20	Mengede	112.70
Aplerbeck	136.00	Huckarde	89.10

5 Problems with the Data Base

The Dortmund police press office who was finally confronted with the results of our estimation warns about over-interpreting the study. They stress that one should have in mind that the published press releases are chosen solely for media. This means it is decided how interesting the publications of the incidents are, e.g., for the newspapers in Dortmund. Smaller offences like pick-pocketing are barely mentioned. Therefore, the evaluation of the releases could draw a wrong picture of the offence distribution. Also, according to the press office some colleagues on the

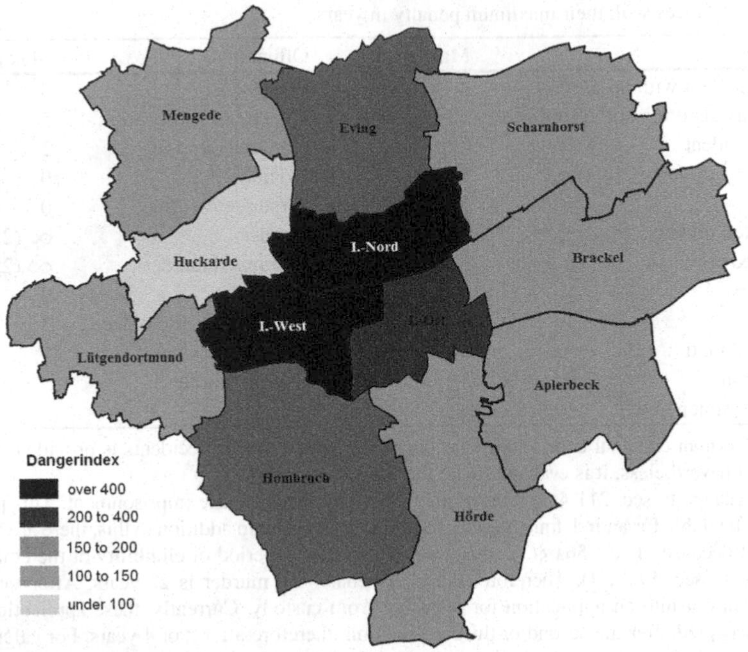

Fig. 3 Danger index in the city of Dortmund

Table 5 Regression results for the danger index after backward selection

Factor of influence	$\hat{\beta}$	$\lvert t \rvert$	p-value	$\hat{\beta}_s$	$\lvert t_s \rvert$	p-value$_s$
Intercept	1,000.867	2.676	0.0440	1,241.062	8.932	< 0.0001
Population density	−2.382	−0.997	0.3646			
Traffic density	0.085	2.844	0.0361	0.051	2.784	0.0238
Unemployment rate	−11.001	−1.042	0.3452			
Share of migrants	7.160	1.878	0.1192			
Old rate	−25.585	−3.647	0.0148	−30.062	−10.230	< 0.0001
Share of right-wing	−72.994	−3.207	0.0238	−87.738	−4.465	0.0021

various police stations have to be encouraged to submit their press releases while others lavish with releases on the press office. This effect may be reflected in the ranking.

6 Summary and Outlook

This study examined the danger in the districts of Dortmund in the year 2011. Offences are weighted by means of the corresponding maximum penalty in the German criminal code. Different regression models estimate whether some influential factors increase or decrease the risk of an offence. The general conclusion might be that more crimes occur where lots of people meet together, so when the traffic density is high. In contrast to this, districts with higher old rates reduce the risk of robberies, thefts and bodily harm, but increase the risk of accidents significantly.

A major problem within this project is the small sample size. Although sufficient offences (1,053) are identified, which are spread on just 12 districts. Because of this, it is difficult to analyze the factors of influence properly. In particular, the sensitivity with respect to outliers is large. It would be advisable to extend the study to North-Rhine-Westphalia or to use the statistical sub-districts of Dortmund to enlarge the sample size. In the latter case it would be necessary to observe the offences for a longer time period to obtain a sufficient number of observations. Perhaps this also could decrease the correlation problem. Furthermore, other influential variables could be examined, e.g. social welfare data or an indicator which classifies a shopping district. This could clarify the results overall.

References

R Development Core Team. (2011). *A language and environment for statistical computing*. Wien, Österreich: R Foundation for Statistical Computing. URL http://www.R-project.org/. ISBN 3-900051-07-0.

Sierau, U. (2006). *Dortmunderinnen und Dortmunder unterwegs - Ergebnisse einer Befragung von Dortmunder Haushalten zu Mobilität und Mobilitätsverhalten, Ergebnisbericht*. Dortmund-Agentur/Graphischer Betrieb Dortmund 09/2006.

Benchmarking Classification Algorithms on High-Performance Computing Clusters

Bernd Bischl, Julia Schiffner, and Claus Weihs

Abstract Comparing and benchmarking classification algorithms is an important topic in applied data analysis. Extensive and thorough studies of such a kind will produce a considerable computational burden and are therefore best delegated to high-performance computing clusters. We build upon our recently developed R packages `BatchJobs` (Map, Reduce and Filter operations from functional programming for clusters) and `BatchExperiments` (Parallelization and management of statistical experiments). Using these two packages, such experiments can now effectively and reproducibly be performed with minimal effort for the researcher. We present benchmarking results for standard classification algorithms and study the influence of pre-processing steps on their performance.

1 Introduction

Assessing the performance of (supervised) classification methods by means of benchmark experiments is common practice. For example, a well-known study of such a kind was conducted in the StatLog project (King et al. 1995). Benchmark studies often require large computational resources and are therefore best executed on high-performance computing clusters. Bischl et al. (2012) have recently developed two R packages `BatchJobs` and `BatchExperiments` that allow to comfortably control a batch cluster within R. An interesting problem that can be investigated by means of a benchmark study is the impact of data pre-processing operations on the performance of classification methods. Questions of interest are for example: "How often does pre-processing lead to

B. Bischl (✉) · J. Schiffner · C. Weihs
Chair of Computational Statistics, Department of Statistics, TU Dortmund, Germany
e-mail: bischl@statistik.tu-dortmund.de; schiffner@statistik.tu-dortmund.de; weihs@statistik.tu-dortmund.de

M. Spiliopoulou et al. (eds.), *Data Analysis, Machine Learning and Knowledge Discovery*, Studies in Classification, Data Analysis, and Knowledge Organization, DOI 10.1007/978-3-319-01595-8_3,

a considerably increased/decreased performance?" or "Are there pre-processing steps that work well with certain classification methods?". There are many case studies available which report that certain pre-processing options work well for the classification problem at hand. We have found also some studies that compare several pre-processing options (e.g. Pechenizkiy et al. 2004), but many of them consider only very few classifiers and/or classification problems. To our knowledge there are no larger studies that systematically investigate the usefulness of several pre-processing options and their combinations for several classification methods and a larger number of data sets (cp. e.g. Crone et al. 2006).

We investigate the effect of three common steps, *outlier removal*, *principal component analysis* and *variable selection* and their combinations on the performance of eight standard classification methods based on 36 benchmark data sets. Data pre-processing is discussed in Sect. 2. In Sect. 3 the design of the benchmark study is described. Section 4 addresses some technical details concerning the execution of the study by means of the R packages `BatchJobs` and `BatchExperiments`. The results are given in Sect. 5. Section 6 summarizes our findings and provides an outlook to future research.

2 Data Pre-processing

In supervised classification we are given a training data set $\{(\boldsymbol{x}_i, y_i), i = 1, \ldots, n\}$, where $\boldsymbol{x}_i \in \mathbb{R}^p$, $i = 1, \ldots, n$, are realizations of p random variables. We suppose that there are p_{num} numerical and p_{cat} categorical variables ($p_{\text{num}} + p_{\text{cat}} = p$), and write $\boldsymbol{x}_i = (\boldsymbol{x}_{i,\text{num}}, \boldsymbol{x}_{i,\text{cat}})$. Each observation has a class label $y_i = k \in \{1, \ldots, K\}$. The number of training observations from class k is denoted by n_k.

In the following we describe the three pre-processing steps, *outlier removal*, *principal component analysis* and *variable selection*, investigated in our study.

It is well known that if classical classification methods are applied to data containing outliers their performance can be negatively affected. A universal way to deal with this problem is to remove the outliers in an initial pre-processing step. In our study the identification of outliers is based on the numerical variables only and every class is considered separately. We use a common approach based on a robust version of the Mahalanobis distance. For each class k we calculate

$$RM_i^k = \sqrt{(\boldsymbol{x}_{i,\text{num}} - \hat{\boldsymbol{\mu}}_{k,\text{MCD}})' \hat{\boldsymbol{\Sigma}}_{k,\text{MCD}}^{-1} (\boldsymbol{x}_{i,\text{num}} - \hat{\boldsymbol{\mu}}_{k,\text{MCD}})} \tag{1}$$

for all i with $y_i = k$. $\hat{\boldsymbol{\mu}}_{k,\text{MCD}} \in \mathbb{R}^{p_{\text{num}}}$ and $\hat{\boldsymbol{\Sigma}}_{k,\text{MCD}} \in \mathbb{R}^{p_{\text{num}} \times p_{\text{num}}}$ are the class-specific *minimum covariance determinant* (MCD) estimates of location and scatter of the numerical variables computed by the Fast MCD algorithm (Rousseeuw and van Driessen 1999). The i-th observation is regarded as outlier and $(\boldsymbol{x}_i, y_i)$ is removed from the training set if $RM_i^k > \chi^2_{p_{\text{num}},0.975}$ (the 0.975-quantile of the χ^2-distribution with p_{num} degrees of freedom). For MCD estimation only the

Table 1 Misclassification rates of CART on the threenorm problem obtained by using the p original variables and all p principal components

p	2	4	6	8	10	12	14	16	18	20
Original variables	0.04	0.07	0.11	0.12	0.16	0.14	0.19	0.20	0.22	0.21
Principal components	0.03	0.02	0.02	0.01	0.02	0.02	0.02	0.02	0.02	0.01

$h_k < n_k$ observations whose covariance matrix has the lowest determinant are used. h_k is calculated by

$$h_k = \begin{cases} m_k & \text{if } \alpha = 0.5 \\ \lfloor 2m_k - n_k + 2(n_k - m_k)\alpha \rfloor & \text{if } 0.5 < \alpha < 1 \end{cases} \tag{2}$$

with $m_k = \lfloor (n_k + p_{\text{num}} + 1)/2 \rfloor$ (Rousseeuw et al. 2012). In our study, for each class the same α-value is used and α is tuned as described in Sect. 3.

Principal Component Analysis (PCA) converts a set of variables via an orthogonal transformation into a set of uncorrelated variables that are called principal components. In our study a PCA is conducted for the numerical variables, which are scaled to zero mean and unit variance first, based on all training observations. The original observations $\boldsymbol{x}_{i,\text{num}}$ can be replaced by the PCA scores $z_i \in \mathbb{R}^{p_{\text{num}}}$ resulting in a training set with elements $((z_i, \boldsymbol{x}_i^{\text{cat}}), y_i)$. Usually, PCA is used for dimension reduction and just the first few principal components that explain a fixed large percentage of the total variance are selected. Alternatively, the most important principal components can be chosen via some variable selection method. But even if no dimension reduction is done, the rotation of the original data may have a positive effect on the classification performance. As an illustration we consider the *threenorm* problem of Breiman (1996), an artificial binary classification problem. The data for the first class are drawn with equal probability from two p-dimensional standard normal distributions with mean $(a, a, \ldots, a)' \in \mathbb{R}^p$ and $(-a, -a, \ldots, -a)'$ respectively. The second class is drawn from a multivariate normal with mean $(a, -a, a, -a, \ldots)'$ where $a = 2/\sqrt{p}$. The dimension p was varied from 2 to 20 and for each p we generated a training and a test data set of size 1,000. A classification tree (CART) was fitted to the training data and predicted on the test data. As Table 1 shows, the misclassification rate of CART obtained on the original variables increases with the dimension p. Moreover, a PCA was conducted on the training data and all p principal components were used in place of the original variables. In this case the error rate of CART is nearly zero, even for large values of p. However, since PCA does not take class labels into account, it is not guaranteed that the principal components are helpful for discriminating the classes. Moreover, PCA captures only linear relationships in the data. For this reason kernel PCA or (kernel) Fisher Discriminant Analysis are also in use. But since we found that PCA is regularly applied and that, to our knowledge, there are only few studies that assess the impact of PCA pre-processing in classification (e.g. Pechenizkiy et al. 2004),

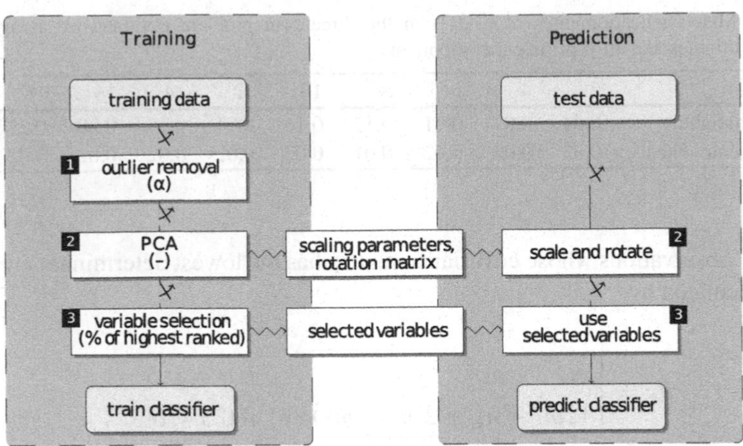

Fig. 1 Pre-processing steps done before training and predicting a classifier

PCA is investigated in our study. As described above we conduct a PCA for the numerical variables based on all training observations. We either use all components in place of the original variables or choose only some of them by applying a variable selection method in the next step.

For variable selection we consider a filter approach. Filter methods rank the variables according to some importance measure. In order to reduce the number of variables based on this ranking, it is common to select either all variables with an importance value larger than a fixed threshold or a certain percentage of highest ranked variables (cp. Guyon and Elisseeff 2003). We employ the latter approach and use the mean decrease in accuracy in random forests as importance criterion. In contrast to the first two pre-processing steps, numerical as well as categorical variables are taken into account. The percentage of selected variables is tuned as described in Sect. 3.

Figure 1 summarizes the pre-processing steps conducted every time when training a classifier (left-hand side) and when making predictions (right-hand side). The three pre-processing operations are applied in the displayed order from top to bottom. Outlier removal, PCA and variable selection are conducted solely on the data used for training. The scaling parameters and the rotation matrix determined by PCA, as well as the names of the selected variables, are stored and the test data are transformed accordingly when making predictions. The switch symbols between the individual pre-processing steps in Fig. 1 indicate that every single step can be activated or deactivated. Thus, there exist $2^3 = 8$ possible pre-processing variants (including the case where no pre-processing is done at all).

Table 2 Classification methods and pre-processing steps under consideration

Method		Hyper-parameters	Box constraints	R package
ro	Outlier removal	α	$[0.5, 1]$	`robustbase`
pca	PCA	–	–	`stats`
fil	Filter	Percentage	$[0.7, 1]$	`FSelector`
lda	Linear discriminant analysis	–	–	`MASS`
multinom	Multinomial regression	–	–	`nnet`
qda	Quadratic discriminant analysis	–	–	`MASS`
naiveBayes	Naive Bayes	–	–	
rbfsvm	Support vector machine	`C`	$[2^{-10}, 2^{10}]$	`kernlab`
	with RBF kernel	`sigma`	$[2^{-10}, 2^{10}]$	
nnet	Neural networks	`decay`	$[0.001, 0.1]$	`nnet`
rpart	CART decision tree	`cp`	$[0.001, 0.1]$	`rpart`
		`minsplit`	$\{5, \ldots, 50\}$	
randomForest	Random forest	`ntree`	$\{100, \ldots, 2000\}$	`randomForest`

3 Study Design

In order to assess the impact of the pre-processing operations described in Sect. 2 on the performance of classification methods we have used the following experimental setup: For every classifier we considered all eight possible pre-processing variants. Prediction performance was evaluated by using nested resampling and measuring the misclassification error rate. In the outer loop, the complete data set was subsampled 30 times, producing 30 (outer) training data sets of size 80 % and 30 (outer) test sets of size 20 %. If a classifier (with pre-processing) had $q > 0$ associated hyperparameters, these were tuned on the training data set by measuring the misclassification rate via threefold cross-validation. Tuning was performed by choosing an effective sequential model-based optimization approach, in which the true relation between the parameters and the performance is approximated by a kriging regression model in each iteration (Jones et al. 1998; Koch et al. 2012). In every iteration the so called expected improvement was maximized to generate a new promising design point to visit subsequently. The budget for the optimization process was $10q$ evaluations for an initial latin hypercube design and $40q$ evaluations for sequential improvements. After tuning, the best parameter combination was selected, the model was trained on the complete (if necessary pre-processed) outer training data set and the (pre-processed) outer test set was predicted. Table 2 shows the pre-processing steps and classification methods under consideration and displays the box constraints for the optimized hyperparameters.

We used 36 data sets from the UCI Machine Learning Repository. Table 3 provides a survey of basic properties of the data sets.

Table 3 Data sets taken from the UCI repository. Displayed are the number of observations and the number of numerical and categorical variables

Data	obs	num	cat	Data	obs	num	cat
BalanceScale	625	4	0	LiverDisorders	345	6	0
BloodTransfusion	748	4	0	MolecularBiologyPromoters	106	0	57
BreastCancer	699	0	9	Monks3	122	6	0
BreastCancerUCI	286	0	9	Parkinsons	195	22	0
BreastTissue	106	9	0	Phoneme	4,509	256	0
Cmc	1,473	2	7	PimaIndiansDiabetes	768	8	0
CoronaryHeartSurgery	1,163	2	20	SAheart	462	8	1
Crabs	200	5	1	Segment	2,310	19	0
Dermatology	366	1	33	Shuttle	256	0	6
Glass2	163	9	0	Sonar	208	60	0
GermanCredit	1,000	7	13	Spambase	4,601	57	0
Haberman	306	2	1	Spect	80	0	22
HayesRoth	132	4	0	Spectf	80	44	0
HeartCleveland	303	6	7	Splice	3,190	0	60
HeartStatlog	270	13	0	TicTacToe	958	0	9
Ionosphere	351	32	1	Vehicle	846	18	0
Iris	150	4	0	VertebralColumn	310	6	0
KingRook_vs_KingPawn	3,196	0	36	Waveform5000	5,000	40	0

4 BatchExperiments and Parallelization Scheme

Bischl et al. (2012) have recently published two R packages `BatchJobs` and `BatchExperiments` for parallelizing arbitrary R code on high-performance batch computing clusters. The former enables the basic parallelization of Map and Reduce operations from functional programming on batch systems. In this study we have used `BatchExperiments`, as it is especially constructed for evaluating arbitrary algorithms on arbitrary problem instances. A problem instance in our case is a classification data set, while an algorithm application is one run of tuning, model-fitting and test set prediction for one classifier with pre-processing operations. This leads to 36 datasets $\times$ 30 iterations of outer subsampling $\times$ 8 classifiers $\times$ 8 preprocessing variants = 69,120 jobs. It should be noted that one computational job is already quite complicated as it contains up to 200 iterations of tuning via sequential model-based optimization. Due to space limitations we cannot go into more technical details how the code is structured, but refer the reader to Bischl et al. (2012), who demonstrate the parallelization of a simple classification experiment for the well-known iris data set. Job runtimes were quite diverse and ranged from a few seconds to more than 18 h, depending on the classifier and data set, summing up to more than 750 days of sequential computation time.

5 Results

In order to analyze the results we have used the non-parametric Friedman test as a means of comparing the locations of the 30 misclassification rates per classifier (0.05 level of significance). Significant differences were detected by post-hoc analysis on all pairs of considered classifiers. We controlled the family-wise error rate through the usual procedure for multiple comparisons for this test as outlined in Hollander and Wolfe (1999). We have performed comparisons in two different ways: First, we compared in the group of basic classifiers without any pre-processing (to figure out which classifiers worked best in their basic form), then we compared in 8 groups of the 8 pre-processing variants of each classifier (to figure out which pre-processing operations worked best for which classifiers). In Table 4 the main aggregated results of this study are presented: The first row displays how often each basic classifier was among the best basic methods for each of the 36 considered data sets. "Among the best" here means that it was not significantly outperformed by another basic method. The rows labeled "ro" (outlier removal), "pca" (PCA) and "fil" (variable selection) count how often a classifier was significantly improved by adding only the respective pre-processing operation. The number in parentheses indicates how often the classifier was significantly worsened by doing this. The last line counts how often a classifier was significantly improved by comparing it to the best of the seven pre-processing variants of itself.

It is in line with theoretical considerations of the eight basic classifiers that (a) non-robust methods like lda and naiveBayes benefit from outlier removal, (b) a method like naiveBayes, which assumes independent variables given the class, benefits from decorrelating the variables by PCA and (c) that the performance of methods like naiveBayes and qda can deteriorate with an increasing number of variables and therefore a filter method might be helpful. Table 5 displays some additional selected results, where extremely large absolute error reductions were observed.

Unfortunately not every (tuned) pre-processing operation will always either improve the model or result in comparable performance. The filtering operation is an exception here (see Table 4). This is due to the fact that setting the percentage parameter of the filter operator to 1 results in the basic classifier with all variables, and our tuning process is apparently able to detect this for the data sets where this is appropriate. Actually, this should be the preferable behavior of the operator for outlier removal as well: When it is best to remove no data point, tuning should detect this and fall back to the basic model. The reason that this does not perfectly work in all of our experiments, points to the fact that the quantile value of the χ^2- distribution for outlier removal should have been included in tuning as well. In summary: If one is interested in the absolute best model for a given data set, we recommend to tune across the whole model space of all reasonable pre-processing variants. This is time-consuming, but can again be sped up by parallelization (and the use of our packages).

Table 4 Main aggregated results, for details see text in this section

	rbfsvm	lda	multinom	naiveBayes	nnet	qda	randomForest	rpart
Basic	32 (–)	14 (–)	17 (–)	13 (–)	13 (–)	5 (–)	25 (–)	13 (–)
ro	0 (4)	3 (4)	2 (7)	3 (4)	0 (2)	1 (8)	0 (4)	0 (3)
pca	0 (3)	1 (2)	0 (0)	7 (7)	6 (0)	1 (0)	2 (8)	3 (8)
fil	2 (0)	1 (0)	1 (0)	5 (0)	3 (0)	5 (0)	1 (0)	0 (0)
any	2 (–)	6 (–)	4 (–)	14 (–)	9 (–)	10 (–)	5 (–)	3 (–)

Table 5 Some selected results where strong improvements occurred

Data	Learner	Pre-processing	Error reduction
BreastTissue	nnet	pca	0.226
Crabs	naiveBayes	pca	0.361
Crabs	randomForest	ro+pca+fil	0.083
Haberman	qda	pca+fil	0.125
HayesRoth	lda	ro	0.122
HayesRoth	multinom	ro+fil	0.086
Segment	nnet	pca+fil	0.221
Spectf	lda	ro+fil	0.121
Spectf	nnet	pca+fil	0.081

6 Summary and Outlook

In this article we have applied the R packages of Bischl et al. (2012) to perform a large scale experimental analysis of classification algorithms on a high-performance batch computing cluster. In this study, our goal was to analyze the influence of various pre-processing operations on eight different classifiers. It appears that it is possible to considerably improve the performance by data pre-processing in some cases. However, for the majority of the investigated classification problems pre-processing did not result in improvements. We can also see that for different classifiers different pre-processing options are beneficial and that some classifiers profit much more from the pre-processing steps in this investigation than others. It was especially hard to improve upon the best performing basic method per data set. Here, sometimes improvements around 1–2 % could be observed but as none of these were significant we were reluctant to report these. We also think that it would be useful for the community as a whole, if a digital repository would exist, where results and descriptions of experiments, such as the ones conducted in this paper, are stored in digital form.

References

Breiman, L. (1996). *Bias, variance, and arcing classifiers*. Technical Report 460, Statistics Department, University of California at Berkeley, Berkeley, CA.

Bischl, B., Lang, M., Mersmann, O., Rahnenführer, J., & Weihs, C. (2012). *Computing on high performance clusters with* R*: Packages* `BatchJobs` *and* `BatchExperiments`. SFB 876, TU Dortmund University. http://sfb876.tu-dortmund.de/PublicPublicationFiles/bischl_etal_2012a.pdf

Crone, S. F., Lessmann, S., & Stahlbock, R. (2006). The impact of preprocessing on data mining: An evaluation of classifier sensitivity in direct marketing. *European Journal of Operational Research, 173*, 781–800.

Guyon, I., & Elisseeff, A. (2003). An introduction to variable and feature selection. *Journal of Machine Learning Research, 3*, 1157–1182.

Hollander, M., & Wolfe, D. A. (1999). *Nonparametric statistical methods* (2nd ed.). New York: Wiley.

Jones, D. R., Schonlau, M., & Welch, W. J. (1998). Efficient global optimization of expensive black-box functions. *Journal of Global Optimization, 13*(4), 455–492.

King, R. D., Feng, C., & Sutherland, A. (1995). StatLog: Comparison of classification algorithms on large real-world problems. *Applied Artificial Intelligence, 9*(3), 289–333.

Koch, P., Bischl, B., Flasch, O., Bartz-Beielstein, T., Weihs, C., & Konen, W. (2012). Tuning and evolution of support vector kernels. *Evolutionary Intelligence, 5*(3), 153–170.

Pechenizkiy, M., Tsymbal, A., & Puuronen, S. (2004). PCA-based feature transformation for classification: Issues in medical diagnostics. In *Proceedings of the 17th IEEE Symposium on Computer-Based Medical Systems*. Silver Spring: IEEE Computer Society.

Rousseeuw, P., Croux, C., Todorov, V., Ruckstuhl, A., Salibian-Barrera, M., Verbeke, T., Koller, M., & Maechler, M. (2012). *Robustbase: Basic Robust Statistics. R package version 0.9–2*. URL http://CRAN.R-project.org/package=robustbase.

Rousseeuw, P. J., & Van Driessen, K. (1999). A fast algorithm for the minimum covariance determinant estimator. *Technometrics, 41(3)*, 212–232.

Visual Models for Categorical Data in Economic Research

Justyna Brzezińska

Abstract This paper is concerned with the use of visualizing categorical data in qualitative data analysis (Friendly, Visualizing categorical data, SAS Press, 2000. ISBN 1-58025-660-0; Meyer et al., J. Stat. Softw., 2006; Meyer et al., vcd: Visualizing Categorical Data. R package version 1.0.9, 2008). Graphical methods for qualitative data and extension using a variety of R packages will be presented. This paper outlines a general framework for visual models for categorical data. These ideas are illustrated with a variety of graphical methods for categorical data for large, multi-way contingency tables. Graphical methods are available in R software in `vcd` and `vcdExtra` library including mosaic plot, association plot, sieve plot, double-decker plot or agreement plot. These R packages include methods for the exploration of categorical data, such as fitting and graphing, plots and tests for independence or visualization techniques for log-linear models. Some graphs, e.g. mosaic display plots are well-suited for detecting patterns of association in the process of model building, others are useful in model diagnosis and graphical presentation and summaries. The use of log-linear analysis, as well as visualizing categorical data in economic research, will be presented in this paper.

1 Introduction

This paper fully illustrates the use of modern graphical methods and visualization techniques for categorical data in economic research. Categorical data are usually presented in multi-way tables, giving the frequencies of observations cross-classified by more that two variables (Friendly 2000; Meyer et al. 2006, 2008).

J. Brzezińska (✉)
Department of Statistics, University of Economics in Katowice, 1 Maja 50, 40–287 Katowice, Poland
e-mail: justyna.brzezinska@ue.katowice.pl

M. Spiliopoulou et al. (eds.), *Data Analysis, Machine Learning and Knowledge Discovery*, Studies in Classification, Data Analysis, and Knowledge Organization, DOI 10.1007/978-3-319-01595-8_4,

Nowadays the dynamic development of computer software allows to analyse any number of variables in multi-way tables. Visual representation of data depends on an appropriate visual scheme for mapping numbers into patterns (Bertin 1983). Discrete data often follow various theoretical probability models. Graphical displays are used to visualize goodness-of-fit, to diagnose an appropriate model, and to determine the impact of individual observations on estimated parameters (Friendly 2000). Many of the graphical displays for qualitative data are not readily available in standard software and they are not widely used compared to plots for quantitative data. Mosaic displays (two- and three-dimensional) and a doubledecker plot, will be presented for the analysis of two- and multi-way tables. This paper describes visual models and graphical techniques for categorical data in economic research and presents their application with the use of R software in `vcd` and `vcdExtra` packages.

2 Log-Linear Models for Contingency Tables

Most statistical methods are concerned with understanding the relationships among variables. For categorical data, these relationships are usually studied from a contingency table (two- or multi-way). Log-linear analysis provides a comprehensive scheme to describe the association for categorical variables in contingency table. Log-linear models are a standard tool to analyze structures of dependency in a multi-way contingency tables. The criteria to be analyzed are the expected cell frequencies as a function of all the variables. There are several types of log-linear models depending on the number of variables and interactions included (the saturated model, conditional independence model, homogeneous association model, complete independence). Stepwise procedures are used for model selection. The aim of a researcher is to find a reduced model containing only few parameters. A reduced model is a more parsimonious model with fewer parameters and thus fewer dependencies and effects. The hierarchy principle used in the analysis reveals that a parameter of lower order cannot be removed when there is still a parameter of higher order that concerns at least one of the same variable. By using this approach, log-linear models are called hierarchical. The overall goodness-of-fit of the model is assessed by comparing the expected frequencies to the observed cell frequencies for each model. The goodness of fit of a log-linear model for a three-way table $H \times J \times K$ $(h = 1, 2, \ldots, H, j = 1, 2, \ldots, J, k = 1, 2, \ldots, K)$ can be tested using either the Pearson chi-square test statistic or the likelihood ratio statistic:

$$G^2 = 2 \sum_{h=1}^{H} \sum_{j=1}^{J} \sum_{k=1}^{K} n_{hjk} \cdot log(\frac{n_{hjk}}{m_{hjk}}), \tag{1}$$

where: n_{hjk}—observed cell frequency, m_{hjk}—expected cell frequency.

In order to find the best model from a set of possible models, additional measurements should be considered. The Akaike information criterion (Akaike 1973) refers to the information contained in a statistical model according to the equation:

$$AIC = G^2 - 2 \cdot df. \tag{2}$$

Another information criterion is the Bayesian Information Criterion (Raftery 1986):

$$BIC = G^2 - df \cdot log(n). \tag{3}$$

The model that minimizes AIC and BIC will be chosen.

The maximum likelihood method incorporating iterative proportional fitting is used to estimate the parameters, and estimated parameter values may be used in identifying which variables are of great importance in predicting the observed values. However, estimation is not the main topic of this paper, and thus only model selection and visual representation of log-linear models will be presented.

3 Problem of Unemployment in Poland

With the rising unemployment rate in recent years, unemployment is one of the most important economic and social problems in Poland. According to the Central Statistical Office of Poland the unemployment rate in October 2012 was 12,5 %. A strong differentiation is observed in the unemployment rates for various parts of Poland, especially for the young and university graduates, as well as for males and females.

The data on the number of the unemployed by the level of education in 2010 in 16 voivodeships in Poland come from Demographic Yearbook of Poland published by the Central Statistical Office of Poland. Three variables are considered: Voivodeship (V) (lodzkie 1, mazowieckie 2, malopolskie 3, slaskie 4, lubelskie 5, podkarpackie 6, podlaskie 7, swietokszyskie 8, lubuskie 9, wielkopolskie 10, zachodniopomorskie 11, dolnoslaskie 12, opolskie 13, kujawsko–pomorskie 14, pomorskie 15, warminsko–mazurskie 16), Education level (E) (higher, higher vocational, high-school, lower vocational), Sex (S) (Male M, Female F). Voivodeships (provinces) considered in the research are 16 administrative subdivisions of Poland (from 1999). The sample size is 1,436,814.

A three-way table with Voivodeship, Education level and Sex will be analysed. All possible log-linear models for a three-way table were built and the likelihood ratio with degrees of freedom were computed. Two mosaic plots are presented to display the residuals. The Goodman's bracket notation is used to express the model equation (Goodman 1970). For example [VE][VS] model allows association

Table 1 Goodness of fit statistics for models with three variables

Model	χ^2	G^2	*df*	*p-value*	AIC	BIC
[V][S][E]	77,280.71	79,283.84	108	0	79,067.84	77,752.62
[ES][V]	18,239.83	18,314.23	105	0	18104.23	16,825.55
[VE][S]	66,999.68	67,562.55	63	0	67,436.55	66,669.34
[VS][E]	75,805.59	76,031.78	93	0.014	75,845.78	74,713.23
[VS][ES]	14,948.51	15,062.17	90	0.291	14,882.17	13,786.16
[VE][ES]	6,593.30	6,592.94	60	0.014	6,472.94	5,742.26
[VE][VS]	63,838.36	64,310.49	48	0	64,214.49	63,629.95
[VE][VS][ES]	1,005.21	1,005.52	45	0	915.52	367.51
[VSE]	0	0	0	1	0.000	0.000

between Voivodeship (V) and Education level (E), as well as the interaction between Voiodeship (V) and Sex (S), but it denies a direct link between Education level (E) and Sex (S). All possible models for these three variables were tested.

The best fitting models, except the saturated model, are the homogeneous association model [VE][VS][ES] and conditional independence model [VE][ES]. For the homogeneous association model [VE][VS][ES] the conditional odds ratios for each pair of variables ([VE],[VS] and [ES]) given the third are identical at each level of the third variable. For the homogeneous association model [VE][VS][ES] $G^2=1005.52, df=45, AIC=915.52, BIC=367.51$. The model [VE][VS][ES] allows association between all possible two-way interaction terms: Voivodeship (V) and Education level (E), Voivodeship (V) and Sex (S) as well as Education level (E) and Sex (S). It means that in this model for each of the partial odds ratio for the two variables given level of the third depends only on the corresponding two-way interaction terms and does not depend on levels of the third variable. The second best-fitting model is the conditional independence model [VE][ES]. For this model the goodness-of-fit statistics are: $G^2=6592.94$, $df=60$, $AIC=6472.94$, $BIC=5742.26$. This model allows the association between Voivodeship (V) and Education level (E) as well as between Education level (E) and Sex (S), but it denies a link between Voivodeship (V) and Sex (S).

4 Visual Models for Categorical Data in R

Many statistical graphs for the exploration and modeling of continuous data have been developed in statistics, however graphs for the interpretation and modeling of categorical data are still rarely to be found (Anderson and Vermunt 2000; Rosmalen et al. 1999). Several schemes for representing contingency tables graphically are based on the fact that when the row and column variables are independent, the expected frequencies are a product of the row and column totals (divided by the grand total). Then, each cell can be represented by a rectangle whose area shows the cell frequency or deviation from the independence (Friendly 1995). Visualizing tools for categorical data are available in R in `vcd` and `vcdExtra` packages.

Mosaic plot was first introduced by Hartigan and Kleiner (1981, 1984) and Theus and Lauer (1999) and is one of the most popular and useful method for log-linear modeling. Mosaic plots generalize readily to multi-way tables (Friendly 1992, 1994, 1995, 1999, 2000) extended the use of the mosaic plots for fitting log-linear models. A mosaic represents each cell of the table by a rectangle (or tile) whose area is proportional to the cell count. The mosaic is constructed by dividing a unit square vertically by one variable, then horizontally by the other. Further variables are introduced by recursively subdividing each tile by the conditional proportions of the categories of the next variable in each cell, alternating on the vertical and horizontal dimensions of the display (Meyer et al. 2006). This scheme allows an arbitrary number of variables to be represented and helps in understanding the structure of log-linear models themselves and can be used to display the residuals from any log-linear model. For a two-way table the tiles are obtained by recursive partitioning splits of squares, where the area is proportional to the observed cell frequencies n_{hj}, width is proportional to one set of margins $n_{h.}$, height is the relative proportion of other variable $\frac{n_{hj}}{n_{h.}}$. In an extended mosaic plot shading is a sign and magnitude of Pearson's residuals: $d_{hj} = \frac{n_{hj}-m_{hj}}{\sqrt{m_{hj}}}$. For a given model, the expected frequencies are found with the use of iterative proportional fitting (Deming and Stephen 1940). The mosaic plot display generalizes readily to multi-way table, displays the deviations (residuals) from a given log-linear model that has been fit to the table and provides an illustration of the relationship among variables that are fitted by various log-linear models. In the example the fit of two models is compared.

The negative signs are red, the positive signs are blue, and the magnitude is reflected in the intensity of shading ($|d_{hj}| > 0, 2, 4, \ldots$). The purpose of highlighting the cells is to draw attention to the pattern of departures of the data from the assumed model. Positive values indicate the cells whose observed frequency is substantially greater than that to be found under independence; the negative values indicate the cells that occur less often than under independence (Friendly 2000). When the cells are empty then variables are independent. Mosaic displays can be easily extended to multi-way tables. The relative frequencies of the third variable are used to subdivide each two-way cell, and so on, recursively. Mosaic plots can be done in R with the use of `mosaic()` function in library `vcd` (Fig. 1).

Positive values indicate the cells whose observed frequency is substantially greater that to be found under independence; negative ones indicate the cells that occur less often than under independence.

A much better fit can be seen for the model [VE][VS][ES] as most of the squares are light red and blue which means small differences between the observed and expected cells. For the model [VE][ES] most of the cells show very high Pearson's value ($|d_{hj}| > 4$). Three-dimensional plots are also available in `vcdExtra` package in R with the use of `mosaic3d()` command (Fig. 2).

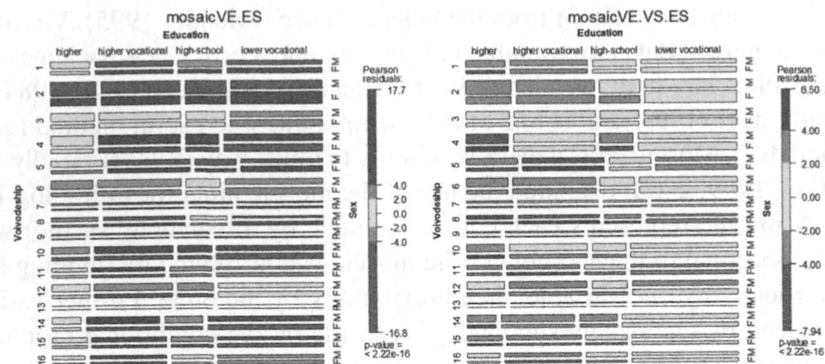

Fig. 1 Extended mosaic plots for [VE][ES] and [VE][VS][ES] models

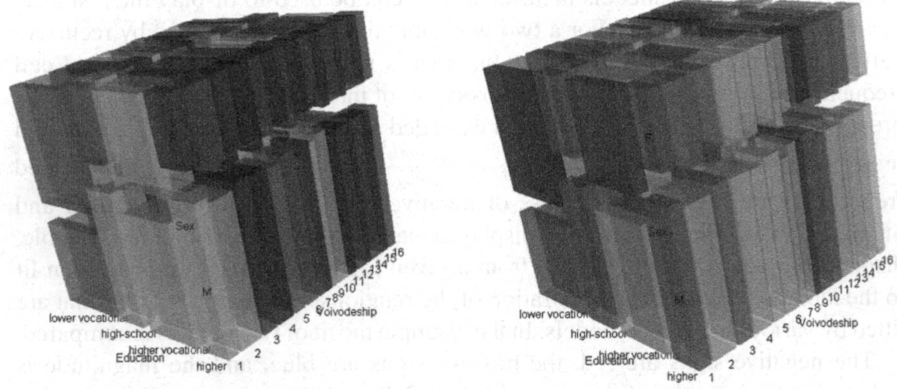

Fig. 2 Three-dimensional mosaic plots for [VE][ES] and [VE][VS][ES] models

For those two models G^2 is relatively large which indicates lack of fit. The first mosaic plot has light residuals which means that the difference between empirical and theoretical cell counts is smaller than for the second model. The deviations, displayed by shading, often suggest terms to be added to an explanatory model that achieved a better fit (Friendly 2000). In this case mosaic plots can be used for testing goodness of fit.

A doubledecker plot is similar to the mosaic, however a doubledecker plot uses only a single horizontal split. It can be seen that the highest level of unemployed females in all voivodeships is observed among women with lower vocational education level, whereas higher education level for men. This disproportion indicated differences in this sector in terms of sex and the education level of the respondents (Fig. 3). A doubledecker plot can be done in R with the use of the `doubledecker()` function in `vcd` package.

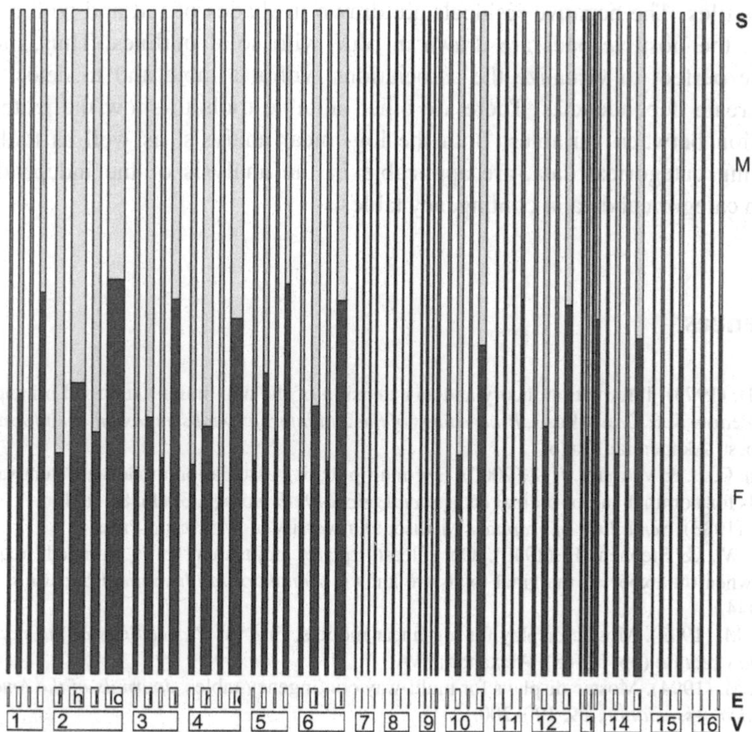

Fig. 3 Doubledecker plot

5 Conclusions

Graphical methods for categorical data are still poorly developed and little used in comparison to methods for qualitative data. However, they can constitute a powerful tool not only for seeing, but also for thinking. It can show how to think outside the box. The categorical data analysis and log-linear analysis have rapidly become major statistical tools for deciphering multidimensional tables arising through Poisson distribution. The purpose of this paper was to provide an overview of log-linear models and visualizing categorical data, and their place in economic research. The `vcd` and `vcdExtra` packages in R provide very general visualization methods via the strucplot framework (the mosaic and association plot, sieve diagram, doubledecker plot) that can be applied to any contingency table. These plots are used to display the deviations (residuals) from the various log-linear models to enable interpretation and visualization. Mosaic graph represents a contingency table, each cell corresponding to a piece of the plot, whose size is proportional to the cell entry. Extended mosaic displays show the standardized residuals of a log-linear model of the counts by the colour and outline of the mosaic's tiles. The negative residuals are drawn in shades of red and with broken outlines; the positive ones are drawn in blue with solid outlines. Thus, mosaic plots are perfect to visualize the associations within a table and to detect cells which create dependencies. There are other suitable tools to show the pattern of association between variables. Both the log-linear analysis, as well as tools for visualizing categorical data, are applicable in the analysis of the independence between categorical data in contingency tables.

References

Akaike, H. (1973). Information theory and an extension of the maximum likelihood principle. In B. N. Petrow & F. Czaki (Eds.), *Proceedings of the 2nd international symposium on information*. Budapest: Akademiai Kiado.

Anderson, C. J., & Vermunt, J. K. (2000). Log-multiplicative association models as latent variable models for nominal and/or ordinal data. *Sociological Methodology, 30*(1), 81–121.

Bertin, J. (1983). *Semiology of graphics*. Madison: University of Wisconsin Press.

Deming, W., & Stephan, F. (1940). On a least squares adjustment of a sampled frequency table when the expected marginal totals are known. *Annals of Mathematical Statistics, 11*(4), 427–444.

Friendly, M. (1992). Mosaic display for log-linear models. In *ASA, Proceedings of the statistical graphics section* (pp. 61–68), Alexandria, VA.

Friendly, M. (1994). Mosaic displays for multi-way contingency tables. *Journals of the American Statistical Association, 89*, 190–200.

Friendly, M. (1995). Conceptual and visual models for categorical data. *The Amercian Statistician, 49*(2), 153–160.

Friendly, M. (1999). Extending mosaic displays: Marginal, conditional, and partial views of categorical data. *Journal of Computational and Graphical Statistics, 8*(3), 373–395.

Friendly, M. (2000). *Visualizing categorical data*. Cary, NC: SAS Press. ISBN 1-58025-660-0.

Goodman, L. A. (1970). The multivariate analysis of qualitative data: Interaction among multiple classifications. *Journal of the American Statistical Association, 65*, 226–256.

Hartigan, J. A., & Kleiner, B. (1981). Mosaics for contingency tables. In *Computer Science and Statistics: Proceedings of the 13th Symposium on the Interface*.

Hartigan, J. A., & Kleiner, B. (1984). A mosaic of television ratings. *The Amercian Statistician, 38*(1), 32–35.

Knoke, D., & Burke, P. J. (1980). *Log-linear models*. Beverly Hills: Sage.

Mayer, D., Hornik, K., & Zeileis, A. (2006). The strucplot framework: Visualizing multi-way contingency tables with vcd. *Journal of Statistical Software, 17*(3), 1–48.

Meyer, D., Zeileis, A., & Hornik, K. (2008). *VCD: Visualizing categorical data, R package*. http:// CRAN.R-project.org.

Raftery, A. E. (1986). Choosing models for cross-classification. *American Sociological Review, 51*, 145–146.

Van Rosmalen, J. M., Koning, A. J., & Groenen, P. J. F. (1999). Optimal scaling of interaction effects in generalized linear modelling. *Multivariate Behavioral Research, 44*, 59–81.

Theus, M., & Lauer, R. W. (1999). Technical report 12, visualizing loglinear models. *Journal of Computational and Graphical Methods, 8*(3), 396–412.

How Many Bee Species? A Case Study in Determining the Number of Clusters

Christian Hennig

Abstract It is argued that the determination of the best number of clusters k is crucially dependent on the aim of clustering. Existing supposedly "objective" methods of estimating k ignore this. k can be determined by listing a number of requirements for a good clustering in the given application and finding a k that fulfils them all. The approach is illustrated by application to the problem of finding the number of species in a data set of Australasian tetragonula bees. Requirements here include two new statistics formalising the largest within-cluster gap and cluster separation. Due to the typical nature of expert knowledge, it is difficult to make requirements precise, and a number of subjective decisions is involved.

1 Introduction

Determining the number of clusters is a notoriously hard key problem in cluster analysis. There is a large body of literature about it (for some references beyond those given below see Jain 2010).

One of the reasons why the problem is so hard is that most of the literature is based on the implicit assumption that there is a uniquely best or "true" clustering for given data or a given underlying statistical model assumed to be true without defining unambiguously what is meant by this. This obscures the fact that there are various ways of defining a "best clustering" which may lead to different solutions for the same data set or model. Which of these definitions is appropriate depends on the meaning of the data and the aim of analysis. Therefore there is no way to find a uniquely best clustering considering the data (or a model assumption) alone.

C. Hennig
Department of Statistical Science, University College London, Gower St, London, WC1E 6BT, UK
e-mail: c.hennig@ucl.ac.uk

M. Spiliopoulou et al. (eds.), *Data Analysis, Machine Learning and Knowledge Discovery*, Studies in Classification, Data Analysis, and Knowledge Organization,
DOI 10.1007/978-3-319-01595-8_5,

For example, the "true clusters" to be counted could correspond to, among others,

- Gaussian mixture components,
- density modes,
- connected data subsets that are strongly separated from the rest of the data set,
- intuitively clearly distinguishable patterns,
- the smallest number of data subsets with a given maximum within-cluster distance.

It is clear that these definitions can lead to different "true" numbers of clusters for the same data. For example it is well known that a mixture of several Gaussians can be unimodal or have more than two modes, two density modes are not necessarily separated by deep gaps, connected and well separated data subsets may include very large within-cluster distances etc. Note further that finding the "true" number of density modes or Gaussian mixture components is an ill posed problem, because it is impossible to distinguish models with k density modes or Gaussian mixture components from models with arbitrarily more of them based on finite data for which a model with k modes/mixture components fits well. A well known implication of this (Hennig 2010) is that the BIC, a consistent method for estimating the number of Gaussian mixture components, will estimate a k tending to infinity for $n \to \infty$ (n being the number of observations) because of the fact that the Gaussian mixture model does not hold precisely for real data, and therefore more and more mixture components will fit real data better and better if there are only enough observations to fit a large number of parameters.

Different concepts to define the number of clusters are required for different applications and different research aims. For example, in social stratification, the poorest people with the lowest job status should not be in the same cluster (social stratum) as the richest people with the highest job status, regardless of whether there is a gap in the data separating them, or whether these groups correspond to different modes, i.e., large within-cluster dissimilarities should not occur. On the other hand, in pattern recognition on images one often wants to only separate subsets with clear gaps between them regardless of whether there may be large distances or even multiple weak modes within the clusters.

In the present paper I suggest a strategy to determine the number of clusters depending on the research aim and the researcher's cluster concept, which requires input based on an expert's subject matter knowledge. Subject matter knowledge has already been used occasionally in the literature to determine the number of clusters, see e.g., Chaturvedi et al. (2001), Morlini and Zani (2012), but mostly informally.

Section 2 introduces a number of methods to estimate the number of clusters. The new approach is illustrated in Sect. 3 by applying it to the problem of determining the number of species in a data set of tetragonula bees. Some limitations are discussed in Sect. 4.

2 Some Methods to Determine the Number of Clusters

Here are some standard approaches from the literature to determine the number of clusters. Assume that the data $x_1, \ldots, x_n$ in some space $\mathcal{S}$ are to be partitioned into exhaustive and non-overlapping sets $C_1, \ldots C_k$, and that there is a dissimilarity measure d defined on $\mathcal{S}^2$.

Calinski and Harabasz (1974) index. k_{CH} maximises $\frac{B(k)(n-k)}{W(k)(k-1)}$, where

$$W(k) = \sum_{h=1}^{k} \frac{1}{|C_h|} \sum_{x_i, x_j \in C_h} d(x_i, x_j)^2, \text{ and}$$
$$B(k) = \frac{1}{n} \sum_{i,j=1}^{n} d(x_i, x_j)^2 - W(k).$$

Note that k_{CH} was originally defined for Euclidean distances and use with k-means, but the given form applies to general distances.

Average silhouette width (Kaufman and Rousseeuw 1990). k_{ASW} maximises $\frac{1}{n} \sum_{i=1}^{n} s(i,k)$, where

$$s(i,k) = \frac{b(i,k) - a(i,k)}{\max(a(i,k), b(i,k))},$$
$$a(i,k) = \frac{1}{|C_j| - 1} \sum_{x \in C_j} d(x_i, x), \; b(i,k) = \min_{x_i \notin C_l} \frac{1}{|C_l|} \sum_{x \in C_l} d(x_i, x),$$

C_j being the cluster to which x_i belongs.

Pearson-version of Hubert's Γ (Halkidi et al. 2001). k_{PG} maximises the Pearson correlation between a vector of all dissimilarities and the corresponding binary vector with 0 for a pair of observations in the same cluster and 1 for a pair of observations in different clusters.

Bootstrap stability selection (Fang and Wang 2012). This is one of a number of stability selection methods in the literature. For each number of clusters k of interest, B pairs of standard nonparametric bootstrap subsamples are drawn from the data. For each pair, both subsamples are clustered, and observations not occurring in any subsample are classified to a cluster in both clusterings in a way adapted to the used clustering method. For example, in Sect. 3, average linkage clustering is used and unclustered points are classified to the cluster to which they have the smallest average dissimilarity. For each pair of clusterings the relative frequency of point pairs in the same cluster in one of the clusterings but not in the other is computed, these are averaged over the B bootstrap samples, and k_{BS} is the k that minimised the resulting instability measure.

As many methods in the literature, the former three methods all try to find a compromise between within-cluster homogeneity (which generally improves with increasing k) and between-cluster separation (which usually is better for smaller k). The terms "within-cluster homogeneity" and "between-cluster separation" are meant here in a general intuitive sense and admit various ways of measuring them, which are employed by the various different criteria. The k optimising these indexes may differ. For example, experiments indicate that k_{ASW} may lump together relatively weakly separated data subsets if their union is strongly separated from what is left, whereas k_{CH} may leave them separated if putting them together makes the resulting cluster too heterogeneous. k_{PG} tends less than the two former methods to integrate single outliers in clusters.

A general remark on stability selection is that although *good* stability is a reasonable requirement in many applications, *optimal* stability is more difficult to motivate, because there is no reason why "bad" clusterings cannot be stable.

3 Analysis of the Tetragonula Bees Data Set

Franck et al. (2004) published a data set giving genetic information about 236 Australasian tetragonula bees, in which it is of interest to determine the number of species. The data set is incorporated in the package "fpc" of the software system R (www.r-project.org). Bowcock et al. (1994) defined the "shared allele dissimilarity" formalising genetic dissimilarity appropriately for species delimitation, which is used for the present data set. It yields values in [0, 1].

In order to apply the approach taken here and in fact also in order to choose an appropriate clustering method, it is important to specify formal requirements of species delimitation. The following list was compiled with help of the species expert Bernhard Hausdorf, museum of zoology, University of Hamburg.

- Large within-cluster gaps should be avoided, because genetic gaps are essential for the species concept. Some caution is needed, though, because gaps could be caused by incomplete sampling and by regional separation within a species.
- Species need to be well separated for the same reason. Experts would normally speak of different species even in case of rather moderate separation among regionally close individuals, so to what extent separation is required depends on the location of the individuals to some extent.
- In order to count as species, a group of individuals needs to have a good overall homogeneity, which can be measured by the average within-species dissimilarity.
- Cluster stability is needed in order to have confidence that the clustering is not a random structure, although there is no specific reason why the best clustering needs to have maximum stability.

The third criterion motivates the average linkage hierarchical clustering, which is applied here, see Fig. 1 (it is beyond the scope of the paper to give a more conclusive justification). Determining the number of species amounts to finding the best height at which the dendrogram is cut. Values of k between 2 and 15 were examined.

The criteria introduced in Sect. 2 do not yield a consistent decision about the number of clusters, with $k_{CH} = k_{ASW} = 10$, $k_{PG} = 9$, $k_{BS} = 5$. Note that for $k > 3$ all instability values are smaller than 0.08, so all clusterings are rather stable and fulfil the fourth requirement. k_{BS} may generally be rather low, because splitting up somewhat ambiguous data subsets may harm stability. Just taking the general

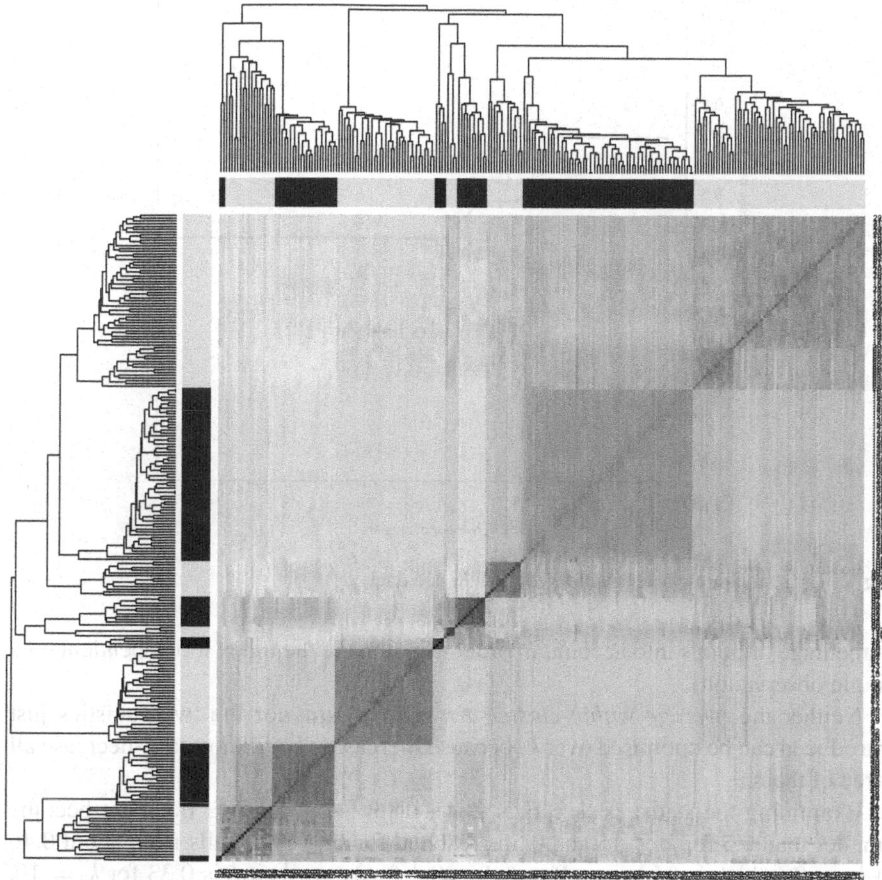

Fig. 1 Heatplot and average linkage clustering for tetragonula bee data. Colour bars at the *left side* and on *top* indicate the clustering with $k = 10$

behaviour of the criteria into account, k_{ASW} with its strong emphasis on separation looks closest to the listed requirements.

An approach driven stronger by the aim of clustering is to find a number of clusters that fulfils all listed requirements separately instead of using a criterion that aggregates them without caring about specific details.

To this end, the *largest within-cluster gap wg* of a clustering can be defined as the maximum over all clusters of the dissimilarity belonging to the the last connecting edge of the minimum spanning tree within each cluster.

Cluster separation se of a clustering can be measured by computing, for all observations, the distance to the closest cluster to which the observation does not belong. Then the average of the minimum 10 % of these distances is taken in order to consider only points close to the cluster borders (one should take a certain

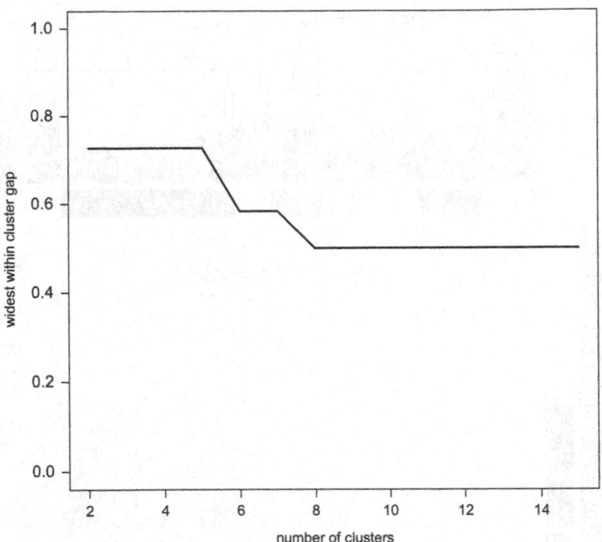

Fig. 2 Largest within-cluster gap for tetragonula bee data

percentage of points into account in order to not make the index too dependent on a single observation).

Neither the *average within-cluster dissimilarity* ad nor the two statistics just introduced can be optimised over k, because increasing k will normally decrease all three of them.

Examining the statistics (see Figs. 2–4), it turns out that wg does not become smaller than 0.5 for $k \leq 15$ and 0.5 is reached for $k \geq 8$. se falls from about 0.46 at $k = 9$ (which is fairly good) to below 0.4 for larger k. ad is 0.33 for $k = 10$, does not improve much for larger k, and is much higher for $k \leq 9$; 0.46 for $k = 9$. Overall this means that $k = 9$ and $k = 10$ can be justified, with $k = 9$ much better regarding separation and $k = 10$ much better regarding ad, i.e., homogeneity.

Automatic aggregation of these aspects by formal criteria such as k_{ASW} or k_{PG} obscures the fact that in this situation the decision which of the requirements is more important really must come from subject matter expertise and cannot be determined from the data alone.

For the given problem, the importance of separation depends on how closely together the sampled individuals actually were taken geographically. From existing information on the sampling of individuals it can be seen that the two clusters merged going from $k = 10$ to 9 consist of individuals that are rather close together, in which case according to B. Hausdorf one would accept a weaker separation and demand more homogeneity. This favours the solution with $k = 10$. This solution is illustrated in Fig. 1 by the colour bars on the left side and above (for $k = 9$, the two clusters in the upper right are merged).

Note that for this data set an expert decision about the existing species exists (cf. Franck et al. 2004; we did not use this in order to define criteria make decisions

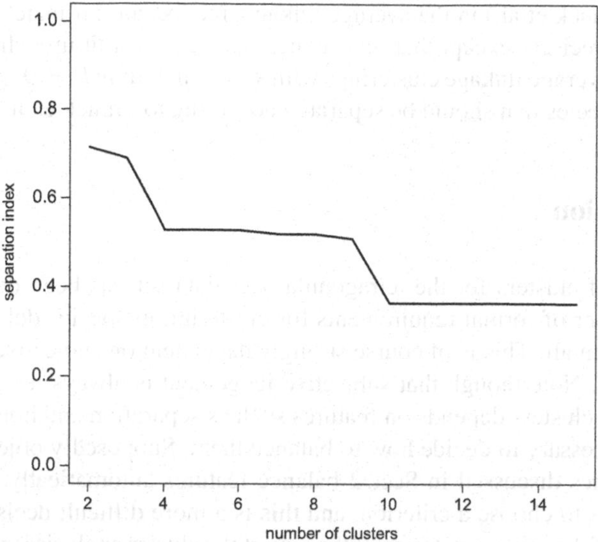

Fig. 3 Cluster separation for tetragonula bee data

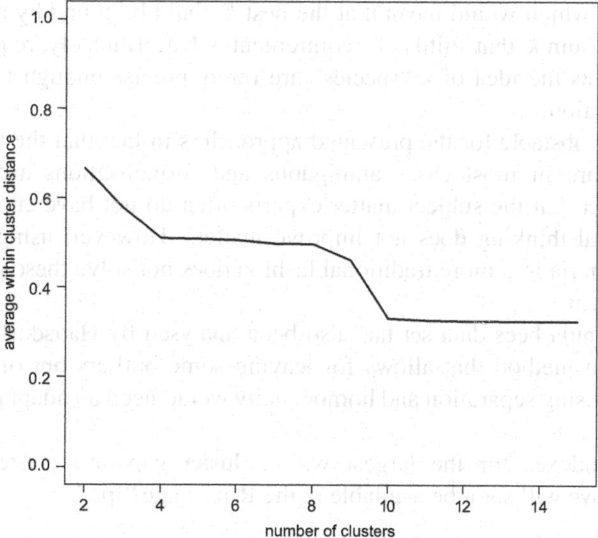

Fig. 4 Average within-cluster dissimilarity for tetragonula bee data

here), using information beyond the analysed data. This could be taken as a "ground truth" but one needs to keep in mind that there is no precise formal definition of a "species". Therefore experts will not always agree regarding species delimitation. According to the expert assessment there are nine species in the data, but in fact the solution with $k = 10$ is the best possible one (in the sense of matching the species decided by Franck et al.) in the average linkage tree, because it matches the expert delimitation precisely except that one "expert species" is split up, which is in fact split up in all average linkage clusterings with $k \geq 6$ including $k = 9$, which instead merges two species that should be separated according to Franck et al. (2004).

4 Conclusion

The number of clusters for the tetragonula bees data set has been determined by listing a number of formal requirements for clustering in species delimitation and examining them all. This is of course strongly dependent on subjective judgements by the experts. Note though that subjective judgement is always needed if in fact the number of clusters depends on features such as separation and homogeneity, of which it is necessary to decide how to balance them. Supposedly objective criteria such as the ones discussed in Sect. 2 balance features automatically, but then the user still needs to choose a criterion, and this is a more difficult decision, because the meaning of the criteria in terms of the aim of the cluster analysis is more difficult to understand than statistics that formalise the requirements directly.

Optimally the statistician would like the expert to specify precise cutoff values for all criteria, which would mean that the best k could be found by a formal rule (e.g., the minimum k that fulfils all requirements). Unfortunately, required cluster concepts such as the idea of a "species" are rarely precise enough to allow such exact formalisation.

The biggest obstacle for the presented approach is in fact that the requirements of clustering are in most cases ambiguous and formalisations are difficult to obtain. The fact that the subject matter experts often do not have enough training in mathematical thinking does not improve matters. However, using supposedly "objective" criteria in a more traditional fashion does not solve these problems but rather hides them.

The tetragonula bees data set has also been analysed by Hausdorf and Hennig (2010), using a method that allows for leaving some outliers out of all clusters. Indexes formalising separation and homogeneity would need an adaptation for such methods.

The new indexes for the largest within-cluster gap and cluster separation introduced above will soon be available in the R-package "fpc".

References

Bowcock, A. M., Ruiz-Linares, A., Tomfohrde, J., Minch, E., Kidd, J. R., & Cavalli-Sforza, L. L. (1994). High resolution of human evolutionary trees with polymorphic microsatellites. *Nature, 368*, 455–457.

Calinski, R. B., & Harabasz, J. (1974). A dendrite method for cluster analysis. *Communications in Statistics, 3*, 1–27.

Chaturvedi, A. D., Green, P. E., & Carrol, J. D. (2001). K-modes clustering. *Journal of Classification, 18*, 35–55.

Fang, Y., & Wang, J. (2012). Selection of the number of clusters via the bootstrap method. *Computational Statistics and Data Analysis, 56*, 468–477.

Franck, P., Cameron, E., Good, G., Rasplus, J.-Y., & Oldroyd, B. P. (2004). Nest architecture and genetic differentiation in a species complex of Australian stingless bees. *Molecular Ecology, 13*, 2317–2331.

Halkidi, M., Batistakis, Y., & Vazirgiannis, M. (2001). On clustering validation techniques. *Journal of Intelligent Information Systems 17*, 107–145.

Hausdorf, B., & Hennig, C. (2010). Species delimitation using dominant and codominant multilocus markers. *Systematic Biology, 59*, 491–503.

Hennig, C. (2010). Methods for merging Gaussian mixture components. *Advances in Data Analysis and Classification, 4*, 3–34.

Jain, A. K. (2010). Data clustering: 50 years beyond K-means. *Pattern Recognition Letters, 31*, 651–666.

Kaufman, L., & Rousseeuw, P. J. (1990). *Finding Groups in Data*. New York: Wiley.

Morlini, I., & Zani, S. (2012). A new class of weighted similarity indices using polytomous variables. *Journal of Classification, 29*, 199–226.

Two-Step Linear Discriminant Analysis for Classification of EEG Data

Nguyen Hoang Huy, Stefan Frenzel, and Christoph Bandt

Abstract We introduce a multi-step machine learning approach and use it to classify electroencephalogram (EEG) data. This approach works very well for high-dimensional spatio-temporal data with separable covariance matrix. At first all features are divided into subgroups and linear discriminant analysis (LDA) is used to obtain a score for each subgroup. Then LDA is applied to these scores, producing the overall score used for classification. In this way we avoid estimation of the high-dimensional covariance matrix of all spatio-temporal features. We investigate the classification performance with special attention to the small sample size case. We also present a theoretical error bound for the normal model with separable covariance matrix, which results in a recommendation on how subgroups should be formed for the data.

1 Introduction

Fisher's classical linear discriminant analysis (LDA) is still one of the most widely used techniques for data classification. For two normal distributions with common covariance matrix Σ and different means μ_1 and μ_2, LDA classifier achieves the minimum classification error rate. The LDA score or discriminant function δ of an observation X is given by

$$\delta(X) = (X - \mu)^T \Sigma^{-1} \alpha \text{ with } \alpha = \mu_1 - \mu_2 \text{ and } \mu = \frac{1}{2}(\mu_1 + \mu_2).$$

N.H. Huy (✉) · S. Frenzel · C. Bandt
Department of Mathematics and Computer Science, University of Greifswald, Greifswald, Germany
e-mail: nhhuy@hua.edu.vn

M. Spiliopoulou et al. (eds.), *Data Analysis, Machine Learning and Knowledge Discovery*, Studies in Classification, Data Analysis, and Knowledge Organization,
DOI 10.1007/978-3-319-01595-8_6,

In practice we do not know Σ and μ_i, and have to estimate them from training data. This worked well for low-dimensional examples, but estimation of Σ for high-dimensional data turned out to be really difficult. When the sample size n of the training data is smaller than the number d of features, then the empirical $d \times d$ covariance matrix $\hat{\Sigma}$ is not invertible. The pseudoinverse can be used but this will impair the classification. Even when n is larger, but of the same magnitude as d, the aggregated estimation error over many entries of the sample covariance matrix will significantly increase the error rate of LDA. For our EEG data, where $200 \leq n \leq 3500$ and $160 \leq d \leq 1280$, these facts will be discussed below in Sect. 5 and Fig. 3.

One possible solution of the estimation problem is regularized LDA, where a multiple of the unity matrix I is added to the empirical covariance. $\hat{\Sigma} + rI$ is invertible for each $r > 0$. The most useful regularization parameter r has to be determined by time-consuming optimization, however.

Bickel and Levina (2004) recommended a simpler solution: to neglect all correlations of the features and use the diagonal matrix D_Σ of Σ instead of Σ. This is called the independence rule. Its discriminant function δ_I is defined by

$$\delta_I(X) = (X - \mu)^T D_\Sigma^{-1} \alpha.$$

In this paper, we present another solution, which uses some but not all correlations of the features and which worked very well for the case of spatio-temporal data, in the context of an experiment with a brain-computer interface.

2 Two-Step Linear Discriminant Analysis

We introduce multi-step linear discriminant analysis which applies LDA in several steps instead of applying it to all features at one time. Here we consider the case of two steps (two-step LDA). All d features of an observation $X \in \mathbb{R}^d$ are divided into disjoint subgroups

$$X = \left[X_1^T, \cdots, X_q^T\right]^T,$$

where $X_j \in \mathbb{R}^p$, and $pq = d$. LDA is applied to obtain a score for each subgroup of features. In the second step, LDA is again applied to these scores which gives the overall score used for classification. Thus the discriminant function of two-step LDA is

$$\delta^\star(X) = \delta(\delta(X_1), \cdots, \delta(X_q)),$$

where δ denotes the LDA function. Figure 1a illustrates the two-step LDA procedure. The assumption of normality which is needed for LDA will be fulfilled in the second step. The distribution of scores can be calculated applying basic linear algebra and the properties of the multivariate normal distribution.

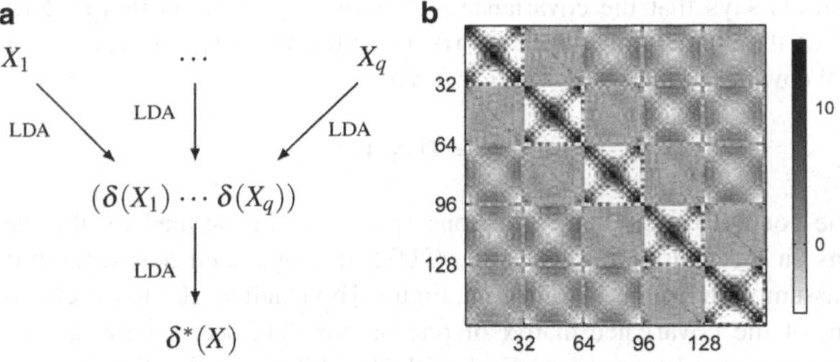

Fig. 1 (a) Schematic illustration of two-step LDA. (b) Sample covariance matrix of a single dataset estimated from 5 time points and 32 locations

Proposition 1. *Suppose X is normally distributed with known μ_1, μ_2 and Σ. Let $\mu_2 - \mu_1 = (\alpha_1^T, \ldots, \alpha_q^T)^T$ and $\Sigma_{ij} \in \mathbb{R}^{p\times p}$ denote the submatrix of Σ corresponding to subgroups i and j such that $\Sigma = (\Sigma_{ij})_{i,j=1}^q$. The scores $(\delta(X_1), \ldots, \delta(X_q))^T$ are then normally distributed with common covariance matrix Θ and means $\pm(1/2)m$ given by*

$$\Theta_{ij} = \alpha_i^T \Sigma_{ii}^{-1} \Sigma_{ij} \Sigma_{jj}^{-1} \alpha_j, \quad m_i = \Theta_{ii}, \quad \text{with } i, j = 1, \ldots, q.$$

3 Separable Models

Statistical modelling of spatio-temporal data often is based on separable models which assume that the covariance matrix of the data is a product of spatial and temporal covariance matrices. This greatly reduces the number of parameters in contrast to unstructured models. Genton (2007) argues that separable approximations can be useful even when dealing with non-separable covariance matrices.

A spatio-temporal random process $X(\cdot,\cdot) : S\times T \to \mathbb{R}$ with time domain $T \subset \mathbb{R}$ and space domain $S \subset \mathbb{R}^3$ is said to have a separable covariance function if, for all $s_1, s_2 \in S$ and $t_1, t_2 \in T$, it holds

$$\text{Cov}(X(s_1,t_1), X(s_2,t_2)) = u(t_1,t_2)\cdot v(s_1,s_2), \tag{1}$$

where u and v is the temporal and spatial covariance function, respectively. Suppose that the data from $X(\cdot,\cdot)$ is only selected at a finite set of locations $s_1, \ldots, s_p$ and time points $t_1, \ldots, t_q$. An observation for classification is obtained by concatenation of spatial data vectors at times $\{t_1, \cdots, t_q\}$

$$X = \left[X(s_1;t_1)\cdots X(s_p;t_1)\cdots X(s_1;t_q)\cdots X(s_p;t_q)\right]^T. \tag{2}$$

Equation (1) says that the covariance matrix of X can be written as Kronecker product of the spatial covariance matrix V with entries $v_{ij} \equiv v(s_i, s_j)$ and the temporal covariance matrix U with $u_{ij} \equiv u(t_i, t_j)$,

$$\Sigma = U \otimes V.$$

In the context of EEG, the locations $s_1, \ldots, s_p$ are defined by the electrode positions on the scalp. Huizenga et al. (2002) demonstrated that separability is a proper assumption for this kind of data. Figure 1b visualizes the Kronecker product structure of the covariance matrix of one of our data sets. There are $p = 32$ electrodes and $q = 5$ time points. Each of the five blocks on the diagonal represents the covariance between the electrodes for a single time point. The other blocks represent covariance for different time points.

4 An Error Bound for Two-Step LDA

In this section we derive a theoretical error estimate for two-step LDA in the case of separable models. The following theorem, illustrated in Fig. 2a, shows that the loss in efficiency of two-step LDA in comparison to ordinary LDA even in the worst case is not very large when the condition number of the temporal correlation matrix is moderate. The assumption that the means and covariance matrices are known may seem a bit unrealistic, but it is good to have such a general theorem. The numerical results in Sect. 5 will show that the actual performance of two-step LDA for finite samples is much better. To compare the error rate of δ and $\delta^{\star}$, we use the technique of Bickel and Levina (2004) who compared independence rule and LDA in a similar way.

Theorem 1. *Suppose that mean vectors μ_1, μ_2 and common separable covariance matrix $\Sigma = U \otimes V$ are known. Then the error rate e_2 of the two-step LDA fulfils*

$$e_1 \leq e_2 \leq \Phi\left(\frac{2\sqrt{\kappa}}{1+\kappa}\,\Phi^{-1}(e_1)\right), \tag{3}$$

where e_1 is the LDA error rate, $\kappa = \kappa(U_0)$ denotes the condition number of the temporal correlation matrix $U_0 = D_U^{-1/2} U D_U^{-1/2}$, $D_U = diag(u_{11}, \cdots, u_{qq})$, and Φ is the Gaussian cumulative distribution function.

Proof. $e_1 \leq e_2$ follows from the optimality of LDA. To show the other inequality, we consider the error $\bar{e}$ of the two-step discriminant function $\bar{\delta}$ defined by

$$\bar{\delta}(X) = \delta_I(\delta(X_1), \cdots, \delta(X_q)),$$

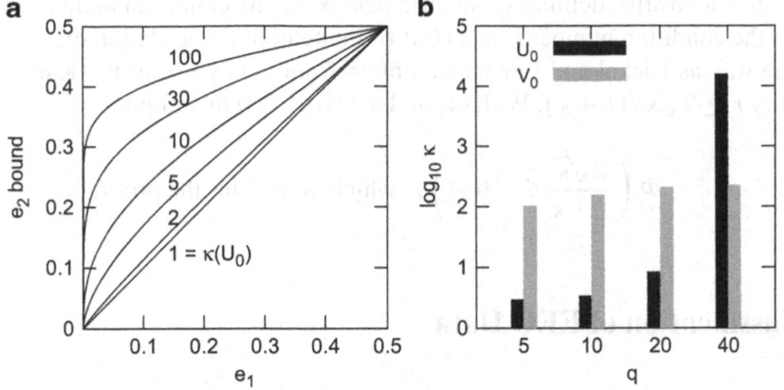

Fig. 2 **(a)** The error bound of two-step LDA as function of the LDA error rate. **(b)** Condition numbers of U_0 and V_0 estimated from a single dataset for different number of time points q

where δ_I is the discriminant function of the independence rule. The relation $e_2 \leq \bar{e}$ again follows from the optimality of LDA and Proposition 1. We complete the proof by showing that $\bar{e}$ is bounded by the right-hand side of (3), by the technique of Bickel and Levina (2004). We repeat their argument in our context, demonstrating how U_0 comes up in the calculation. We rewrite the two-step discriminant function $\bar{\delta}$ applied to the spatio-temporal features X with $\alpha = \mu_1 - \mu_2$ and $\mu = (\mu_1 + \mu_2)/2$

$$\bar{\delta}(X) = (X-\mu)^T \bar{\Sigma}^{-1}\alpha, \quad \text{where} \quad \bar{\Sigma} = D_U \otimes V = \begin{bmatrix} u_{11}V & \cdots & 0 \\ \vdots & \ddots & \vdots \\ 0 & \cdots & u_{qq}V \end{bmatrix}$$

The errors e_1 of $\delta(x)$ and $\bar{e}$ of $\bar{\delta}(x)$ are known, see Bickel and Levina (2004):

$$e_1 = \Phi\left(\frac{-(\alpha^T \Sigma^{-1}\alpha)^{1/2}}{2}\right), \quad \bar{e} = \Phi\left(\frac{-\alpha^T \bar{\Sigma}^{-1}\alpha}{2(\alpha^T \bar{\Sigma}^{-1}\Sigma\bar{\Sigma}^{-1}\alpha)^{1/2}}\right).$$

Writing $\alpha_0 = \bar{\Sigma}^{-1/2}\alpha$, we determine the ratio

$$r = \frac{\Phi^{-1}(\bar{e})}{\Phi^{-1}(e_1)} = \frac{(\alpha_0^T\alpha_0)}{[(\alpha_0^T\tilde{\Sigma}\alpha_0)(\alpha_0^T\tilde{\Sigma}^{-1}\alpha_0)]^{1/2}}, \tag{4}$$

where

$$\begin{aligned}\tilde{\Sigma} = \bar{\Sigma}^{-1/2}\Sigma\bar{\Sigma}^{-1/2} &= (D_U^{-1/2} \otimes V^{-1/2})(U \otimes V)(D_U^{-1/2} \otimes V^{-1/2}) \\ &= (D_U^{-1/2}UD_U^{-1/2}) \otimes (V^{-1/2}VV^{-1/2}) = U_0 \otimes I.\end{aligned}$$

Clearly $\tilde{\Sigma}$ is a positive definite symmetric matrix and its condition number $\kappa(\tilde{\Sigma})$ is equal to the condition number $\kappa = \kappa(U_0)$ of the temporal correlation matrix U_0. In the same way as Bickel and Levina we obtain from (4) by use of the Kantorovich inequality $r \geq 2\sqrt{\kappa}/(1+\kappa)$. With (4) and $\Phi^{-1}(e_1) < 0$ this implies

$$\bar{e} \leq \Phi\left(\frac{2\sqrt{\kappa}}{1+\kappa}\,\Phi^{-1}(e_1)\right), \quad \text{which completes the proof.}$$

5 Classification of EEG Data

To check the performance of two-step LDA, we use the data of a brain-computer interface experiment by Frenzel et al. (2011). A mental typewriter was established using 32-electrode EEG. Users sit in front of a screen which presents a matrix of characters. They are instructed to concentrate on one target character by performing a mental count. Then characters are highlighted many times in random order. About 300 ms after highlighting the target character, a so-called event-related potential should appear in the EEG signal. This potential should not appear for the other characters.

The experiment intended to control the effect of eye view. Users were told to concentrate their eyes on a specific character. When this is the target character, the condition is described as overt attention, and the expected potential is fairly easy to identify. However, most of the time users looked at a different character and counted the target character in their visual periphery. This is referred to as covert attention, see Treder and Blankertz (2010). Controlling a brain-computer interface by covert attention is particularly difficult.

Detection of target characters from the EEG data is a binary classification problem. Each time interval where a character was highlighted is considered as a sample. Class labels are defined according to whether the target character is presented or not. Our data consists of nine datasets of $m = 7,290$ samples measured with $p = 32$ electrodes. For each sample, data of the time interval of about 600 ms were downsampled from the acquisition rate of the hardware to a predefined sampling frequency. For typical values of 8, 16, 32, 64 Hz one obtains $q = 5, 10, 20, 40$ time points and thus $d = pq = 160, 320, 640, 1280$ spatio-temporal features in total.

Defining the Feature Subgroups of Two-step LDA

LDA is invariant with respect to reordering of features whereas two-step LDA is only when reordering is performed within the subgroups. For the latter we saw that it is preferable to define the subgroups such that the statistical dependencies between them are smaller than within. This is reflected in the influence of condition number of U_0 in the bound of the error rate (3).

In Sect. 3 we defined the features to be ordered according to their time index, see (2), and X_i to contain all features at time point i. In other words, in the first step LDA was applied to the spatial features. However, it also seems natural to order the features according to their spatial index and to assign all features from electrode i to X_i, thus interchanging the role of space and time. In this case the covariance matrix becomes $V \otimes U \neq U \otimes V$ and we have to replace $\kappa(U_0)$ by $\kappa(V_0)$ in (3). We argue that the decision between both approaches should be based on a comparison of both condition numbers using the data. This done in the following.

Our EEG data is, rather typical, normalized such that the means over all time points and all electrodes are zero. This implies U and V both to have a single zero eigenvalue. Maximum-likelihood estimation of both in general requires their inverses to exist, see Mitchell et al. (2006). We bypassed this problem by using the simple average-based estimator

$$\hat{V} = \frac{1}{q} \sum_{i=1}^{q} \hat{\Sigma}_{ii} ,$$

where $\hat{\Sigma}_{ii}$ is the sample covariance matrix of the i-th subgroup. It can be shown that $\hat{V}$ is an unbiased and consistent estimator of $\bar{\lambda} V$ with $\bar{\lambda}$ being the average eigenvalue of U. Since the correlation matrix corresponding to $\bar{\lambda} V$ is V_0 we estimated $\kappa(V_0)$ by $\kappa(\hat{V}_0)$, ignoring the single zero eigenvalue. Estimation of $\kappa(U_0)$ was done in the same way.

Figure 2b shows the condition numbers estimated from a single dataset for different number of time points q. Except for $q = 40$ the condition numbers of U_0 were much smaller than those of V_0. This also applied for the corresponding error bounds, see Fig. 2a. It is thus likely that the actual error rates are smaller. Indeed, we never encountered a single case where first applying LDA to the temporal features gave better results for our data. For $q = 40$ the upper bounds were too loose to draw any conclusions. This could be observed in all nine datasets and gives rise to the following recommendation.

Remark 1. When two-step LDA is applied to EEG data it is preferable to define the feature subgroups such that X_i contains all features at time point i.

Learning Curves

We investigated the classification performance using $p = 32$ electrodes and $q = 40$ time points and hence $d = 1280$ features in total. Two-step LDA was compared to ordinary and regularized LDA. For each dataset classifiers were trained using the first n samples, with $200 \leq n \leq 3500$. Scores of the remaining $m - n$ samples were calculated and classification performance was measured by the AUC value, i.e. the relative frequency of target trials having a larger score than non-target ones.

Figure 3 shows the learning curves for all nine datasets. The prominent dip in the learning curves of LDA around d is due to the use of the pseudoinverse

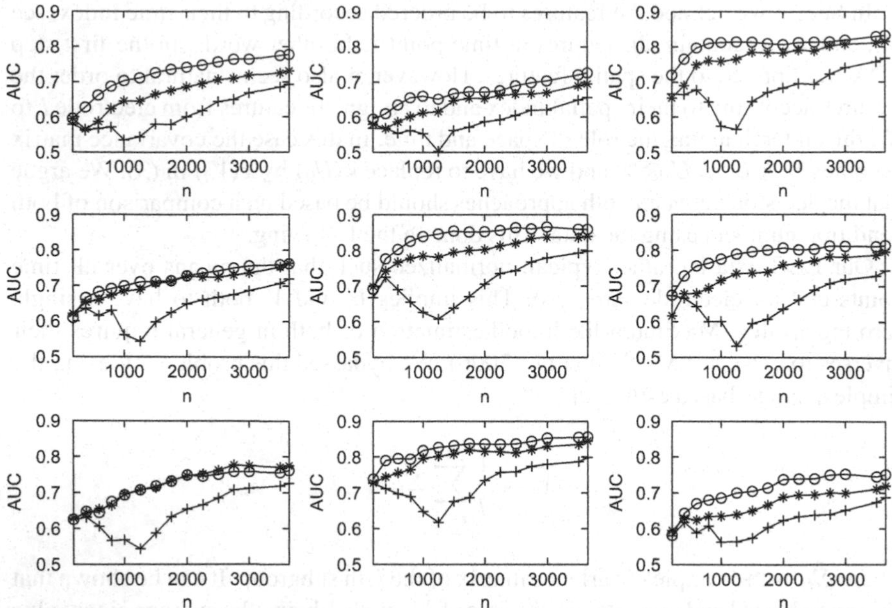

Fig. 3 Learning curves of two-step LDA (*open circle*), regularized LDA (*asterisk*) and LDA (*plus*) for all nine datasets

for $n < d + 2$. Regularized LDA for $n \approx d$ performed much better than LDA, supporting the findings of Blankertz et al. (2011).

Two-step LDA showed similar or slightly better performance than both regularized and ordinary LDA. For large n the difference was rather small. For some datasets, however, it showed much faster convergence, i.e. it needed less training samples to achieve a certain classification performance. Sample size $n = 3500$ corresponds to a training period of approximately 20 min. Since two-step LDA gave reasonable performance even with short training periods, it might offer an practically relevant advantage. Although all three classifiers are computationally cheap, it should be noted that two-step LDA does not involve the inversion of the full sample covariance matrix.

6 Conclusion

When linear discriminant analysis is applied to high-dimensional data, it is difficult to estimate the covariance matrix. We introduced a method which avoids this problem by applying LDA in two steps for the case of spatio-temporal data. For our EEG data, the two-step LDA performed better than regularized LDA.

References

Bickel, P. J. & Levina, E. (2004). Some theory of Fisher's linear discriminant function, 'naive Bayes', and some alternatives when there are many more variables than observations. *Bernoulli, 10*, 989–1010.

Blankertz, B., Lemm, S., Treder, M., Haufe, S., & Müller, K. R. (2011). Single-trial analysis and classification of ERP components – A tutorial. *NeuroImage, 56*, 814–825.

Frenzel, S., Neubert, E., & Bandt, C. (2011). Two communication lines in a 3 × 3 matrix speller. *Journal of Neural Engineering, 8*, 036021.

Genton, M. G. (2007). Separable approximation of space-time covariance matrices. *Environmetrics, 18*, 681–695.

Huizenga, H. M., De Munck, J. C., Waldorp, L. J., & Grasman, R. P. P. P. (2002). Spatiotemporal EEG/MEG source analysis based on a parametric noise covariance model. *IEEE Transactions on Biomedical Engineering, 49*, 533–539.

Mitchell, M. W., Genton, M. G., & Gumpertz, M. L. (2006). A likelihood ratio test for separability of covariances. *Journal of Multivariate Analysis, 97*, 1025–1043.

Treder, M. S. & Blankertz, B. (2010). (C)overt attention and visual speller design in an ERP-based brain-computer interface. *Behavioral and Brain Functions, 6*, 28.

Predictive Validity of Tracking Decisions: Application of a New Validation Criterion

Florian Klapproth, Sabine Krolak-Schwerdt, Thomas Hörstermann, and Romain Martin

Abstract Although tracking decisions are primarily based on students' achievements, distributions of academic competences in secondary school strongly overlap between school tracks. However, the correctness of tracking decisions usually is based on whether or not a student has kept the track she or he was initially assigned to. To overcome the neglect of misclassified students, we propose an alternative validation criterion for tracking decisions. We applied this criterion to a sample of $N = 2{,}300$ Luxembourgish 9th graders in order to identify misclassifications due to tracking decisions. In Luxembourg, students in secondary school attend either an academic track or a vocational track. Students' scores of academic achievement tests were obtained at the beginning of 9th grade. The test-score distributions, separated by tracks, overlapped to a large degree. Based on the distributions' intersection, we determined two competence levels. With respect to their individual scores, we assigned each student to one of these levels. It turned out that about 21 % of the students attended a track that did not match their competence level. Whereas the agreement between tracking decisions and actual tracks in 9th grade was fairly high ($\kappa = 0.93$), the agreement between tracking decisions and competence levels was only moderate ($\kappa = 0.56$).

1 Introduction

Tracking in school refers to the ability-based selection of students and their assignment to different school tracks. In educational systems with hierarchical tracks in secondary school (like in Germany or Luxembourg), tracking decisions are

F. Klapproth (✉) · S.Krolak-Schwerdt · T. Hörstermann · R. Martin
University of Luxembourg, Route de Diekirch, 7220 Walferdange, Luxembourg
e-mail: florian.klapproth@uni.lu; sabine.krolak@uni.lu;
thomas.hoerstermann001@student.uni.lu; romain.martin@uni.lu

M. Spiliopoulou et al. (eds.), *Data Analysis, Machine Learning and Knowledge Discovery*, Studies in Classification, Data Analysis, and Knowledge Organization,
DOI 10.1007/978-3-319-01595-8_7,

mainly based on students' achievements in primary school. Outcomes of tracking decisions determine the type of future schooling students will be permitted to attend, and therefore the type of education they will receive. Measuring the correctness of tracking decisions (hence, their predictive validity) usually is based on whether or not a student has kept the track to that she or he has been initially assigned. According to this criterion, a high amount of predictive validity will be achieved if the quality of the tracking decision is high, that is, if teachers are able to validly predict students' future achievements in school. However, predictive validity will also reach high values if the school system's permeability is rather low. In the latter case, changing tracks is impeded because schools try to keep their students (regardless of their achievements). Thus, estimation of predictive validity of tracking decisions may be biased if it is based on the frequencies of students who keep the track after some years of schooling.

Although tracking decisions are primarily based on students' achievements, the distributions of their academic competences in secondary school have been shown to strongly overlap between different school tracks (e.g., Bos and Gröhlich 2010). That is, students with high academic competences are often assigned to lower school tracks, whereas students with low academic competences often attend higher school tracks. Hence, students can be regarded as being misclassified if their academic competences do not match their actual track.

To overcome the neglect of misclassified students, we propose an alternative validation criterion for tracking decisions which is based on standardized achievement test scores rather than on keeping the track. In the present study, we applied this validation criterion to a sample of $N = 2{,}300$ Luxembourgish 9th graders in order to examine the degree of misclassification due to tracking decisions.

2 Development of the Validation Criterion

The starting point for the development of the validation criterion is the distribution of test scores that were obtained from students' results in standardized academic achievement tests in secondary school (see Fig. 1).

In Fig. 1, distributions of test scores of students coming from a lower track (the vocational track) and from a higher track (the academic track) are displayed. As can be seen in the figure, both distributions overlap. The test score that is assigned to the distributions' intersection shall be called x_{inter}. This test score divides the test score continuum into two parts. Each part of the test-score continuum represents one of two competence levels that will later on serve as the validation criterion. We shall call one competence level the *VT* competence level (in analogy to the vocational track) and the other competence level the *AT* competence level (in analogy to the academic track). Students' test scores of the *VT* competence level occur more frequently within the vocational track than within the academic track. Correspondingly, test scores of the *AT* competence level occur more frequently within the academic track than within the vocational track.

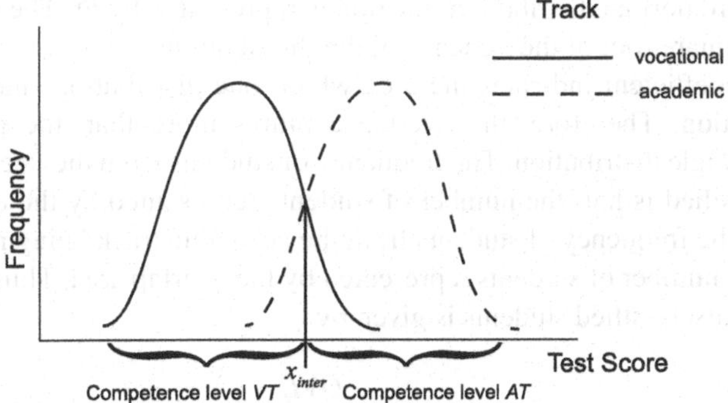

Fig. 1 Hypothetical distributions of test scores with equal variances, separated for different tracks. *Continuous line*: Test score distribution obtained from students of the vocational track. *Intermittent line*: Test score distribution obtained from students of the academic track. x_{inter} marks the intersection of the distributions. The distributions' intersection defines the competence levels

If the variances of the test-score distributions are equal, the intersection of the distributions can be estimated as

$$x_{inter} = \frac{\mu_1 + \mu_2}{2}, \tag{1}$$

with μ_1 being the mean of the distribution of the vocational track, μ_2 being the mean of the distribution of the academic track, and $\mu_1 < \mu_2$.

Students with test scores $x_i < x_{inter}$ are assigned to the *VT* competence level, whereas students with test scores $x_i > x_{inter}$ are assigned to the *AT* competence level. We define students being correctly classified as those students whose actual track matches their competence level. Correspondingly, students are considered misclassified if they obtained test scores that fall into a competence level that does not fit their actual track. That is, students of the vocational track are misclassified if their test scores are captured by the *AT* competence level, and vice versa, students of the academic track are misclassified if they obtained test scores falling into the *VT* competence level. The number of misclassifications can be expressed formally by the area under the distributions that is defined by the overlap and the intersection. According to Inman and Bradley (1989), the overlap *OVL* of two normal distributions obtained from independent random samples with equal variances is given by

$$OVL = 2\Phi\left(-\frac{1}{2}abs\left(\delta\right)\right), \tag{2}$$

with

$$\delta = \frac{\mu_1 - \mu_2}{\sigma}, \tag{3}$$

and the standard normal distribution function is represented by Φ. The value of σ is given by the square root of the variance of the distributions.

The *OVL* coefficient indicates the area which one distribution shares with the other distribution. Therefore, the overlap captures more than the misclassified students of a single distribution. The frequency of students from the vocational track being misclassified is half the number of students represented by the overlap area. Accordingly, the frequency of students from the academic track being misclassified is also half the number of students represented by the overlap area. Thus, the overall frequency of misclassified students is given by

$$f_{misclass} = \frac{OVL}{2}. \tag{4}$$

So far, we have shown how the number of misclassified students can be estimated in case their test scores are distributed with equal variances. However, this approach is also feasible with distributions of unequal variances. If the variances are unequal, there will result two points of intersection. These points are defined by

$$x_{inter_{1,2}} = -\frac{p}{2} \pm \sqrt{\left(\frac{p}{2}\right)^2 - q}, \tag{5}$$

with

$$p = \frac{2\left(\mu_2\sigma_1^2 - \mu_1\sigma_2^2\right)}{\sigma_2^2 - \sigma_1^2} \tag{6}$$

and

$$q = \frac{\mu_1^2\sigma_2^2 - \mu_2^2\sigma_1^2 - 2\ln\left(\frac{\sigma_2}{\sigma_1}\right)\sigma_1^2\sigma_2^2}{\sigma_2^2 - \sigma_1^2}, \tag{7}$$

with μ_1 and σ_1 as parameters of the distribution of test scores obtained from students of the vocational track, and μ_2 and σ_2 as parameters of the distribution of test scores obtained from students of the academic track.

As can be seen in Fig. 2, both intersections define three (instead of two) competence levels: Competence level AT_1 where test scores are more likely for students from the academic track, competence level VT where test scores are more likely for students from the vocational track, and competence level AT_2 where again test scores are more likely for students from the academic track. However, the competence level AT_2 which contains the smallest test scores is of no relevance for our purpose since surely all students of the academic track falling into this category would be considered as being misclassified since their test scores are below x_{inter_2}. According to Inman and Bradley (1989), the overlap of two normal distributions with unequal variances is given by

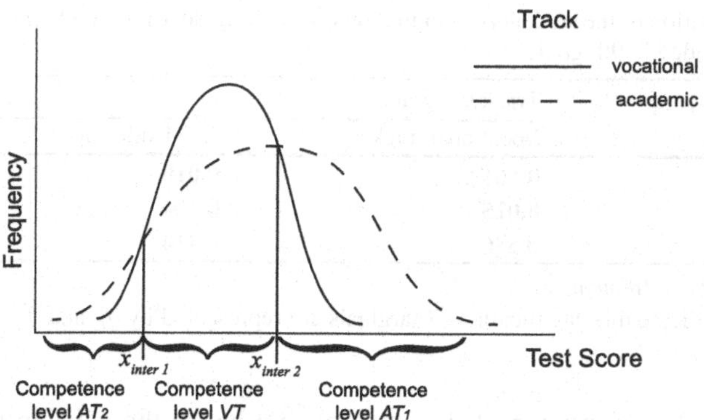

Fig. 2 Hypothetical distributions of test scores with unequal variances, separated for different tracks. *Continuous line*: Test score distribution obtained from students of the vocational track. *Intermittent line*: Test score distribution obtained from students of the academic track. x_{inter_1} and x_{inter_2} mark the intersections of the distributions. The distributions' intersections define the competence levels VT, AT_1 and AT_2

$$OVL = \Phi\left(\frac{x_{inter_1} - \mu_1}{\sigma_1}\right) + \Phi\left(\frac{x_{inter_2} - \mu_2}{\sigma_2}\right) - \Phi\left(\frac{x_{inter_1} - \mu_2}{\sigma_2}\right) - \Phi\left(\frac{x_{inter_2} - \mu_1}{\sigma_1}\right) + 1. \quad (8)$$

However, the entire overlap captures not all misclassified students, since students from the academic track whose test scores are assigned to competence level AT_2 are not part of the overlap. Equation (9) provides an estimate of the proportion of students being misclassified if the variances of the distributions differ.

$$f_{misclass} = \frac{n_2}{n_1 + n_2}\Phi\left(\frac{x_{inter_2} - \mu_2}{\sigma_2}\right) + \frac{n_1}{n_1 + n_2}\left(1 - \Phi\left(\frac{x_{inter_2} - \mu_1}{\sigma_1}\right)\right), \quad (9)$$

with n_1 representing the sample size of the distribution of students from the vocational track, and n_2 representing the sample size of the distribution of students from the academic track. The parameters μ and σ are estimated by the distributions' means and standard deviations.

3 Application of the Validation Criterion to a Representative Sample of Luxemburgish Secondary-School Students

In Luxembourg, students were oriented to one of three tracks after 6th grade of primary school. These tracks are one academic track and two vocational tracks. For the sake of simplicity, we treated students of both vocational tracks as being

Table 1 Distribution of the students depending on the tracking decision in 6th grade and the track the students attended in 9th grade

Tracking decision	Track 9th grade		
6th grade	Vocational track	Academic track	$f_{i.}$
Vocational track	0.561	**0.010**	0.571
Academic track	**0.025**	0.404	0.429
$f_{.j}$	0.586	0.414	1.000

Numbers are relative frequencies
Bold numbers indicate misclassifications. Marginals are represented by $f_{i.}$ and $f_{.j}$

in a single track. Orientation of the students and hence the split-up of the whole student population was done according to the tracking decisions at the end of primary school. These decisions were primarily based on the students' academic achievements in the main curricular fields. On average, students assigned to the vocational track show lower achievements than students assigned to the academic track. In our sample, $n_1 = 1{,}346$ students (59 %) attended the vocational track, whereas $n_2 = 954$ students (41 %) attended the academic track in 9th grade. Table 1 shows the distribution of the students depending on the initial tracking decision in 6th grade and the track they attended in 9th grade.

Of all students, 3.5 % attended tracks for which they were not recommended. This figure represents the number of misclassified students according to the usual validation criterion. As an indicator of academic competence, we resorted to results obtained from standardized academic achievement tests that were administered in all 9th grades of Luxemburgish secondary school in 2011. The test items were constructed with respect to standards proposed by the Luxemburgish Ministry of Education, and they covered contents of the main curricular fields (French, German, Mathematics) of all tracks (Ministere de l'Education nationale et de la Formation professionelle 2010). Although different versions of the test were administered depending on the different tracks, common scales of the versions were established after items response theory (e.g., Drasgow and Hulib 1990). Thus, according to the curricular fields, three competence scales resulted, with a mean of each competence scale that was set to $M = 500$, and a standard deviation of $SD = 100$. For the purpose of this study, test scores of the three competence scales were averaged. Hence, the averaged test scores represented the global academic competence of the students. Figure 3 shows the distribution of the students' mean test scores plotted against the different tracks.

The means and standard deviations were $M_1 = 480.79$ ($SD_1 = 61.03$) for the distribution of test scores obtained from the vocational track, and $M_2 = 580.21$ ($SD_2 = 62.59$) for the distribution of test scores obtained from the academic track. The difference between both means was significant, $p < 0.001$. According to Kolmogorov–Smirnov tests, both distributions conformed to normality ($Z_1 = 0.62$, $p = 0.838$; $Z_2 = 1.11$, $p = 0.170$). Estimation of the distributions' intersection was done after Equation (1) since the variances of both distributions did not differ significantly, $F(1312, 986) = 0.951$, $p > 0.21$. The test score belonging to the

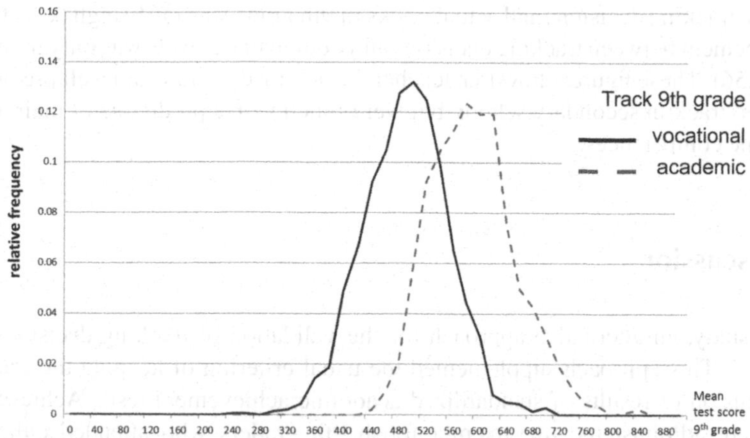

Fig. 3 Distributions of the students' mean test scores plotted against the different tracks

Table 2 Distribution of the students depending on the tracking decision in 6th grade, the track the students attended in 9th grade, and the competence level

Tracking decision 6th grade	Competence level	Track 9th grade: Vocational track	Academic track	$f_{i.}$
Vocational	VT	0.456	**0.006**	0.571
Track	AT	**0.105**	0.004	
Academic	VT	0.015	**0.091**	0.429
Track	AT	**0.010**	0.313	
$f_{.j}$		0.586	0.414	1.000

Numbers are relative frequencies
Bold numbers indicate misclassifications. Marginals are represented by $f_{i.}$ and $f_{.j}$

intersection was $x_{inter} = 530.8$. All students who obtained test scores that were below x_{inter} were assigned to competence level VT, and all students who obtained test scores above x_{inter} were assigned to competence level AT. Next, the area of overlap of both distributions was estimated after Equation (2). For estimation of δ, the mean of the distributions' standard deviations was used. The overlap was $OVL = 0.422$, from which followed that the number of misclassified students was $f_{misclass} = 0.211$. That is, about 21 % of the students were on the wrong track. Table 2 shows the number of students of each competence level, plotted against the tracking decision and the track they attended in 9th grade. As can be seen, the frequency of the misclassified students estimated by Equation (4) matched the number of misclassified students given in Table 2.

Finally, we estimated the predictive validity of the tracking decision. We did this first by measuring the level of agreement between the initial tracking decision and the actual track the students attended in 9th grade, and second by measuring the level of agreement between the tracking decision and the competence level. Cohen's kappa provided an indicator of agreement (Cohen 1960). Whereas the agreement between tracking decisions and actual tracks in 9th grade was fairly high ($\kappa = 0.93$), the agreement between tracking decisions and competence levels was only moderate ($\kappa = 0.56$). These figures show that teachers' tracking decisions did well predict the students' track in secondary school, but were much worse predictors of their actual academic competences.

4 Discussion

In this study, an alternative approach for the validation of tracking decisions was proposed. This approach supplemented the usual criterion of keeping the track by the inclusion of results of standardized academic achievement tests. Achievement tests were administered to Luxembourgish 9th graders who attended either the lower (vocational) track or the higher (academic) track. Test scores were then classified into two categories according to the intersection of the distributions of test scores obtained from students of either track. Since the categories represented the competence of students that was more likely for students from one track compared to students from the opposite track, we treated these categories as competence levels pertinent to a certain track. We presented a formula according to which the frequency of misclassified students can be estimated, if the test-score distributions are normal and the parameters (μ, σ, n) of the test-score distributions are known. When applied to the sample of Luxembourgish 9th graders, it turned out that about 21 % of the students were on the wrong track.

The construction of standardized scholastic achievement tests is often guided by the aim of achieving normality of the test score distributions. Provided that these distributions are normal, the formal approach allows for a more precise estimation of the intersection of the distributions and hence of the distributions' overlap than if one tries to gauge the intersection by visual inspection of the curves. Our approach is feasible if the distributions are normal and completely specified. However, alternative approaches relaxing this assumption might also be developed.

The discrepancy between the tracks students actually attend, and their academic competence demonstrates that the validation of tracking decisions must go beyond the assessment of the level of agreement between tracking decisions and actual tracks. Therefore, we additionally estimated the validity of tracking decisions as the level of agreement between tracking decisions and the competence level a student was assigned to according the intersection of the test-score distributions. Since the latter was much lower than the former, we suppose that success and failure of students in secondary school partly depend on factors that are not related to student's actual academic achievement. This supposition is confirmed by empirical evidence. For instance, it has been shown that teacher's evaluations of students are affected by their expectations regarding prior achievements of the students (Jussim and Eccles 1992). Moreover, teachers frequently stick to their evaluations of students even if students' achievements obviously have changed (e.g., Schlechty 1976).

Of course, test results in 9th grade do not only reflect the actual competence of the students, but are also affected by the impact a single track had on the students' individual development. However, the comparison of the test results obtained from students of different tracks is nevertheless useful since it indicates whether at the time of testing students exhibit achievements which are below or above the general level, or fit well the general level, despite the impact of their corresponding track.

We conclude that the discrepancy between tracking and students' academic competences reflects both a limited ability of teachers to validly predict students academic achievements in secondary school as well as a low permeability of the Luxembourgish school system. Both factors may prevent the students' academic advancement and therefore decrease their chances for occupational success.

Acknowledgements This study was supported by grant F3R-LCM-PFN-11PREV from the Fonds National de la Recherche Luxembourg. We are grateful to Martin Brunner who provided us with the test results.

References

Bos, W., & Gröhlich, C. (2010). *Kompetenzen und Einstellungen von Schülerinnen und Schülern am Ende der Jahrgangsstufe* (Vol. 8). Münster: Waxmann.

Cohen, J. (1960). A coefficient of agreement for nominal scales. *Educational and Psychological Measurement, 20*, 37–46.

Drasgow, F., & Hulib, C. L. (1990). Item response theory. In M. D. Dunnette, & L. M. Hough (Eds.), *Handbook of industrial and organizational psychology* (Vol. 1). Mountain View: Davies-Black Pub.

Inman, H. F., & Bradley, E. L. (1989). The overlapping coefficient as a measure of agreement between probability distributions and point estimation of the overlap of two normal densities. *Communications in Statistics–Theory and Methods, 18*, 3851–3874.

Jussim, L., & Eccles, J. S. (1992). Teacher expectations II: Construction and reflection of student achievement. *Journal of Personality and Social Psychology, 63*, 947–961.

Ministere de l'Education Nationale et de la Formation Professionelle (2010). *Nationaler Bericht Epreuves Standardisees (EpStan)*. Luxembourg: MENFP.

Schlechty, P. C. (1976) *Teaching and social behavior: Toward an organizational theory of instruction.* Boston: Ally and Bacon.

DDα-Classification of Asymmetric and Fat-Tailed Data

Tatjana Lange, Karl Mosler, and Pavlo Mozharovskyi

Abstract The DDα-procedure is a fast nonparametric method for supervised classification of d-dimensional objects into $q \geq 2$ classes. It is based on q-dimensional depth plots and the α-procedure, which is an efficient algorithm for discrimination in the depth space $[0, 1]^q$. Specifically, we use two depth functions that are well computable in high dimensions, the zonoid depth and the random Tukey depth, and compare their performance for different simulated data sets, in particular asymmetric elliptically and t-distributed data.

1 Introduction

Classical procedures for supervised learning like LDA or QDA are optimal under certain distributional settings. To cope with more general data, nonparametric methods have been developed. A point may be assigned to that class in which it has maximum depth (Ghosh and Chaudhuri 2005; Hoberg and Mosler 2006). Moreover, data depth is suited to reduce the dimension of the data and aggregate their geometry in an efficient way. This is done by mapping the data to a depth-depth (DD) plot or, more generally, to a DD-space: a unit cube of dimension $q \geq 2$, where each axis indicates the depth w.r.t. a certain class (e.g. see Fig. 1, left and middle). A proper classification rule is then constructed in the DD-space, see Li et al. (2012). In Lange et al. (2012) the DDα-classifier is introduced, which employs a

T. Lange (✉)
Hochschule Merseburg, Geusaer Straße, 06217 Merseburg, Germany
e-mail: tatjana.lange@hs-merseburg.de

K. Mosler · P. Mozharovskyi
Universität zu Köln, 50923 Köln, Germany
e-mail: mosler@statistik.uni-koeln.de; mozharovskyi@statistik.uni-koeln.de

M. Spiliopoulou et al. (eds.), *Data Analysis, Machine Learning and Knowledge Discovery*, Studies in Classification, Data Analysis, and Knowledge Organization, DOI 10.1007/978-3-319-01595-8_8,

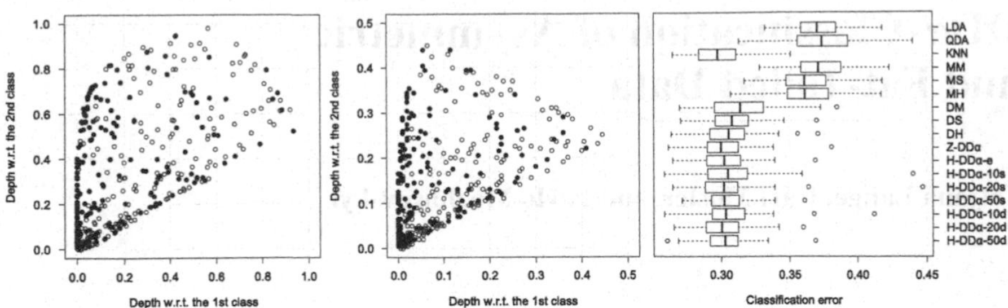

Fig. 1 DD-plots using zonoid (*left*) and location (*middle*) depth with *black* and *open circles* denoting observations from the two different classes and combined box-plots (*right*) for Gaussian location alternative

modified version of the α-procedure (Vasil'ev 1991, 2003; Vasil'ev and Lange 1998; Lange and Mozharovskyi 2012) for classification in the DD-space. For other recent depth-based approaches, see Dutta and Ghosh (2012a), Dutta and Ghosh (2012b), and Paindaveine and Van Bever (2013); all need intensive computations.

However, when implementing the DDα-classifier a data depth has to be chosen and the so called "outsiders" have to be treated in some way. (An *outsider* is a data point that has depth 0 in each of the classes.) The present paper also addresses the question which notion of data depth should be employed. To answer it we consider two depth notions that can be efficiently calculated in higher dimensions and explore their sensitivity to fat-tailedness and asymmetry of the underlying class-specific distribution. The two depths are the zonoid depth (Koshevoy and Mosler 1997; Mosler 2002) and the location depth (Tukey 1975). The zonoid depth is always exactly computed, while the location depth is either exactly or approximately calculated. In a large simulation study the average error rate of different versions of the DDα-procedure is contrasted with that of standard classifiers, given data from asymmetric and fat-tailed distributions. Similarly the performance of different classifiers is explored depending on the distance between the classes, and their speed both at the training and classification stages is investigated. We restrict ourselves to the case $q = 2$, see Lange et al. (2012) for $q > 2$. Outsiders are randomly assigned with equal probabilities; for alternative treatments of outsiders, see Hoberg and Mosler (2006) and Lange et al. (2012).

The rest of the paper is organized as follows. Section 2 shortly surveys the depths notions used in the DDα-procedure and their computation. In Sect. 3 the results of the simulation studies are presented and analyzed, regarding performance, performance dynamics and speed of the proposed classifiers. Section 4 concludes.

2 Data Depths for the DDα-Classifier

Briefly, a data depth measures the centrality of a given point $\mathbf{x}$ in a data set X in $\mathbb{R}^d$; see e.g. Zuo and Serfling (2000) and Mosler (2002) for properties.

2.1 *Zonoid* vs. *Location Depth*

In the sequel we employ two data depths that can be efficiently computed for high-dimensional data ($d = 20$ and higher): the zonoid depth and the location depth. The computational aspect plays a significant role here as the depths have to be calculated for each point of each class at the training stage, and they still have to be computed for a new point w.r.t. each class at the classification stage.

Zonoid depth. For a point $\mathbf{x}$ and a set $X = \{\mathbf{x}^1, \dots, \mathbf{x}^n\} \in \mathbb{R}^d$ the zonoid depth is defined as

$$ZD(\mathbf{x}, X) = \begin{cases} \sup\{\alpha : \mathbf{x} \in ZD_\alpha(X)\} & \text{if } \mathbf{x} \in ZD_\alpha(X) \text{ for some } 0 < \alpha \le 1, \\ 0 & \text{otherwise,} \end{cases} \tag{1}$$

with the zonoid region (2)

$$ZD_\alpha(X) = \left\{ \sum_{i=1}^{n} \lambda_i \mathbf{x}^i \ : 0 \le \lambda_i \le \frac{1}{n\alpha}, \sum_{i=1}^{n} \lambda_i = 1 \right\} ; \tag{2}$$

see Koshevoy and Mosler (1997) and Mosler (2002) for properties.
Location depth, also known as halfspace or Tukey depth, is defined as

$$HD(\mathbf{x}, X) = \frac{1}{n} \min_{\mathbf{u} \in S^{d-1}} \sharp\{i : \langle \mathbf{x}^i, \mathbf{u} \rangle \ge \langle \mathbf{x}, \mathbf{u} \rangle\}, \tag{3}$$

where $\langle \cdot, \cdot \rangle$ denotes the inner product. The location depth takes only discrete values and is robust (having a large breakdown point), while the zonoid depth takes all values in the set $\{[\frac{1}{n}, 1] \cup \{0\}\}$, is maximum at the mean $\frac{1}{n} \sum_i \mathbf{x}^i$, and therefore less robust. For computation of the zonoid depth we use the exact algorithm of Dyckerhoff et al. (1996).

2.2 *Tukey Depth* vs. *Random Tukey Depth*

The location depth can be exactly computed or approximated. *Exact computation* is described in Rousseeuw and Ruts (1996) for $d = 2$ and in Rousseeuw and Struyf (1998) for $d = 3$. For bivariate data we employ the algorithm of Rousseeuw and Ruts (1996) as implemented in the R-package "depth". In higher dimensions exact computation of the location depth is possible (Liu and Zuo 2012), but the algorithm involves heavy computations. Cuesta-Albertos and Nieto-Reyes (2008) instead propose to approximate the location depth, using (3), by minimizing the univariate location depth over randomly chosen directions $u \in S^{d-1}$. Here we explore two different settings where the set of randomly chosen u is either *generated once and for all* or *generated instantly* when computing the depth of a given point. By construction, the random Tukey depth is always greater or equal to the exact location depth. Consequently, it yields fewer outsiders.

3 Simulation Study

A number of experiments with simulated data is conducted. Firstly, the error rates of 17 different classifiers (see below) are evaluated on data from asymmetric t- and exponential distributions in $\mathbb{R}^2$. Then the performance dynamics of selected ones is visualized as the classification error in dependence of the Mahalanobis distance between the two classes. The third study explores the speed of solutions based on the zonoid and the random Tukey depth.

3.1 Performance Comparison

While the usual multivariate t-distribution is elliptically symmetric, it can be made asymmetric by conditioning its scale on the angle from a fixed direction, see Frahm (2004). For each degree of freedom, ∞ (= Gaussian), 5 and 1 (= Cauchy), two alternatives are investigated: one considering differences in location only (with $\mu_1 = \left[\begin{smallmatrix}0\\0\end{smallmatrix}\right]$, $\mu_2 = \left[\begin{smallmatrix}1\\1\end{smallmatrix}\right]$ and $\Sigma_1 = \Sigma_2 = \left[\begin{smallmatrix}1&1\\1&4\end{smallmatrix}\right]$) and one differing in both location and scale (with the same μ_1 and μ_2, $\Sigma_1 = \left[\begin{smallmatrix}1&1\\1&4\end{smallmatrix}\right]$, $\Sigma_2 = \left[\begin{smallmatrix}4&4\\4&16\end{smallmatrix}\right]$), skewing the distribution with reference vector $v_1^+ = (\cos(\pi/4), \sin(\pi/4))$, see Frahm (2004) for details. Further, the bivariate Marshall–Olkin exponential distribution (BOMED) is looked at: $(\min\{Z_1, Z_3\}, \min\{Z_2, Z_3\})$ for the first class and $(\min\{Z_1, Z_3\} + 0.5, \min\{Z_2, Z_3\} + 0.5)$ for the second one with $Z_1 \sim Exp(1)$, $Z_2 \sim Exp(0.5)$, and $Z_3 \sim Exp(0.75)$. Each time we generate a sample of 400 points (200 from each class) to train a classifier and a sample containing 1,000 points (500 from each class) to evaluate its performance (= error rate).

DD-plots of a training sample for *the Gaussian location alternative* using zonoid (left) and location (middle) depth are shown in Fig. 1. For each classifier, training and testing is performed on 100 simulated data sets, and a box-plot of error rates is drawn; see Fig. 1 (right). The first group of non-depth classifiers includes linear (LDA) and quadratic (QDA) discriminant analysis and k-nearest-neighbors classifier (KNN). Then the maximal depth classifiers (MM, MS and MH; cf. Ghosh and Chaudhuri 2005) and the DD-classifiers (DM, DS and DH; cf. Li et al. 2012) are regarded. Each triplet uses the Mahalanobis (Mahalanobis 1936; Zuo and Serfling 2000), simplicial (Liu 1990) and location depths, respectively. The remaining eight classifiers are DDα-classifiers based on zonoid depth (Z-DDα), exactly computed location depth (H-DDα-e), random Tukey depth for once-only (H-DDα-$\sharp$s) and instantly (H-DDα-$\sharp$d) generated directions, each time using $\sharp = 10, 20, 50$ random

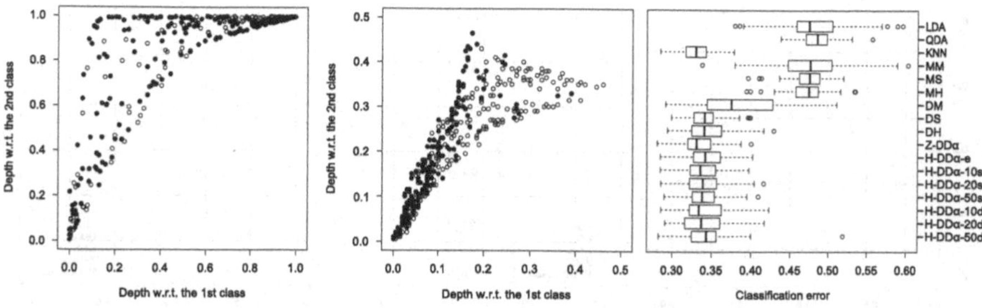

Fig. 2 DD-plots using zonoid (*left*) and location (*middle*) depth and combined box-plots (*right*) for Cauchy location-scale alternative

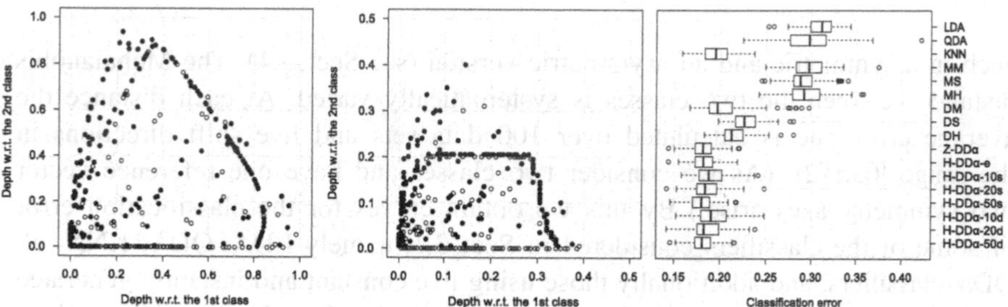

Fig. 3 DD-plots using zonoid (*left*) and location (*middle*) depth and combined box-plots (*right*) for BOMED location alternative

directions, respectively. The combined box-plots together with corresponding DD-plots using zonoid and location depth are presented for *the Cauchy location-scale alternative* (Fig. 2) and *the BOMED location alternative* (Fig. 3).

Based on these results (and many more not presented here) we conclude: In many cases DDα-classifiers, both based on the zonoid depth and the random Tukey depth, are better than their competitors. The versions of the DDα-classifier that are based on the random Tukey depth are not outperformed by the exact computation algorithm. There is no noticeable difference between the versions of the DDα-classifier based on the random Tukey depth using same directions and an instantly generated direction set. The statement "the more random directions we use, the better classification we achieve" is not necessarily true with the DDα-classifier based on the random Tukey depth, as the portion of outsiders and their treatment are rather relevant.

3.2 *Performance Dynamics*

To study the performance dynamics of the various DDα-classifiers in contrast with existing classifiers we regard t-distributions with ∞, 5 and 1 degrees of freedom,

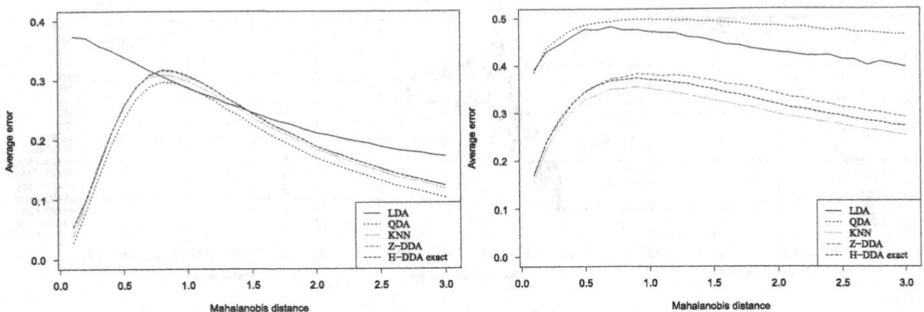

Fig. 4 Performance dynamic graphs for Gaussian (*left*) and asymmetric conditional scale Cauchy (*right*) distributions

each in a symmetric and an asymmetric version (see Sect. 3.1). The Mahalanobis distance between the two classes is systematically varied. At each distance the average error rate is calculated over 100 data sets and five shift directions in the range $[0, \pi/2]$. (As we consider two classes and have one reference vector two symmetry axes arise.) By this we obtain curves for the classification error of some of the classifiers considered in Sect. 3.1, namely LDA, QDA, KNN, all DDα-classifiers, and additionally those using five constant and instantly generated random directions. The results for two extreme cases, Gaussian distribution (left) and asymmetric conditional scale Cauchy distribution (right) are shown in Fig. 4.

Under bivariate elliptical settings (Fig. 4, left) QDA, as expected from theory, outperforms other classifiers and coincides with LDA when the Mahalanobis distance equals 1. DDα-classifiers suffering from outsiders perform worse but similarly, independent of the number of directions and the depth notion used; they are only slightly outperformed by KNN for the "upper" range of Mahalanobis distances. (Note that KNN does not have the "outsiders problem".) But when considering an asymmetric fat-tailed distribution (Fig. 4, right), neither LDA nor QDA perform satisfactorily. The DDα-classifiers are still outperformed by KNN (presumably because of the outsiders). They perform almost the same for different numbers of directions. The DDα-classifier based on zonoid depth is slightly outperformed by that using location depth, which is more robust.

3.3 The Speed of Training and Classification

The third task tackled in this paper is comparing the speed of the DDα-classifiers using zonoid and random Tukey depth, respectively. (For the latter we take 1,000 random directions and do not consider outsiders.) Two distributional settings are investigated: $N(\mathbf{0}_d, \mathbf{I}_d)$ *vs.* $N(0.25 \cdot \mathbf{1}_d, \mathbf{I}_d)$ and $N(\mathbf{0}_d, \mathbf{I}_d)$ *vs.* $N((0.25\mathbf{0}'_{d-1})', 5 \cdot \mathbf{I}_d)$, $d = 5, 10, 15, 20$. For each *pair of classes* and *number of training points* and *dimension* we train the classifier 100 times and test each of them using 2,500 points. Average times in seconds are reported in Table 1.

Table 1 The average speed of training and classification (in parentheses) using the random Tukey depth, in seconds

	$N(\mathbf{0}_d, \mathbf{I}_d)$ vs $N(0.25 \cdot \mathbf{1}_d, \mathbf{I}_d)$				$N(\mathbf{0}_d, \mathbf{I}_d)$ vs $N((0.25\mathbf{0}'_{d-1})', 5 \cdot \mathbf{I}_d)$			
	d = 5	d = 10	d = 15	d = 20	d = 5	d = 10	d = 15	d = 20
$n = 200$	0.1003	0.097	0.0908	0.0809	0.0953	0.0691	0.0702	0.0699
	(0.00097)	(0.00073)	(0.00033)	(0.00038)	(0.00098)	(0.0005)	(0.00034)	(0.00025)
$n = 500$	0.2684	0.2551	0.2532	0.252	0.2487	0.2049	0.1798	0.1845
	(0.00188)	(0.00095)	(0.00065)	(0.00059)	(0.0019)	(0.00096)	(0.00065)	(0.00049)
$n = 1000$	0.6255	0.6014	0.5929	0.5846	0.5644	0.5476	0.4414	0.4275
	(0.00583)	(0.00289)	(0.00197)	(0.00148)	(0.0058)	(0.00289)	(0.00197)	0.00148

Table 1 and the distributional settings correspond to those in Lange et al. (2012), where a similar study has been conducted with the zonoid depth. We also use the same PC and testing environment. Note firstly that the DDα-classifier with the random Tukey depth requires substantially less time to be trained than with the zonoid depth. The time required for training increases almost linearly with the cardinality of the training set, which can be traced back to the structure of the algorithms used for the random Tukey depth and for the α-procedure. The time decreases with dimension, which can be explained as follows: The α-procedure takes most of the time here; increasing d but leaving n constant increases the number of points outside the convex hull of one of the training classes, that is, having depth = 0 in this class; these points are assigned to the other class without calculations by the α-procedure.

4 Conclusions

The experimental comparison of the DDα-classifiers, using the zonoid depth and the random Tukey depth, on asymmetric and fat-tailed distributions shows that in general both depths classify rather well, the random Tukey depth performs not worse than the zonoid depth and sometimes even outperforms it (cf. Cauchy distribution), at least in two dimensions. Though both depths can be efficiently computed, also for higher dimensional data, the random Tukey depth is computed much faster. Still when employing the random Tukey depth the number of random directions has to be selected; this as well as a proper treatment of outsiders needs further investigation.

References

Cuesta-Albertos, J. A., & Nieto-Reyes, A. (2008). The random Tukey depth. *Computational Statistics and Data Analysis, 52*, 4979–4988.

Dutta, S., & Ghosh, A. K. (2012). On robust classification using projection depth. *Annals of the Institute of Statistical Mathematics, 64*, 657–676.

Dutta, S., & Ghosh, A. K. (2012). *On classification based on L_p depth with an adaptive choice of p*. Technical Report Number R5/2011, Statistics and Mathematics Unit, Indian Statistical Institute, Kolkata, India.

Dyckerhoff, R., Koshevoy, G., & Mosler, K. (1996). Zonoid data depth: Theory and computation. In A. Prat (Ed.), *COMPSTAT 1996 - Proceedings in computational statistics* (pp. 235–240). Heidelberg: Physica.

Frahm, G. (2004). *Generalized elliptical distributions: Theory and applications*. Doctoral thesis. University of Cologne.

Ghosh, A. K., & Chaudhuri, P. (2005). On maximum depth and related classifiers. *Scandinavian Journal of Statistics, 32*, 327–350.

Hoberg, R., & Mosler, K. (2006). Data analysis and classification with the zonoid depth. In R. Liu, R. Serfling, & D. Souvaine (Eds.), *Data depth: Robust multivariate analysis, computational geometry and applications* (pp. 49–59). Providence: American Mathematical Society.

Koshevoy, G., & Mosler, K. (1997). Zonoid trimming for multivariate distributions. *Annals of Statistics, 25*, 1998–2017.

Lange, T., Mosler, K., & Mozharovskyi, P. (2012). *Fast nonparametric classification based on data depth*. Statistical papers (to appear).

Lange, T., & Mozharovskyi, P. (2012). The Alpha-Procedure - a nonparametric invariant method for automatic classification of d-dimensional objects. In *36th Annual Conference of the German Classification Society, Hildesheim*.

Li, J., Cuesta-Albertos, J. A., & Liu, R. Y. (2012). DD-classifier: Nonparametric classification procedure based on DD-plot. *Journal of the Americal Statistical Association, 107*, 737–753.

Liu, R. (1990). On a notion of data depth based on random simplices. *Annals of Statistics, 18*, 405–414.

Liu, X., & Zuo, Y. (2012). *Computing halfspace depth and regression depth*. Mimeo.

Mahalanobis, P. (1936). On the generalized distance in statistics. *Proceedings of the National Institute of Science of India, 2*, 49–55.

Mosler, K. (2002). *Multivariate dispersion, central regions and depth: The lift zonoid approach*. New York: Springer.

Paindaveine, D., & Van Bever, G. (2013). Nonparametrically consistent depth-based classifiers. *Bernoulli* (to appear).

Rousseeuw, P. J., & Ruts, I. (1996). Bivariate location depth. *Journal of the Royal Statistical Society. Series C (Applied Statistics), 45*, 516–526.

Rousseeuw, P. J., & Struyf, A. (1998). Computing location depth and regression depth in higher dimensions. *Statistics and Computing, 8*, 193–203.

Tukey, J. W. (1975). Mathematics and the picturing of data. In *Proceedings of the International Congress of Mathematicians* (pp. 523–531), Vancouver.

Vasil'ev, V. I. (1991). The reduction principle in pattern recognition learning (PRL) problem. *Pattern Recognition and Image Analysis, 1*, 23–32.

Vasil'ev, V. I. (2003). The reduction principle in problems of revealing regularities I. *Cybernetics and Systems Analysis, 39*, 686–694.

Vasil'ev, V. I., & Lange, T. (1998). The duality principle in learning for pattern recognition (in Russian). *Kibernetika i Vytschislit'elnaya Technika, 121*, 7–16.

Zuo, Y., & Serfling, R. (2000). General notions of statistical depth function. *Annals of Statistics, 28*, 462–482.

The Alpha-Procedure: A Nonparametric Invariant Method for Automatic Classification of Multi-Dimensional Objects

Tatjana Lange and Pavlo Mozharovskyi

Abstract A procedure, called α-procedure, for the efficient automatic classification of multivariate data is described. It is based on a geometric representation of two learning classes in a proper multi-dimensional rectifying feature space and the stepwise construction of a separating hyperplane in that space. The dimension of the space, i.e. the number of features that is necessary for a successful classification, is determined step by step using two-dimensional repères (linear subspaces). In each step a repère and a feature are constructed in a way that they yield maximum discriminating power. Throughout the procedure the invariant, which is the object's affiliation with a class, is preserved.

1 Introduction

A basic task of pattern recognition consists in constructing a decision rule by which objects can be assigned to one of two given classes. The objects are characterized by a certain number of real-valued properties. The decision rule is based on a trainer's statement that states for a training sample of objects, whether they belong to class V_1 or class V_2. Many procedures are available to solve this task, among them binary regression, parametric discriminant analysis, and kernel methods like the SVM; see, e.g., Hastie et al. (2009).

A large part of nonparametric approaches search for a separating (or rectifying) hyperplane dividing the two training classes in a sufficiently high-dimensional

T. Lange
Hochschule Merseburg, Geusaer Straße, 06217 Merseburg, Germany
e-mail: tatjana.lange@hs-merseburg.de

P. Mozharovskyi
Universität zu Köln, 50923 Köln, Germany
e-mail: mozharovskyi@statistik.uni-koeln.de

M. Spiliopoulou et al. (eds.), *Data Analysis, Machine Learning and Knowledge Discovery*, Studies in Classification, Data Analysis, and Knowledge Organization, DOI 10.1007/978-3-319-01595-8_9,

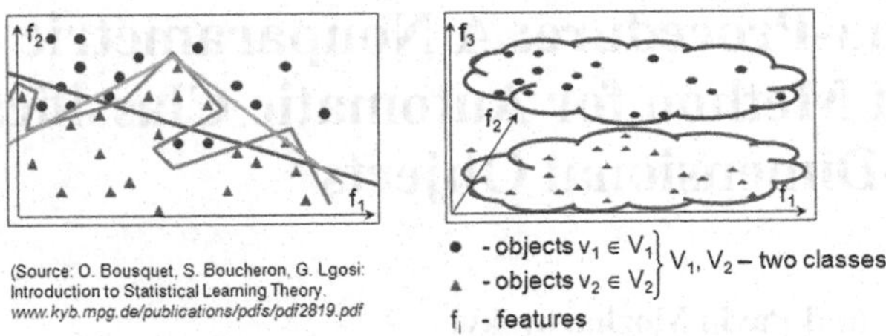

Fig. 1 Instable solution for the separation of the classes (*left*). The selection of an informative property (*right*)

feature space. In doing so we face the problem that the "competencies" of measured properties (forming the axes of the original space) are unknown. Even more, we also do not know the correct scale of a property.

Very often these uncertainties lead to a situation where the complete separation of the patterns (or classes) of a training sample becomes difficult (Fig. 1, left). All these factors can cause a very complex separating surface in the *original* space which correctly divides the classes of the training sample but works rather poorly in case of new measured data. The selection of a more "informative" property (Fig. 1, right) can give less intricate and thus more stable decisions.

The α*-procedure* (Vasil'ev 2003, 2004, 1969, 1996; Vasil'ev and Lange 1998) uses the idea of a *general invariant* for stabilizing the selection of the separating plane. The invariant is the *belonging of an object to a certain class* of the training sample. The α-procedure—using repères—performs a step-by-step search of the direction of a straight line in a given repère that is as near as possible to the trainer's statement, i.e. separates best the training sample. It is completely nonparametric. The properties of the objects which are available for the recognition task are selected in a sequence one by one. With the most powerful properties a new space of "transformed features" is constructed that is as near as possible to the trainer's statement.

2 The α-Procedure

First, we perform some pre-selection, taking into further considerations only those properties $p_q, q = 1, \ldots, m$, whose values are completely separated or have some overlap as shown in Fig. 2. Next, we define the *discrimination power* or *separating power* of a *single* property p_q as

$$F(p_q) = \frac{c_q}{l}, \tag{1}$$

where l is the length of the training sample (= number of objects) and c_q is the number of correctly classified objects.

Fig. 2 Classification by a single property p_q, with $l = 15, c_q = 6$

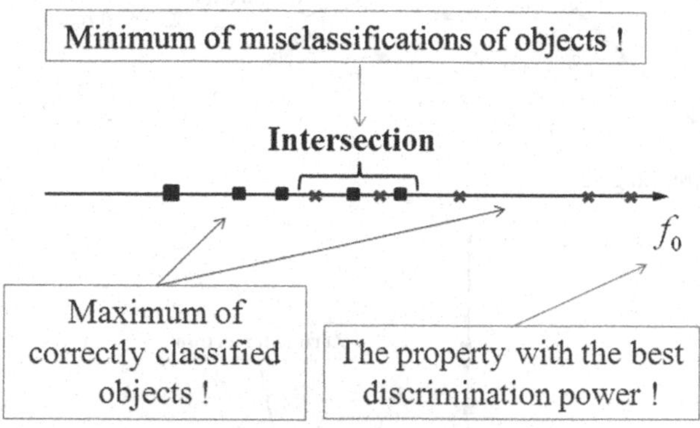

Fig. 3 α-procedure, Stage 1

We set a minimum admissible discrimination power F_{min}, and at the first step select any property as a possible *feature* whose discrimination power exceeds the minimum admissible one:

$$F(p_q) > F_{min} \tag{2}$$

For the synthesis of the space, we select step-by-step those features that have best discrimination power. Each new feature shall increase the number of correctly classified objects. For this, we use the following definition of the *discrimination power* of a feature, selected at step k:

$$F(x_k) = \frac{\omega_k - \omega_{k-1}}{l} = \frac{\Delta\omega_k}{l}, \quad \omega_0 = 0, \tag{3}$$

where ω_{k-1} is the accumulated number of correctly classified objects before the k-th feature was selected and ω_k is the same after it was selected.

At Stage 1 we select a property having best discrimination power as a basis feature f_0 (= first axis) and represent the objects by their values on this axis; see Fig. 3.

At Stage 2 we add a second property p_k to the coordinate system and project the objects to the plane that is spanned by the axes f_0 and p_k. In this plane a ray originating from the origin is rotated up to the point where the *projections* of the objects onto this ray provide the best separation of the objects. The resulting ray, characterized by its rotation angle α, defines a possible new axis. We repeat this procedure for all remaining properties and select the property that gives the best separation of the objects on its rotated axis, which is denoted as $\tilde{f}_1$. This axis is taken as the first *new feature*, and the respective plane as the first *repère*; see Fig. 4.

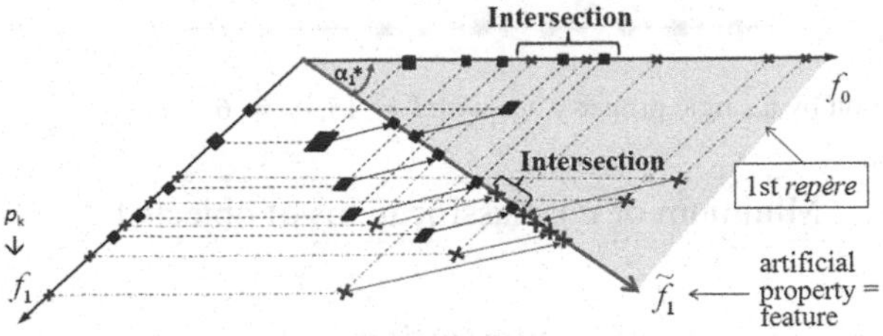

Fig. 4 α-procedure, Stage 2

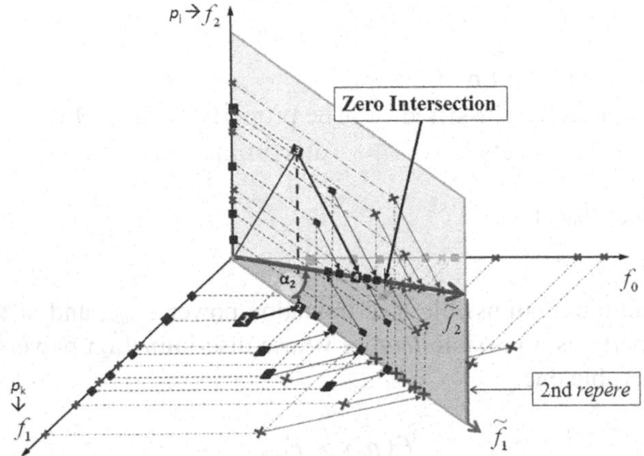

Fig. 5 α-procedure, Stage 3

At Stage 3 we regard another property p_j that has not been used so far and define the position of the objects in a *new plane* that is built by the axes $\tilde{f}_1$ and p_j. Again we consider a ray in this plane and turn it around the origin *by the angle* α until the projections of the objects onto this axis give the best separation. We repeat this procedure for all remaining properties and select the best one, which, together with $\tilde{f}_1$ forms the second repère (Fig. 5). In our simple example this feature already leads to a faultless separation of the objects.

If all properties have been used but no complete separation of all objects reached, a special stopping criterion as described in Vasil'ev (1996) is to be used.

3 Some Formulae

As we see from the description of the idea, the procedure is the same at each step except for the first basic step defining f_0.

Let us assume that we have already selected $k - 1$ features. We will use the symbol $\tilde{x}_{i,(k-1)}, i = 1, \ldots, l$, for the projections of the objects onto the feature f_{k-1}

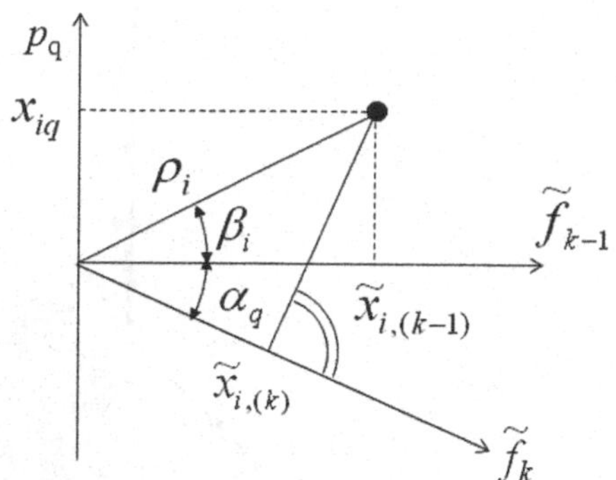

Fig. 6 Calculating the value of feature $k-1$ for object i

and ω_{k-1} as the number of already correctly classified objects (Fig. 6). For the *next step* k we have to compute the projection $\tilde{x}_{i,(k)} = \rho_i \cos(\beta_i + \alpha_q)$ for *all remaining* properties p_q, where $\tilde{x}_{i,(k-1)}$ is the value of feature $k-1$ for object i, x_{iq} is the value of property q for object i, $\rho_i = \sqrt{\tilde{x}^2_{i,(k-1)} + x^2_{iq}}$, $\beta_i = \arctan(x_{iq}/\tilde{x}_{i,(k-1)})$.

After n steps the normal of the separating hyperplane is given by

$$\left(\prod_{k=2}^{n} \cos\alpha_k^0, \sin\alpha_2^0 \prod_{k=3}^{n} \cos\alpha_k^0, \ldots, \sin\alpha_{n-1}^0 \cos\alpha_n^0, \sin\alpha_n^0\right), \tag{4}$$

where α_k^0 denotes the angle α that is best in step k, $k = 2, \ldots, n$. Due to the fact that (4) is stepwise calculated, the underlying features must be assigned *backwards* in practical classification. For example, the separation decision plane and the decomposition of its normal vector are shown in Figs. 7 and 8.

Note: If the separation of objects is not possible in the original space of properties, the space can be extended by building additional properties using products of the type $x_{iq}^s \cdot x_{ir}^t$ for all $q, r \in \{1, \ldots, m\}$ and i and some (usually small) exponents s and t. The solution is then searched in the extended space.

4 Simulations and Applications

To explore the specific potentials of the α-procedure we apply it to simulated data. Besides the application to the original data, the α-procedure can also be applied to properly transformed data; in particular, it has been successfully used to classify data on the basis of their so called DD-plot (DDα-classifier), see Lange et al. (2012a,b). The α-procedure (applied in the original space (α-pr.(1)) and the extended space using polynomials of degree 2 (α-pr.(2)) and 3 (α-pr.(3))) is contrasted with the following nine classifiers: linear discriminant analysis (LDA), quadratic discriminant analysis (QDA), k-nearest neighbors classification (KNN), maximum depth classification based on Mahalanobis (MM), simplicial (MS), and

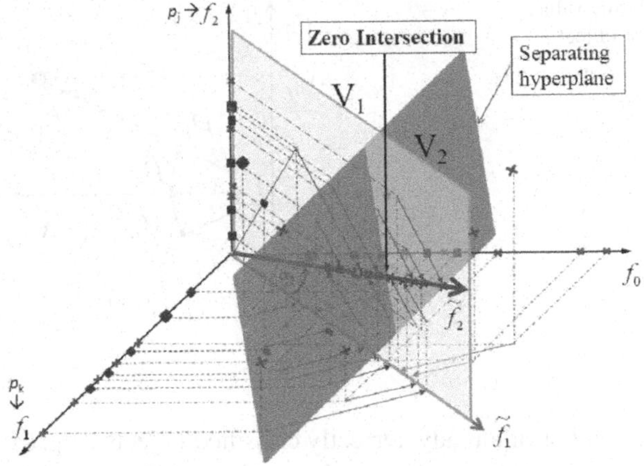

Fig. 7 The separating decision plane

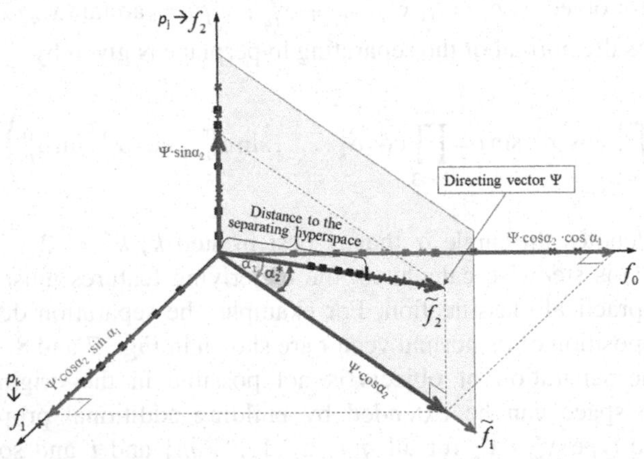

Fig. 8 The separating decision plane with its defining vector

halfspace (MH) depth, and DD-classification with the same depths (DM, DS and DH, correspondingly; see Li et al. 2012 for details), and to the DDα-classifier.

Six simulation alternatives are used; each time a sample of 400 objects (200 from each class) is used as a training sample and 1,000 objects (500 from each class) to evaluate the classifier's performance (= classification error). First, normal location (two classes originate from $N(\begin{bmatrix} 0 \\ 0 \end{bmatrix}, \begin{bmatrix} 1 & 1 \\ 1 & 4 \end{bmatrix})$ and $N(\begin{bmatrix} 1 \\ 1 \end{bmatrix}, \begin{bmatrix} 1 & 1 \\ 1 & 4 \end{bmatrix})$, see Fig. 9, left) and normal location-scale (the second class has covariance $\begin{bmatrix} 4 & 4 \\ 4 & 16 \end{bmatrix}$, see Fig. 9, middle) alternatives are tried.

To test the proposed machinery for robustness properties we challenge it using contaminated normal distribution, where the first class of the training sample of the normal location and location-scale alternatives considered above contains 10 % objects originating from $N(\begin{bmatrix} 10 \\ 10 \end{bmatrix}, \begin{bmatrix} 1 & 1 \\ 1 & 4 \end{bmatrix})$ (see Fig. 9, right and Fig. 10, left

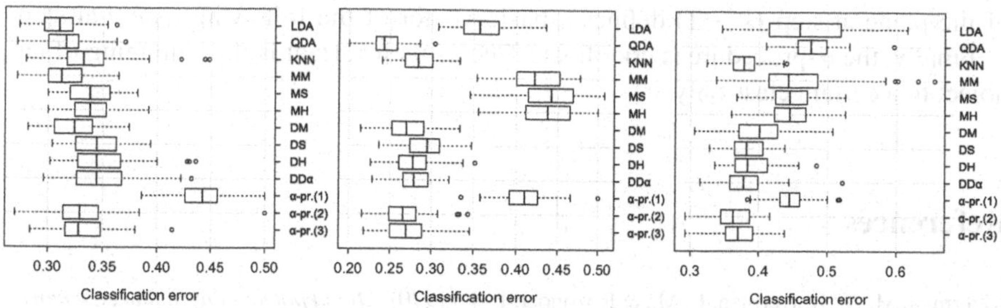

Fig. 9 Boxplots of the classification error for normal location (*left*) and location-scale (*middle*), and normal contaminated location (*right*) alternatives over 100 takes

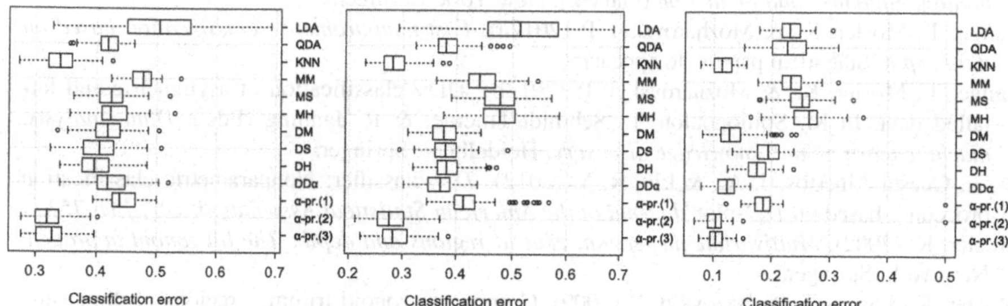

Fig. 10 Boxplots of the classification error for normal contaminated location-scale (*left*), and Cauchy location-scale (*middle*), and exponential (*right*) alternatives over 100 takes

correspondingly). Other robustness aspects are demonstrated with a pair of Cauchy distributions forming a similar location-scale alternative, see Fig. 9, middle. Settings with exponential distributions ($(\text{Exp}(1), \text{Exp}(1))$ *vs.* $(\text{Exp}(1) + 1, \text{Exp}(1) + 1)$, see Fig. 10) conclude the simulation study.

The α-procedure performs fairly well for normal alternatives and shows remarkable performance for the robust alternatives considered. Though it works well on exponential settings as well, its goodness appears to depend on the size of the extended feature space. The last choice can be made by using either external information or cross-validation techniques.

5 Conclusion

The α-procedure calculates a separating hyperplane in a (possibly extended) feature space. In each step a two-dimensional subspace is constructed where, as the data points are naturally ordered, only a circular (that is, linear) search has to be performed. This makes the procedure very fast and stable. The classification task is simplified to a stepwise linear separation of planar points, while the complexity of the problem is coped with by the number of features constructed. The angle α

of the plane at step $(k - 1)$ defines a basic vector of the following *repère* at step k. Finally, the α-procedure is coordinate-free as its invariant is the belonging of an object to a certain class only.

References

Aizerman, M. A., Braverman, E. M., & Rozonoer, L. I. (1970). *Metod potencialnych funkcij v teorii obucenia mashin (The method of potential functions in machine learning theory)* [in Russian]. Moscow: Nauka.

Hastie, T., Tibshirani, R., & Friedman, J. H. (2009). *The elements of statistical learning: Data mining, inference, and prediction* (2nd ed.). New York: Springer.

Lange, T., Mosler, K., & Mozharovskyi, P. (2012a). *Fast nonparametric classification based on data depth*. Statistical papers (to appear).

Lange, T., Mosler, K., & Mozharovskyi, P. (2012b). DDα classification of asymmetric and fat-tailed data. In M. Spiliopoulou, L. Schmidt-Thieme, & R. Janning (Eds.), *Data analysis, machine learning and knowledge discovery*. Heidelberg: Springer.

Li, J., Cuesta-Albertos, J. A., & Liu, R. Y. (2012). DD-classifier: Nonparametric classification procedure based on DD-plot. *Journal of the American Statistical Association, 107*, 737–753.

Mosler, K. (2002). *Multivariate dispersion, central regions and depth: The lift zonoid approach*. New York: Springer.

Mosler, K., Lange, T., & Bazovkin, P. (2009). Computing zonoid trimmed regions of dimension $d > 2$. *Computational Statistics and Data Analysis, 53*(7), 2000–2010.

Novikoff, A. (1962). On convergence proofs for perceptrons. In *Proceedings of the symposium on mathematical theory of automata* (pp. 615–622). Brooklyn, NY: Polytechnic Institute of Brooklyn.

Vapnik, V., & Chervonenkis, A. (1974). *Theory of pattern recognition* (in Russian). Moscow: Nauka.

Vasil'ev, V. I. (1969). *Recognition systems – Reference book* (in Russian). Kyiv: Naukova Dumka.

Vasil'ev, V. I. (1996). The theory of reduction in extrapolation problem (in Russian). *Problemy Upravleniya i Informatiki, 1-2*, 239–251.

Vasil'ev, V. I. (2003). The reduction principle in problems of revealing regularities. Part I (in Russian). *Cybernetics and System Analysis, 39*(5), 69–81.

Vasil'ev, V. I. (2004). The reduction principle in problems of revealing regularities. Part II (in Russian). *Cybernetics and System Analysis, 40*(1), 9–22.

Vasil'ev, V. I., & Lange, T. (1998). The duality principle in learning for pattern recognition (in Russian). *Kibernetika i Vytschislit'elnaya Technika, 121*, 7–16.

Support Vector Machines on Large Data Sets: Simple Parallel Approaches

Oliver Meyer, Bernd Bischl, and Claus Weihs

Abstract Support Vector Machines (SVMs) are well-known for their excellent performance in the field of statistical classification. Still, the high computational cost due to the cubic runtime complexity is problematic for larger data sets. To mitigate this, Graf et al. (Adv. Neural Inf. Process. Syst. 17:521–528, 2005) proposed the Cascade SVM. It is a simple, stepwise procedure, in which the SVM is iteratively trained on subsets of the original data set and support vectors of resulting models are combined to create new training sets. The general idea is to bound the size of all considered training sets and therefore obtain a significant speedup. Another relevant advantage is that this approach can easily be parallelized because a number of independent models have to be fitted during each stage of the cascade. Initial experiments show that even moderate parallelization can reduce the computation time considerably, with only minor loss in accuracy. We compare the Cascade SVM to the standard SVM and a simple parallel bagging method w.r.t. both classification accuracy and training time. We also introduce a new stepwise bagging approach that exploits parallelization in a better way than the Cascade SVM and contains an adaptive stopping-time to select the number of stages for improved accuracy.

1 Introduction

Support vector machines (e.g., Schoelkopf and Smola 2002) are a very popular supervised learning algorithm for both classification and regression due to their flexibility and high predictive power. One major obstacle in their application to larger data sets is that their runtime scales approximately cubically with the

O. Meyer (✉) · B. Bischl · C. Weihs
Chair of Computational Statistics, Department of Statistics, TU Dortmund, Germany
e-mail: meyer@statistik.uni-dortmund.de; bischl@statistik.uni-dortmund.de; weihs@statistik.uni-dortmund.de

M. Spiliopoulou et al. (eds.), *Data Analysis, Machine Learning and Knowledge Discovery*, Studies in Classification, Data Analysis, and Knowledge Organization, DOI 10.1007/978-3-319-01595-8_10,

number of observations in the training set. Combined with the fact that not one but multiple model fits have to be performed due to the necessity of hyperparameter tuning, their runtime often becomes prohibitively large beyond 100.000 or 1 million observations. Many different approaches have been suggested to speed up training time, among these online SVMs (e.g., the LASVM by Border et al. 2005, sampling techniques and parallelization schemes. In this article we will evaluate two quite simple methods of the latter kind. All of our considered approaches break up the original data set into smaller parts and fit individual SVM models on these. Because of the already mentioned cubical time-scaling of the SVM algorithm w.r.t. the number of data points a substantial speed up should be expected.

We state two further reasons for our general approach: (a) Computational power through multiple cores and multiple machines is often cheaply available these days. (b) We would like to keep as much as possible from the original SVM algorithm (use it as a building block in our parallelization scheme) in order to gain from future improvements in this area. It is also of general interest for us, how far we can get with such relatively simple approaches.

In the following sections we will cover the basic SVM theory, describe the considered parallelization approaches, state our experimental setup, report the results and then summarize them with additional remarks for future research.

2 Support Vector Machines

In supervised machine learning, data for classification tasks can be represented as a number of observations $(\boldsymbol{x}_1, y_1), (\boldsymbol{x}_2, y_2), \ldots, (\boldsymbol{x}_n, y_n) \in \mathcal{X} \times \mathcal{Y}$, where the set $\mathcal{X}$ defines the space in which our feature vectors $\boldsymbol{x}_i$ live in (here assumed to be $\mathbb{R}^p$ as we will mainly discuss the Gaussian kernel later on) and $\mathcal{Y} = \{-1, 1\}$ is the set of binary class labels. The support vector machine (SVM) relies on two basic concepts:

(a) Regularized risk minimization: We want to fit a large margin classifier f : $\mathbb{R}^p \to \mathbb{R}$ with a low empirical regularized risk:

$$(\hat{f}, \hat{b}) = \arg \inf_{f \in \mathcal{H}, b \in \mathbb{R}} ||f||^2_{\mathcal{H}} + C \sum_{i=1}^{n} L(y_i, f(\boldsymbol{x}_i) + b) \,. \tag{1}$$

Here, b is the so-called bias term of the classifier and L is a loss function. For classification with the SVM, we usually select the hinge loss $L(y,t) = \max(0, 1 - yt)$. This is a convex, upper surrogate loss for the 0/1-loss $L(y,t) = I[yt < 0]$, which is of primary interest, but algorithmically intractable.

While the second term above (called the empirical risk) measures the closeness of the predictions $f(x_i) + b$ to the true class labels -1 and $+1$, respectively, by means of L, the first term $||f||^2_H$ is called a regularizer, relates to the maximization of the margin and penalizes "non-smooth" functions f.

The balance between these two terms is controlled by the hyperparameter C. For an unknown observation x the class label is predicted by $\mathrm{sign}(\hat{f}(x) + \hat{b})$.

(b) Kernelization: In order to be able to solve non-linear classification problems we "kernelize" (1) by introducing a kernel function $k : \mathcal{X} \times \mathcal{X} \to \mathbb{R}$, which measures the "similarity" of two observations. Formally, k is a symmetric, positive semi-definite Mercer kernel. And $\mathcal{H}$ is now defined as the associated reproducing kernel Hilbert space for k, our generalized inner product in $\mathcal{H}$ is $\langle x, x' \rangle_{\mathcal{H}} = k(x, x')$ and $||x||^2 = \langle x, x \rangle_{\mathcal{H}}$ By using this so-called "kernel trick" we implicitly map our data into a usually higher-dimensional space, enabling us to tackle nonlinear problems with essentially linear techniques. The Gaussian kernel

$$k(\boldsymbol{x}_i, \boldsymbol{x}_j) = \exp\left(-\sigma||\boldsymbol{x}_i - \boldsymbol{x}_j||_2^2\right) \tag{2}$$

is arguably the most important and popular kernel function and we have therefore focused on it in all subsequent experiments. But note that all following parallelization techniques are basically independent of this choice.

The optimization problem (1) is usually solved in its dual formulation and leads to the following quadratic programming problem:

$$\max_{\alpha} \sum_{i=1}^{n} \alpha_i - \frac{1}{2} \sum_{i,j=1}^{n} y_i y_j \langle \boldsymbol{x}_i, \boldsymbol{x}_j \rangle_{\mathcal{H}} \tag{3}$$

$$\text{s.t. } 0 \leq \boldsymbol{\alpha} \leq C \text{ and } \boldsymbol{y}^T \boldsymbol{\alpha} = 0 \, ,$$

where $\boldsymbol{\alpha}$ denotes the vector of Lagrange multipliers.

As we will usually obtain a sparse solution due to the non-differentiability of our hinge loss L, some α_i will be zero, and the observations x_i with $\alpha_i > 0$ shall be called support vectors (SVs). They are the samples solely responsible for the shape of our decision border $f(\boldsymbol{x}) = 0$. This is implied by the fact that if we retrain an SVM on only the support vectors, we will arrive at exactly the same model as with the full data set.

The SVM performance is quite sensitive to hyperparameter settings, e.g., the settings of the complexity parameter C and the kernel parameter σ for the Gaussian kernel. Therefore, it is strongly recommended to perform hyperparameter tuning before the final model fit. Often a grid search approach is used, where performance is estimated by cross-validation, but more sophisticated methods become popular as well (see e.g., Koch et al. 2012).

Multi-class problems are usually solved by either using a multi-class-to-binary scheme (e.g., one-vs-one) or by directly changing the quadratic programming problem in (3) to incorporate several classes.

3 Cascade Support Vector Machine

The Cascade SVM is a stepwise procedure that combines the results of multiple regular support vector machines to create one final model. The main idea is to iteratively reduce a data set to its crucial data points before the last step. This is done by locating potential support vectors and removing all other samples from the data. The method described here is essentially taken from the original paper by Graf et al. (2005):

1. Partition the data into k disjoint subsets of preferably equal size.
2. Independently train an SVM on each of the data subsets.
3. Combine the SVs of, e.g., pairs or triples of SVMs to create new subsets.
4. Repeat steps 2 and 3 for some time.
5. Train an SVM on all SVs that were finally obtained in step 4.

This algorithm (depicted in the right-hand side of Fig. 1) will be called *Cascade SVM* or simply *cascade*. In the original paper, Graf et al. also considered the possibility of multiple runs through the cascade for each data set. After finishing a run through the cascade the subsets for the first step of the next run are created by combining the remaining SVs of the final model with each subset from the first step of the first run. For speed reasons we always only perform one run through the cascade.

4 Bagging-Like Support Vector Machines

Another generic and well-known concept in machine learning is bagging. Its main advantage is that derived methods are usually accurate and very easy to parallelize. Chawla et al. (2003) introduced and analyzed a simple variant, which proved to perform well on large data sets for decision trees and neural networks. Unlike in traditional bagging algorithms, the original data set is randomly split into n disjoint (and not overlapping) subsamples, which all contain $\frac{1}{n}$-th of the data. Then a classification model is trained on each of these subsets. Classification of new data is done by majority voting with ties being broken randomly. Hence, using SVMs means that the training of this bagging-like method is equivalent to the first step of the Cascade SVM. By comparing these two methods we can analyze if the additional steps of the cascade (and the invested runtime) improves the accuracy of the procedure.

Figure 1 shows the structures of a 4-2 Cascade SVM (C-4-2)—with 4 being the number of subsets in the first step and 2 representing the number of models being combined after every single step—and a bagged SVM using three bags.

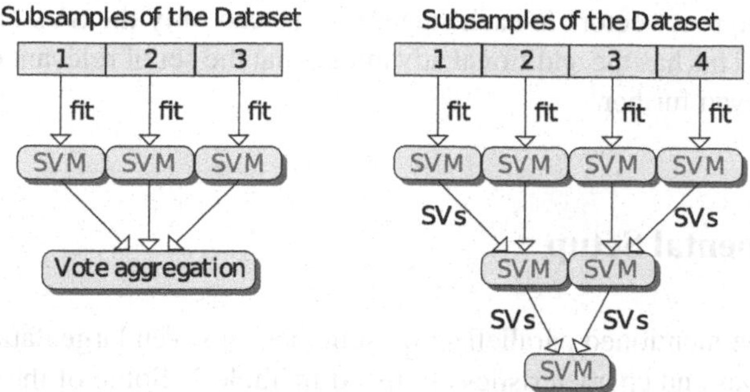

Fig. 1 Schemes for bagged SVM (*left*) and cascade (*right*)

5 Stepwise Bagging

It can easily be seen that the possibility to parallelize the Cascade SVM decreases in every step. This leads to the problem that an increasing number of cores will stay idle during the later stages, and in the last stage only one core can be used. We will also observe in the following experimental results that both algorithms—cascade and bagging—will perform suboptimally in some cases either with regard to runtime or accuracy. We therefore made the following changes to the described cascade algorithm in order to maximally use the number of available cores and to generally improve the algorithm by combining the advantages of both methods:

1. In the first stage, the data is partitioned in k subsets as usual.
2. At beginning of each subsequent stage in the cascade, all remaining vectors are combined into one set and then randomly divided into k overlapping subsets. The size of the subsets is fixed to the size of the subsets of the first stage, but not larger than 2/3 of the current data, if the former cannot be done. Overlapping occurs as vectors are drawn with replacement.
3. In the final stage, a bagged SVM is created instead of a single model.
4. As it is problematic to determine the number of stages of this approach we try to infer the optimal stopping time: At the beginning of the training process we hold out 5 % of the training data as an internal validation set. After each stage we measure the error of the bagged model of k SVMs from the current stage on this validation data. If the accuracy compared to the previous stage decreases, we stop the process and return the bagged model of the previous stage.
5. We have noticed that in some cases of a preliminary version of this stepwise bagging algorithm the performance degraded when the support vectors contained many wrongly classified examples. This happens in situations with high Bayes error/label noise, because all misclassified examples automatically become support vectors and will therefore always be contained in the training set for the next stage. As this seems somewhat counterintuitive, we have opted not to

select the support vectors in each stage, but instead only the SVs on and within the margin. This has the additional advantage that the set of relevant observations is reduced even further.

6 Experimental Setup

We evaluate the mentioned parallelization schemes on seven large data sets.[1] Their respective names and characteristics are listed in Table 1. Some of the data sets are multi-class, but as we want to focus on the analysis of the basic SVM algorithm, which at least in its usual from can only handle two-class problems, we have transformed the multi-class problems into binary ones. The transformation is stated in Table 1 as well. We have also eliminated every feature from every data set which was either constant or for which more than $n-1000$ samples shared the same feature value.

As we are mainly interested in analyzing the possible speedup of the training algorithm we have taken a practical approach w.r.t. the hyperparameter tuning in this article: For all data sets we have randomly sampled 10 % of all observations for tuning and then performed a usual grid search for $C \in 2^{-5}, 2^{-3}, \ldots, 2^{15}$ and $\sigma \in 2^{-15}, 2^{-13}, \ldots, 2^{3}$, estimating the misclassification error by fivefold cross-validation. The whole procedure was repeated five times, for every point (C, σ) the average misclassification rate was calculated, and the optimal configuration was selected. In case of ties, a random one was sampled from the optimal candidates. Table 1 displays the thereby obtained parameterizations and these have been used in all subsequent experiments.

Table 1 Data sets, data characteristics, used hyperparameters, proportional size of smallest class and multi-class binarization

Data set	Size	Features	C	σ	Smaller class	Binarization
covertype	581,012	46	8	0.500	0.488	Class 2 vs. rest
cod	429,030	9	32	0.125	0.333	
ijcnn1	191,681	22	32	3.13e−2	0.096	
miniboone	130,064	50	32768	7.81e−3	0.281	
acoustic	98,528	50	8	3.13e−2	0.500	Class 3 vs. rest
mnist	70,000	478	32	1.95e−3	0.492	Even vs. odd
connect	67,557	94	512	4.88e−4	0.342	Class 1 vs. rest

To compare the speedups of the different parallelization methods, we have run the following algorithms: The basic SVM, the Cascade SVM with C-27-3

[1] Can be obtained either from the LIBSVM web page or the UCI repository.

(27-9-3-1 SVMs), C-27-27 (27-1 SVMs), C-9-3 (9-3-1 SVMs) and C-9-9 (9-1 SVMs), bagging with nine bags (B-9) and the stepwise bagging approach also with nine subsets (SWB-9). For the basic SVM algorithm we have used the implementation provided in the `kernlab` R package by Karatzoglou et al. (2004). For all parallel methods we have used nine cores. For all algorithms we have repeatedly (ten times) split the data into a 3/4 part for training (with the already mentioned hyperparameters) and 1/4 for testing. In all cases we have measured the misclassification rate on the test set and the time in seconds that the whole training process lasted.

7 Results

The results of our experiments are displayed in Fig. 2 and can loosely be separated into three groups. For the data sets *acoustic*, *cod* and *miniboone*, the bagged SVMs lead to a better or at least equally good accuracy as the Cascade SVMs in only a fraction of its training time. Since the bagged SVM is nothing else but the first step of a cascade, this means that the subsequent steps of the cascade do not increase or even decrease the quality of the prediction. This does not mean that the Cascade SVM leads to bad results on all of these sets. In the case of *miniboone* for example it performs nearly as good as the classic SVM in just about 4,000 s compared to 65,000. But bagging does the same in only 350 s. On these data sets the stepwise bagging procedure usually leads to an accuracy that is equal to those of the standard bagging SVM and needs at worst as much time as the cascade.

The second group consists of *connect*, *covertype* and *mnist*. On these datasets the Cascade SVM leads to results that are as accurate as those from the classic SVM but only needs about half of the training time. The bagged SVM on the other hand again performs several times faster but cannot achieve the same accuracy as the other methods. So in these cases, at the cost of an at least ten times higher training time, the further steps of the cascade actually do increase the accuracy. The stepwise bagging SVM produces results that lie between the cascade and the standard bagging SVMs in both accuracy and training time. The actual boost in accuracy varies from data set to data set.

For the last dataset *ijcnn1*, all three methods perform very fast (cascade 15, bagging 25, stepwise bagging 14 times faster) but none of them achieves an accuracy that is as good as the classic SVM. Again the cascade outperforms bagging w.r.t accuracy while SWB lies between the other two methods.

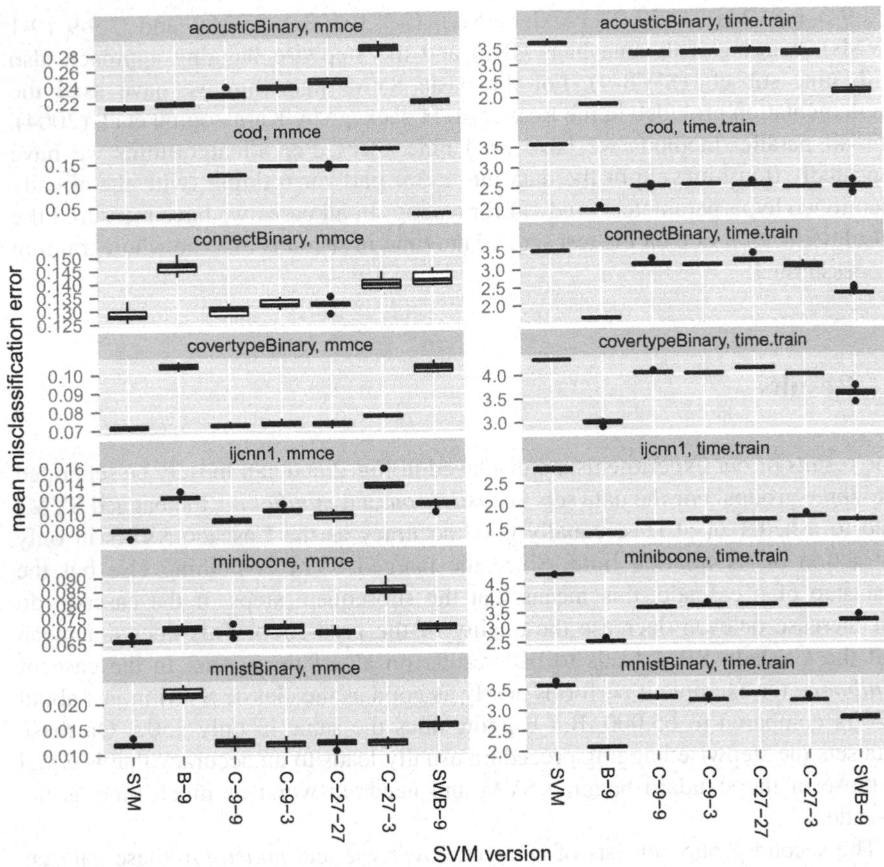

Fig. 2 Misclassification rates and training times (the latter on log10 scale) for normal and parallel SVMs

8 Conclusion and Outlook

We have analyzed simple parallelization schemes for parallel SVMs, namely a bagging-like approach, the Cascade SVM and a new combination of the two. On the considered data sets we could often observe a drastic reduction in training time through the parallelization with only minor losses in accuracy. Especially our new combined approach showed promising results. But still none of the considered algorithms shows optimal results across all considered data sets, and more work has to be done in this regard. One of the major missing features of our method is an efficient procedure for hyperparameter tuning that does not require many evaluations on large subsets of the training data. We have already begun preliminary experiments for this and will continue our research in this direction.

References

Border, A., Ertekin, S., Weston, J., & Bottou, L. (2005). Fast kernel classifiers with online and active learning. *Journal of Machine Learning Research, 6*, 1579–1619.

Chawla, N. V., Moore, T. E., Jr., Hall, L. O., Bowyer, K. W., Kegelmeyer, P., & Springer, C. (2003). Distributed learning with bagging-like performance. *Pattern Recognition Letter, 24*, 455–471.

Graf, H. P., Cosatto, E., Bottou, L., Durdanovic, I., & Vapnik, V. (2005). Parallel support vector machines: The cascade SVM. *Advances in Neural Information Processing Systems, 17*, 521–528.

Karatzoglou, A., Smola, A., Hornik, K., & Zeileis, A. (2004). kernlab - An S4 package for kernel methods in R. *Journal of Statistical Software, 11*(9), 1–20.

Koch, P., Bischl, B., Flasch, O., Bartz-Beilstein, T., & Konen, W. (2012). On the tuning and evolution of support vector kernels. *Evolutionary Intelligence, 5*, 153–170.

Schoelkopf, B., & Smola, A. J. (2002). *Learning with kernels: Support vector machines, regularization, optimization, and beyond*. Cambridge: MIT.

Soft Bootstrapping in Cluster Analysis and Its Comparison with Other Resampling Methods

Hans-Joachim Mucha and Hans-Georg Bartel

Abstract The bootstrap approach is resampling taken with replacement from the original data. Here we consider sampling from the empirical distribution of a given data set in order to investigate the stability of results of cluster analysis. Concretely, the original bootstrap technique can be formulated by choosing the following weights of observations: $m_i = n$, if the corresponding object i is drawn n times, and $m_i = 0$, otherwise. We call the weights of observations masses. In this paper, we present another bootstrap method, called soft bootstrapping, that consists of random change of the "bootstrap masses" to some degree. Soft bootstrapping can be applied to any cluster analysis method that makes (directly or indirectly) use of weights of observations. This resampling scheme is especially appropriate for small sample sizes because no object is totally excluded from the soft bootstrap sample. At the end we compare different resampling techniques with respect to cluster analysis.

1 Introduction

The non-parametric bootstrap approach is resampling taken with replacement from the original distribution (Efron 1979). Asymptotic results and simulations of bootstrapping were presented by many authors like Mammen (1992). Bootstrapping has been established as a statistical method for estimating the sampling distribution

---

H.-J. Mucha (✉)
Weierstrass Institute for Applied Analysis and Stochastics (WIAS), Mohrenstraße 39, 10117 Berlin, Germany
e-mail: mucha@wias-berlin.de

H.-G. Bartel
Department of Chemistry, Humboldt University Berlin, Brook-Taylor-Straße 2, 12489 Berlin, Germany
e-mail: hg.bartel@yahoo.de

M. Spiliopoulou et al. (eds.), *Data Analysis, Machine Learning and Knowledge Discovery*, Studies in Classification, Data Analysis, and Knowledge Organization,
DOI 10.1007/978-3-319-01595-8_11,

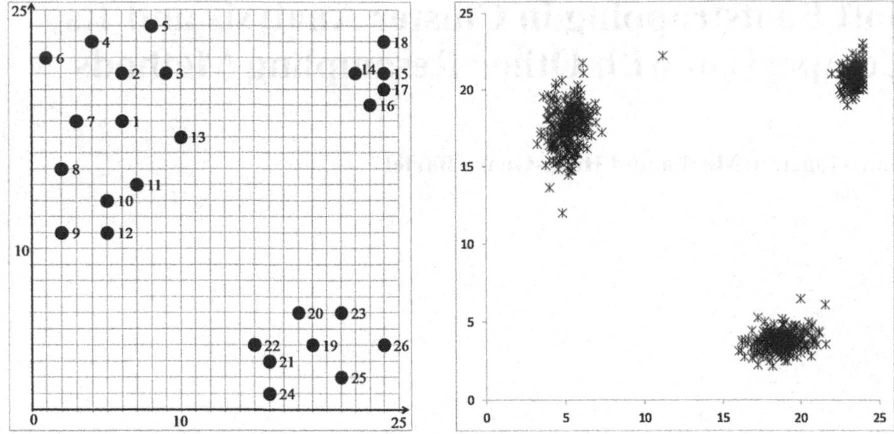

Fig. 1 Plot of the two dimensional toy data set A (on the *left hand side*). On the *right hand side*, the estimates of the locations of clusters are plotted. They are the result of hierarchical clustering of 250 bootstrap samples into three clusters

of an estimator based on the original data (Efron and Tibshrani 1993). This technique allows the estimation of the sampling distribution of almost any statistic using only very simple methods.

Generally, bootstrapping falls in the broader class of resampling methods. Some alternative resampling methods are sub-setting (draw a subsample to a smaller size without replacement) and jittering (add random noise to every measurement of each single observation). These alternatives or a combination of them are often applied in order to investigate the stability of clustering results. This is because no multiple observations disturb the clustering, as it is in bootstrapping where multiple observations can establish "artificial" mini-clusters.

In this paper, we consider non-parametric bootstrap sampling from the empirical distribution of a given data set in order to investigate the stability of results of cluster analysis. In doing so, statistical parameters such as location and variance of individual cluster means can be assessed. Figure 1 introduces a toy data set (plot on the left hand side). Obviously, there are three classes $C_1 = \{1, 2, \ldots, 13\}$, $C_2 = \{14, 15, \ldots, 18\}$, and $C_3 = \{19, 20, \ldots, 26\}$. The data values are integers. They can be taken directly from the plot. Cluster analysis aims at finding groups (clusters) in the data at hand. In addition, Fig. 1 shows estimates of the location parameters that are the result of hierarchical Ward clusterings of 250 non-parametric bootstrap samples of this toy data into three clusters (plot on the right hand side). Here, each bootstrap clustering provides estimates of three cluster centers. So, the plot represents $750 (= 250 * 3)$ estimates. Obviously, almost all estimates of the cluster centers reflect the true classes. The lonely star at the top in the center of the plot presents clearly a mixture of classes C_1 and C_2. Of course, such an estimation of locations is strictly restricted to metric data.

In clustering, the estimation of these parameters is not the main task. The final aim of clustering is the formation of groups either as a partition or a hierarchy of a given set of observations. Therefore, here the focus is on general investigation of stability based on partitions. This covers also hierarchies because they can be considered as a set of partitions. The stability of partitions and hierarchies as well as individual clusters can be assessed by different similarity measures such as the adjusted Rand or Jaccard index (Mucha 2007). Using such measures of comparing partitions or sets, the bootstrap simulations can help to answer the following questions:

- Are there clusters? And if so, how many clusters are most likely?
- What about the stability of individual clusters?
- What about the reliability of cluster membership of each observation?

In doing so, the partition of clustering of the original data set is compared with a bootstrap clustering. Of course, this comparison is repeated with many bootstrap samples. At the end, these comparisons are summarized. Obviously, the objective is the investigation of the stability of a clustering on three levels (Haimerl and Mucha 2007; Mucha 2007):

1. Determine the number of clusters that result in the most stable partition with respect to measures such as the adjusted Rand or the total Jaccard (the latter is the average over the Jaccard values of all individual clusters).
2. Calculate the individual stability for every cluster in a partition with respect to measures such as Jaccard index or Dice index (Hennig 2007).
3. Calculate the value of reliability of cluster membership for every observation in a cluster and test it with a significance level.

In the case of metric data, general stability investigations can be completed by the estimation of statistics of location parameters of clusters as shown in Fig. 1.

2 Resampling by Weighting of the Observations

Let us consider the problem of finding K clusters for I given observations. Without loss of generality, let us consider furthermore the well-known (generalized) sum of squares clustering criterion

$$V_K = \sum_{k=1}^{K} \frac{1}{M_k} \sum_{i \in C_k} m_i \sum_{l \in C_k, l > i} m_l d_{il}, \tag{1}$$

that has to be minimized concerning a partition into the clusters $C_k, k = 1, \ldots, K$, where $M_k = \sum_{i \in C_k} m_i$ and m_i denote the weight of cluster C_k and the weight of observation i, respectively. Here d_{il} are the pair-wise squared Euclidean distances between two observations i and l:

$$d_{il} = d(\mathbf{x}_i, \mathbf{x}_l) = (\mathbf{x}_i - \mathbf{x}_l)^T(\mathbf{x}_i - \mathbf{x}_l), \qquad (2)$$

where $\mathbf{x}_i$ and $\mathbf{x}_l$ are the vectors of measurements of the corresponding observations. Then, for instance, the partitional K means method (Späth 1985 in Chap. 7: "Criteria for given or computed distances not involving centres", Mucha et al. 2002) or the hierarchical Ward's method (Mucha 2009; Späth 1982) find a suboptimum solution of (1). In addition, the latter finds all partitions into $K = 2, 3, \ldots, I - 1$ clusters in one run only, and this result is (usually) unique. Therefore and without loss of generality, we focus here our investigation of stability of clustering on Ward's method (see for more details: Haimerl and Mucha 2007; Mucha and Haimerl 2005).

For simplicity, let us suppose in the following that the original weights of observations are $m_i = 1, i = 1, 2, \ldots, I$ ("unit mass"). Concretely, then the original bootstrap technique based on these masses can be formulated by choosing the following weights of observations:

$$m_i = \begin{cases} n \text{ if observation } i \text{ is drawn } n \text{ times} \\ 0 \text{ otherwise}. \end{cases} \qquad (3)$$

Here $I = \sum_i m_i$ holds in resampling with replacement. Observations with $m_i > 0$ are called selected or active ones, otherwise non-selected or supplementary ones. The latter do not affect the clustering. All statistical methods that make use (directly or indirectly) of masses can do bootstrapping based on (3). By the way, the "centers-free" criterion (1) allows computational efficient K means clusterings because the pairwise distances (2) remain unchanged in simulations based on resampling.

3 Soft Bootstrapping

Bootstrapping generates multiple observations. In clustering, this is often a disadvantage because they can be seen as mini-clusters in itself (Hennig 2007). Otherwise, on average, more than a third of the original observations are absent in the bootstrap sample. In the case of small sample size that can cause another disadvantage. Therefore, here we recommend another bootstrap method, called soft bootstrapping hereafter, that consists of random change of the bootstrap masses (3) to some degree:

$$m_i^* = \begin{cases} n - (qp/(I-q)) \text{ if observation } i \text{ is drawn } n \text{ times} \\ 0 + p \qquad\qquad \text{otherwise}. \end{cases} \qquad (4)$$

Table 1 An example for the toy data sets: Original bootstrap weights m_i and the corresponding soft bootstrap weights m_i^* with the softness parameter $p = 0.1$

	Observations							
Mass	1	2	3	4	5	...	25	26
m_i	1	1	4	0	2	...	3	0
m_i^*	0.956	0.956	3.956	0.100	1.956	...	2.956	0.100

Here q counts the number of observations that are not drawn by bootstrapping and p ($p > 0$) is a softness parameter close to 0. This resampling scheme of assigning randomized masses $m_i^* > 0$ (obviously, guaranteed by construction: $I = \sum_i m_i^*$) is especially appropriate for a small sample size because no object is totally excluded from the soft bootstrap sample. Table 1 shows examples of both the masses (3) coming from original bootstrapping of the toy data and their corresponding soft masses (4). Here the bootstrap sample consists of 18 (individual or multiple) observations, i.e., $q = 8$.

4 Validation of Clustering Using Soft Bootstrapping

Concretely, in (4), each observation gets a quite different degree of importance in the clustering process. The clustering algorithms itself must fulfill the condition that non-selected observations ($m_i^* = p$) do not initialize clusters either by merging two non-selected observations (in case of hierarchical methods) or by setting seed points in an initial partition (in case of K means and other partitional methods). These conditions are easy to realize in the algorithms to make the thing working.

What are stable clusters from a general statistical point of view? These clusters can be confirmed and reproduced to a high degree by using resampling methods. Thus, cluster analysis of a randomly drawn sample of the data should lead to similar results. The Jaccard index is the odds-on favorite to assess cluster stability (Hennig 2007). Both the adjusted Rand index and the total Jaccard index (pooled over all individual clusters) is most appropriate to decide about the number of clusters, i.e. to assess the total stability of a clustering. In a validation study, we compare an original cluster analysis result with many bootstrap clusterings based on these measures.

5 Comparison of Resampling Methods

In clustering, the so-called sub-sampling or sub-setting is another approach introduced by Hartigan (1969):

$$m_i^s = \begin{cases} 1 \text{ if observation } i \text{ is drawn randomly} \\ 0 \text{ otherwise .} \end{cases} \tag{5}$$

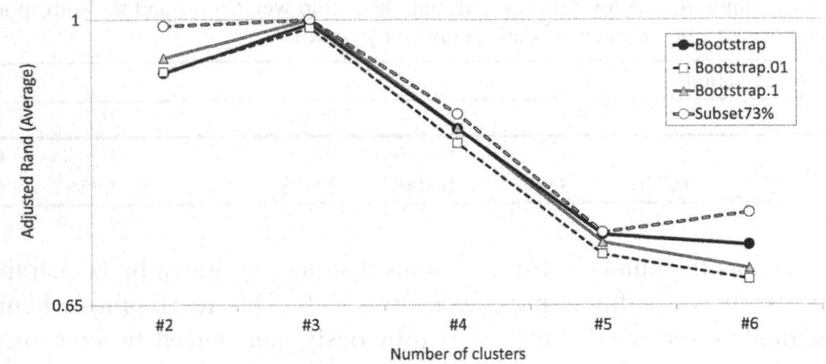

Fig. 2 Ward's clustering of toy data set A of Fig. 1: Comparison of different resampling schemes based on the adjusted Rand index

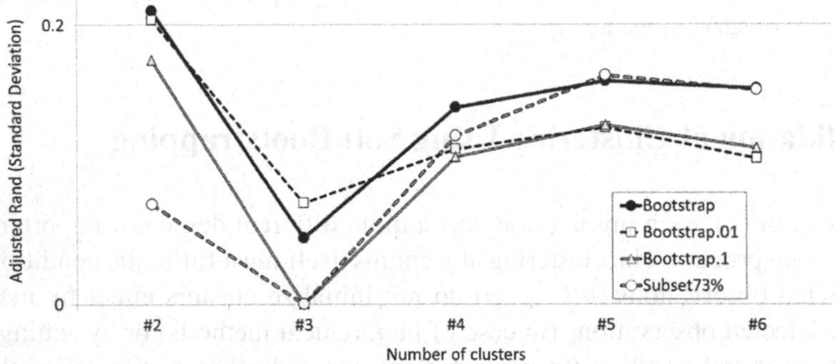

Fig. 3 Ward's clustering of toy data set A: Comparison of different resampling schemes based on the standard deviations of the adjusted Rand index, see Fig. 2

In contradiction to (3) this is resampling taken without replacement from the original data, and a parameter $L < I$ is needed: the cardinality of the drawn sample. As before, observations with $m_i^s = 0$ (non-selected objects) do not affect the cluster analysis in any way.

In the following we compare the applicability of different resampling techniques with respect to cluster analysis:

- original bootstrapping (3),
- soft bootstrapping (4) with different parameters $p = 0.01$, and $p = 0.1$, and
- sub-setting or sub-sampling (5) with $L = 19$ (i.e., about 73 % out of 26 observations of the toy data set are drawn randomly).

Figures 2 and 3 show the averages and standard deviations of the adjusted Rand index based on 250 simulations, respectively. The total Jaccard index behaves very similar to the adjusted Rand, see Fig. 4. Both take their maximum at $K = 3$. Surprisingly, the two cluster solution seem to be likely too. However, the variation of the adjusted Rand index has its distinct minimum at $K = 3$, see Fig. 3.

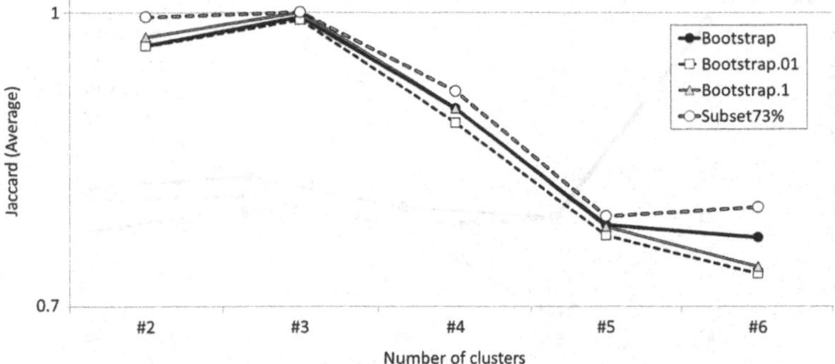

Fig. 4 Ward's clustering of toy data set A: Comparison of different resampling schemes based on the Jaccard index

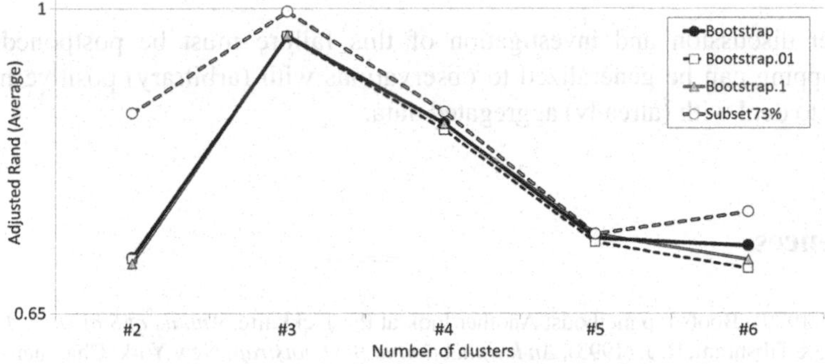

Fig. 5 Ward's clustering of toy data set B: Comparison of different resampling schemes based on the adjusted Rand index

The toy data B consists also of 26 observations in two dimensions as before but the space between the three classes is reduced by four units on each axis. Now it should be more difficult to find the three clusters. Figures 5 and 6 show the averages and standard deviations of the adjusted Rand index, respectively. All of a sudden, the vote for the three cluster solution is much more clear than in Fig. 2. As before in the case of toy data A, the Jaccard index behaves also very similar to the adjusted Rand and votes strictly for $K = 3$.

6 Summary

Soft bootstrapping gives similar simulation results in cluster analysis compared to original bootstrapping. The smaller the softness parameter p the more similar the results are. Surprisingly, for both toy data sets considered here, the sub-setting scheme seems not to be appropriate for detecting the number of clusters. However,

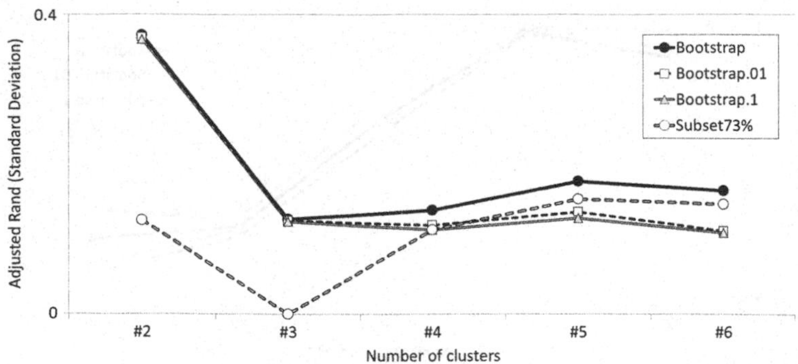

Fig. 6 Ward's clustering of toy data set B: Comparison of different resampling schemes based on the standard deviations of the adjusted Rand index, see Fig. 2

a further discussion and investigation of this failure must be postponed. Soft bootstrapping can be generalized to observations with (arbitrary) positive masses in order to deal with (already) aggregated data.

References

Efron, B. (1979). Bootstrap methods: Another look at the Jackknife. *Annals of Statistics*, *47*, 1–26

Efron, B., & Tibshrani, R. J. (1993). *An Introduction to the bootstrap*. New York: Chapman & Hall.

Haimerl, E., & Mucha, H.-J. (2007). Comparing the stability of different clustering results of dialect data. In R. Decker, & H.-J. Lenz (Eds.), *Advances in data analysis* (pp. 619–626). Berlin: Springer.

Hartigan, J. A. (1969). Using subsample values as typical values. *Journal of the American Statistical Association*, *64*, 1303–1317.

Hennig, C. (2007). Cluster-wise assessment of cluster stability. *Computational Statistics and Data Analysis*, *52*, 258–271.

Mammen, E. (1992). *When does bootstrap work?: Asymptotic results and simulations*. New York: Springer.

Mucha, H.-J. (2007). On validation of hierarchical clustering. In R. Decker, & H.-J. Lenz (Eds.), *Advances in data analysis* (pp. 115–122). Berlin: Springer.

Mucha, H.-J. (2009). ClusCorr98 for Excel 2007: Clustering, multivariate visualization, and validation. In H.-J. Mucha, & G. Ritter (Eds.), *Classification and clustering: Models, software and applications* (pp. 14–40). WIAS, Berlin, Report 26.

Mucha, H.-J., & Haimerl, E. (2005). Automatic validation of hierarchical cluster analysis with application in dialectometry. In C. Weihs, & W. Gaul (Eds.), *Classification - the ubiquitous challenge* (pp. 513–520). Berlin: Springer.

Mucha, H.-J., Simon, U., & Brüggemann, R. (2002). *Model-based cluster analysis applied to flow cytometry data of phytoplankton*. Technical Report No. 5 (http://www.wias-berlin.de/), WIAS, Berlin.

Späth, H. (1982). *Cluster analysis algorithms for data reduction and classification of objects*. Chichester: Ellis Horwood.

Späth, H. (1985). *Cluster dissection and analysis*. Chichester: Ellis Horwood.

Dual Scaling Classification and Its Application in Archaeometry

Hans-Joachim Mucha, Hans-Georg Bartel, and Jens Dolata

Abstract We consider binary classification based on the dual scaling technique. In the case of more than two classes many binary classifiers can be considered. The proposed approach goes back to Mucha (An intelligent clustering technique based on dual scaling. In: S. Nishisato, Y. Baba, H. Bozdogan, K. Kanefuji (eds.) Measurement and multivariate analysis, pp. 37–46. Springer, Tokyo, 2002) and it is based on the pioneering book of Nishisato (Analysis of categorical data: Dual scaling and its applications. The University of Toronto Press, Toronto, 1980). It is applicable to mixed data the statistician is often faced with. First, numerical variables have to be discretized into bins to become ordinal variables (data preprocessing). Second, the ordinal variables are converted into categorical ones. Then the data is ready for dual scaling of each individual variable based on the given two classes: each category is transformed into a score. Then a classifier can be derived from the scores simply in an additive manner over all variables. It will be compared with the simple Bayesian classifier (SBC). Examples and applications to archaeometry (provenance studies of Roman ceramics) are presented.

H.-J. Mucha (✉)
Weierstrass Institute for Applied Analysis and Stochastics (WIAS), Mohrenstraße 39, 10117 Berlin, Germany
e-mail: mucha@wias-berlin.de

H.-G. Bartel
Department of Chemistry at Humboldt University Berlin, Brook-Taylor-Straße 2, 12489 Berlin, Germany
e-mail: hg.bartel@yahoo.de

J. Dolata
Head Office for Cultural Heritage Rhineland-Palatinate (GDKE), Große Langgasse 29, 55116 Mainz, Germany
e-mail: dolata@ziegelforschung.de

M. Spiliopoulou et al. (eds.), *Data Analysis, Machine Learning and Knowledge Discovery*, Studies in Classification, Data Analysis, and Knowledge Organization, DOI 10.1007/978-3-319-01595-8_12,

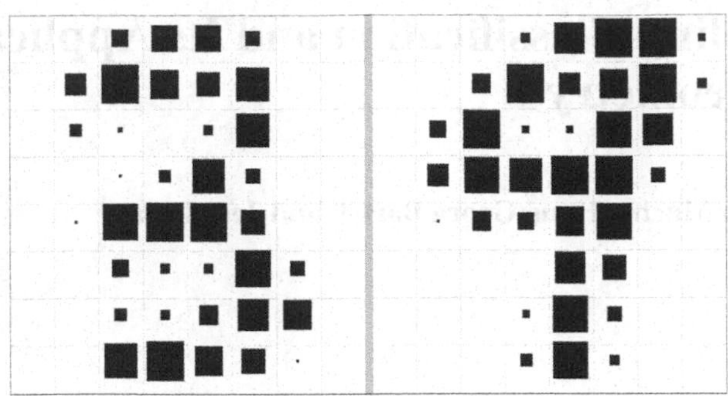

Fig. 1 OCR data: examples of images of the digits 3 and 9 in a 8×8 grid of pixels

1 Introduction

In this paper, we consider binary dual scaling classification (DSC). However, our approach of binary classification is not restricted explicitly to $K = 2$ classes. For $K > 2$ classes, this results in $(K-1)K/2$ binary classifiers. This is denoted as pairwise classification because one has to train a classifier for each pair of the K classes.

The DSC proposed here appears to be motivated by practical problems of analyzing a huge amount of mixed data efficiently. In archaeometry, for instance, both the chemical composition and petrographic characteristics (such as texture) of Roman tiles are investigated in order to find provenances (Giacomini 2005). One way to deal with such problems is down-grading all data to the lowest scale level, that is, downgrading to categories by loosing almost all quantitative information. Another general way is binary coding which is much more expensive in computer space and time, for details see Kauderer and Mucha (1998).

First, an application of the proposed method to optical character recognition (OCR) is presented, see at the website (Frank and Asuncion 2010). The basis of the ORC data are normalized bitmaps of handwritten digits from a preprinted form. From a total of 43 people, 30 contributed to the training set and different 13 to the test set. The resulting 32×32 bitmaps are divided into non-overlapping blocks of 4×4 and the number of pixels are counted in each block. This generates an input matrix of 8×8 where each element is an integer in the range $0 \ldots 16$. This reduces the dimensionality and gives invariance to small distortions. Figure 1 shows examples of digits 3 and 9: the first observation of "3" and "9" of the database, respectively. The area of a square represents the count of the corresponding element of the 8×8 matrix (64 variables).

Usually, these 64 ordinal variables with 17 categories at most can be directly processed by DSC. However, the number of categories is high with respect to the sample size of a few hundred and there still is an ordinal scale.

never	never	never	seldom	often	always	always	seldom
never	never	sometimes	always	sometimes	often	always	seldom
never	sometimes	always	seldom	seldom	often	often	never
never	sometimes	often	often	always	always	sometimes	never
never	seldom	sometimes	sometimes	often	often	never	never
never	never	never	never	often	sometimes	never	never
never	never	never	seldom	often	sometimes	never	never
never	never	never	sometimes	often	seldom	never	never

Fig. 2 The codes of the digit "9" of Fig. 1 after transformation into text data

Therefore (and because there is a general need of categorization in case of metric variables) here the original ORC data is down-graded exemplarily in a quite rough manner into at most five categories: "never" (count = 0), "seldom" (1–4), "sometimes" (5–11), "often" (12–15), and "always" (16). The corresponding transformation to do this is simply: if(c=0;"never";if(c=16;"always"; if(c>11;"often";if(c<5;"seldom";"sometimes")))) (where c stands for count and based on the standard function if(test;then;else)). Figure 2 shows a result of binning into categorical data as described by the simple formula above. The quantitative meaning of the target words is only for illustrative purposes. DSC makes no use of any ordering. However, especially in medical applications, special optimal scaling techniques for ordered categories can be of great interest, see for instance (Pölz 1995).

2 Binary Classification Based on Dual Scaling

The question arises: Can we build a good classifier nevertheless? The starting point is a $(I \times J)$-data table $\mathbf{X} = (x_{ij}), i = 1, 2, \ldots, I, j = 1, 2, \ldots, J$ of I observations and J variables. It is only supposed for DSC that the number of different values (or categories, words, ...) of each variable should be at least two. On the other hand, for reasons of getting a stable result for classifying unseen observations, the number of categories should be "as small as possible" with regard to the number of observations. In addition to $\mathbf{X}$, a binary class membership variable is required. It can be used in order to give categorical data a quantitative meaning. Generally, we want to obtain a new variable $\mathbf{z}_j$ so as to make the derived scores within the given classes as similar as possible and the scores between classes as different as possible (Nishisato 1980, 1994). The basis is a contingency table, which can be obtained by crossing a categorical variable $\mathbf{x}_j$ of M_j categories with the class membership variable. Considering the special case of two classes, dual scaling can be applied without the calculation of orthogonal eigenvectors. A given category

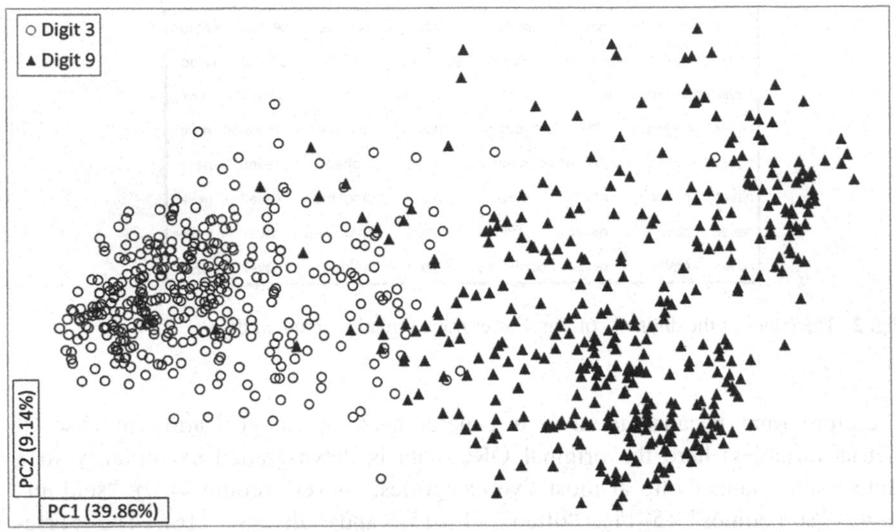

Fig. 3 OCR data: PCA plot of the two classes "digit 3" and "digit 9" based on DSC scores

y_{mj} $(m = 1, 2, \ldots, M_j)$ of a variable j is transformed into an optimally scaled in sense of maximal between classes variances by

$$u_{mj} = \frac{p_{mj}^{(1)}}{p_{mj}^{(1)} + p_{mj}^{(2)}}, \quad j = 1, 2, \ldots, J, m = 1, 2, \ldots, M_j. \tag{1}$$

Here $p_{mj}^{(1)}$ is an estimate of the probability for being a member of class 1 when coming from category m of variable j, whereas on the other side $p_{mj}^{(2)}$ is an estimate of the probability for being a member of class 2 when coming from category m of variable j. For further details, see Kauderer and Mucha (1998) and Mucha (2002). The final result of the transformations (1) is a quantitative data matrix $\mathbf{Z} = (z_{ij})$, $i = 1, 2, \ldots, I$, $j = 1, 2, \ldots, J$. To be concrete, $\mathbf{Z}$ is obtained from $\mathbf{X}$ by replacing each category by its corresponding score (1). Figure 3, that presents a multivariate view on $\mathbf{Z}$ by principal component analysis (PCA), emphasizes the inhomogeneity of the class "digit 9". This fact becomes also visible in density plot in Fig. 4.

The well-known correspondence analysis (Greenacre 1989) is simply related to (1) by a scale factor $b = \sqrt{I_1 I_2 / I}$ and a shift c, where I_1 and I_2 is the number of observations in class 1 and class 2, respectively, and c is a constant (Mucha 2002). Concerning other scaling approaches, see Gebelein (1941), Fahrmeir and Hamerle (1984), and Pölz (1988).

After recoding the categories of $\mathbf{X}$ all data values of $\mathbf{Z} = (z_{ij})$ are in the interval [0,1]. Without loss of generality the hypothetical worst-case $\mathbf{z}_w \equiv \mathbf{0}$ is considered here (this naming comes from credit scoring, see Kauderer and Mucha 1998).

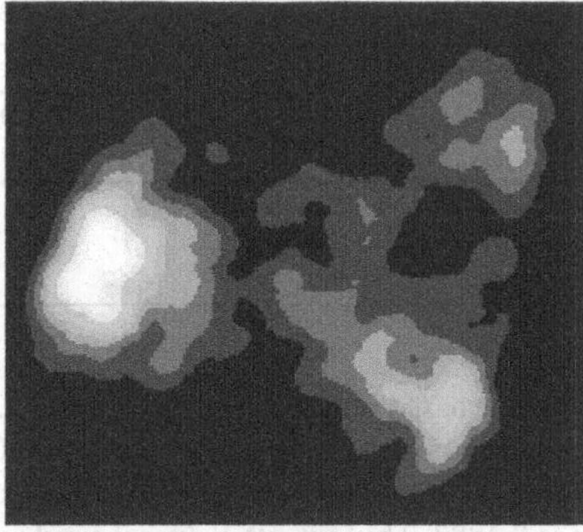

Fig. 4 Several cuts of the nonparametric density estimation based on the first two PC of Fig. 3

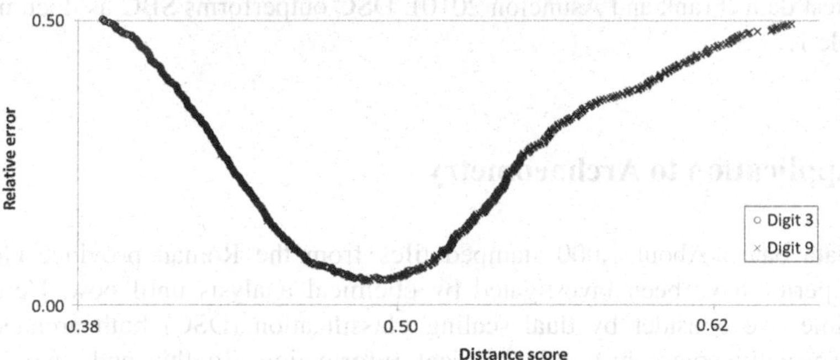

Fig. 5 OCR data: plot of distance scores t_{iw} (2) versus the classification error

Then the Manhattan distance t_{iw} between an observation $\mathbf{z}_i$ and the worst case $\mathbf{z}_w$ has both the suitable and really simple form

$$t_{iw} = t(\mathbf{z}_i, \mathbf{z}_w) = \sum_{j=1}^{J} z_{ij}. \qquad (2)$$

We call t_{iw} the distance score. In contrast, the SBC, sometimes called Naive–Bayes (Good 1965), is based on a conditional independence model of each variable given the class. Once the discrete probabilities $p_{mj}^{(1)}$ and $p_{mj}^{(2)}$ of (1) are estimated, the posterior probabilities can be computed easily as a by-product. Concretely, they have to be computed for each individual class, and the largest one predicts the class.

For the training data, Fig. 5 visualizes the relation between the distance scores and the error rate with respect to the class membership variable. Now the question

Table 1 OCR test data: comparison of results of dual scaling and Naive Bayes classification

Given	Dual scaling classification			Naive Bayes classification		
	Confusion matrix		Error	Confusion matrix		Error
Classes	Class 1	Class 2	rate	Class 1	Class 2	rate
Digit "3"	164	19	10.38 %	160	23	12.57 %
Digit "9"	9	171	5.00 %	7	173	3.89 %
Total	173	190	**7.71 %**	167	196	**8.26 %**

is how to build up a classifier based on the distance scores (2). One simple way is looking for a suitable cut-off-point on the distance score axis. Obviously, in Fig. 5, the optimum error rate is near 0.04 (4.15 %). Here the corresponding cut-off-value of the minimum error rate is 0.4883. This cut-off-values is used as a classifier later on for a decision-making on new observations. For comparison, the SBC, also based on the categorized data, has a slightly lower error rate of 3.24 %. Concerning the OCR test data (Frank and Asuncion 2010), DSC outperforms SBC as documented in Table 1.

3 Application to Archaeometry

Our data basis: About 1,000 stamped tiles from the Roman province Germania Superior have been investigated by chemical analysis until now. Here, the first time, we consider by dual scaling classification (DSC) both archaeological (especially epigraphy) and chemical information. To this end, two of the most important and already confirmed areas for brick-production (provenances), Straßburg–Königshofen ($I_1 = 113$) and Groß–Krotzenburg ($I_2 = 63$), will be investigated. They are located side by side in a PCA plot of eight provenances, see Fig. 19 in Mucha (2009). Altogether, two archaeological categorical variables (type of the tile, name of the military unit) and nine oxides and ten trace elements are the starting point for DCS.

Preparation of the metric data: The chemical elements are measured with X-ray fluorescence analysis (XRF) in quite different scales. Thus, first the metric data matrix $\mathbf{Y} = (y_{ij})$ of these measurements is standardized by:

$$v_{ij} = \frac{y_{ij} - y_j^*}{s_j} \quad i = 1, 2, \ldots, I, \;\; j = 1, 2, \ldots, J,$$

where y_j^* is the minimum and s_j is the standard deviation of variable j. Then the v_{ij} are non-negative values. Second, almost all quantitative information is lost by down-scaling to categories simply by:

No.	S_iO_2	T_iO_2	Al_2O_3	Fe_2O_3	MnO	MgO	CaO	Na_2O	K_2O	V	Cr	Ni	Zn	Rb	Sr	Y	Zr	Nb	Ba	Type	Stamp
F043	c5	c3	c2	c0	c0	c1	c0	c0	c0	c2	c1	c5	c0	c2	c0	c0	c1	c3	c0	tegula	no stamp
F056	c3	c2	c3	c3	c0	c2	c0	c0	c1	c3	c2	c5	c0	c3	c1	c0	c0	c2	c0	unknown	XXII
F092	c2	c2	c4	c2	c0	c3	c0	c0	c1	c4	c1	c6	c2	c4	c3	c0	c0	c2	c0	tegula	K IV
F181	c4	c4	c2	c1	c0	c2	c0	c0	c1	c4	c1	c6	c0	c3	c1	c1	c1	c2	c0	brick-kiln	no stamp
G068	c5	c1	c1	c1	c0	c2	c0	c0	c0	c1	c2	c4	c1	c2	c0	c0	c0	c1	c0	later	K IV
G090	c3	c3	c2	c2	c1	c2	c0	c0	c1	c2	c3	c5	c2	c3	c1	c0	c1	c1	c0	later	K IV
G095	c4	c3	c2	c0	c0	c1	c1	c0	c0	c1	c3	c5	c2	c1	c6	c0	c1	c1	c1	tegula	XXII
G102	c1	c3	c4	c4	c1	c3	c0	c0	c2	c3	c4	c7	c2	c3	c0	c1	c0	c1	c0	later	K IV
H198	c4	c3	c2	c1	c0	c1	c0	c0	c0	c0	c2	c3	c0	c3	c1	c1	c2	c2	c0	later	CONSIUS
H277	c4	c3	c2	c1	c1	c2	c0	c0	c0	c1	c2	c3	c1	c3	c1	c1	c1	c1	c0	tegula	VICCOMA

Fig. 6 Snapshot of the first part of the input matrix $\mathbf{X} = (x_{ij})$ for DSC after down-grading the measurements to characters. The last two archaeological variables were already categorical ones

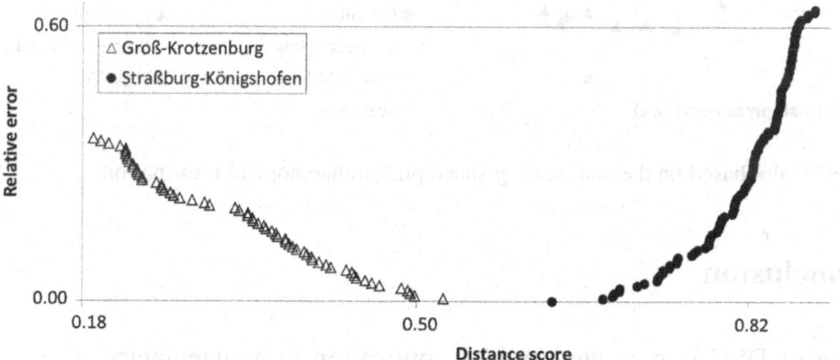

Fig. 7 Result of DSC: the two classes look well-separated and no errors occur

$$x_{ij} = \text{“}c\text{”} \ \& \ \texttt{rounddown}(v_{ij}; 0),$$

where the function `rounddown` makes integer values and "c" makes text data.

Figure 6 shows a part of the result of binning into categorical data as described by the simple formula above. As before in the case of the ORC data, $\mathbf{X} = (x_{ij})$ is the starting point for both DSC and SBC. Figure 7 visualizes the relation between the distance scores and the error rate. Clearly to see, no errors are counted when using a cut-off-value of approximately 0.55. All tiles from Straßburg–Königshofen are located on the right hand side. SBC performs similar: no errors are counted. Finally, Fig. 8 presents the PCA plot based on $\mathbf{Z}$. As before, also this figure shows that DSC does a good job of discriminating between the two provenances Straßburg–Königshofen (on the right hand side: marked by circles) and Groß–Krotzenburg. The latter looks much more inhomogeneous. Also, for both DSC and SBC, no errors are counted by cross validation. This is a helpful result for archaeological investigation of military places of late second and early third century AD. In addition, in Fig. 8, the type of the stamp of each tile is marked by a symbol.

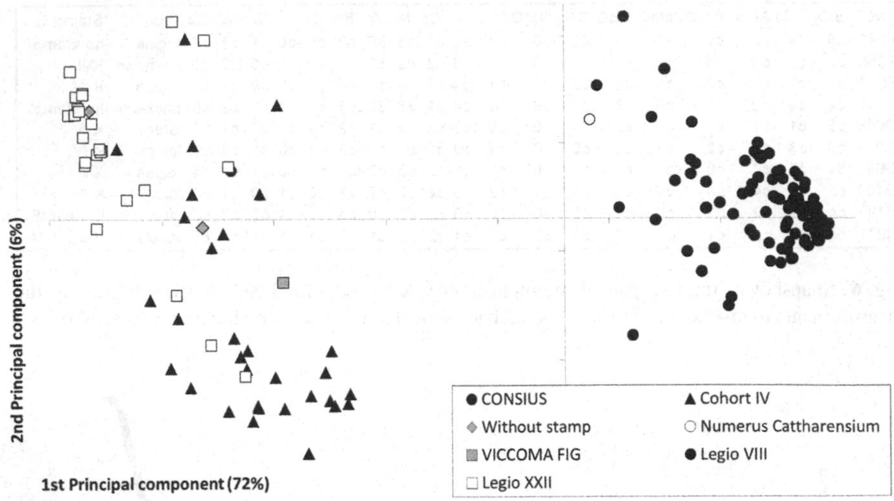

Fig. 8 PCA plot based on the dual scaling space plus archaeological information

4 Conclusion

The idea of DSC is presented and its application to archaeometry. DSC can be applied in case of mixed data. Missing values cause also no problem because they can be seen as an additional category. DSC performs similar to SBC. In addition, DSC allows multivariate graphics based on the scores. It is a computationally efficient classification method. Sure, DSC can be improved by sophisticated categorization methods.

References

Fahrmeir, L., & Hamerle, A. (1984). Multivariate statistische Verfahren. Berlin: Walter de Gruyter.

Frank, A., & Asuncion, A. (2010). *UCI Machine Learning Repository* [http://archive.ics.uci.edu/ml]: [http://archive.ics.uci.edu/ml/datasets/Optical+Recognition+of+Handwritten+Digits]. University of California: School of Information and Computer Science, Irvine.

Gebelein, M. J. (1941). Das statistische Problem der Korrelation als Variations- und Eigenwertproblem und sein Zusammenhang mit der Ausgleichsrechnung. *Zeitschrift für angewandte Mathematik und Mechanik, 21*, 365–375.

Giacomini, F. (2005). The Roman stamped tiles of Vindonissa (1st century AD., Northern Switzerland). *Provenance and technology of production - an archaeometric study. BAR international series* (Vol. 1449), Oxford: Archaeopress.

Good, I. J. (1965). *The estimation of probabilities: An essay on modern Bayesian methods.* Cambridge: MIT.

Greenacre, M. J. (1989). *Theory and applications of correspondence analysis* (3rd edn.). London: Academic.

Kauderer, H., & Mucha, H. J. (1998). Supervised learning with qualitative and mixed attributes. In I. Balderjahn, R. Mathar, & M. Schader (Eds.), *Classification, data analysis, and data highways* (pp. 374–382). Berlin: Springer.

Mucha, H. J. (2002). An intelligent clustering technique based on dual scaling. In S. Nishisato, Y. Baba, H. Bozdogan, & K. Kanefuji (Eds.), *Measurement and multivariate analysis* (pp. 37–46). Tokyo: Springer.

Mucha, H. J. (2009). ClusCorr98 for Excel 2007: Clustering, multivariate visualization, and validation. In H. J. Mucha, & G. Ritter (Eds.), *Classification and clustering: Models, software and applications* (pp. 14–40). WIAS, Berlin, Report No. 26.

Nishisato, S. (1980). *Analysis of categorical data: Dual scaling and its applications*. Toronto: The University of Toronto Press.

Nishisato, S. (1994). *Elements of dual scaling: An introduction to practical data analysis*. Hillsdale: Lawrence Erlbaum Associates Publishers.

Pölz, W. (1988). *Ein dispositionsstatistisches Verfahren zur optimalen Informationsausschöpfung aus Datensystemen mit unterschiedlichem Ausprägungsniveau der Merkmale*. Frankfurt: Peter Lang.

Pölz, W. (1995). Optimal scaling for ordered categories. *Computational Statistics*, *10*, 37–41.

Gamma-Hadron-Separation in the MAGIC Experiment

Tobias Voigt, Roland Fried, Michael Backes, and Wolfgang Rhode

Abstract The MAGIC-telescopes on the canary island of La Palma are two of the largest Cherenkov telescopes in the world, operating in stereoscopic mode since 2009 (Aleksić et al., Astropart. Phys. 35:435–448, 2012). A major step in the analysis of MAGIC data is the classification of observations into a gamma-ray signal and hadronic background. In this contribution we introduce the data provided by the MAGIC telescopes, which has some distinctive features. These features include high class imbalance, unknown and unequal misclassification costs as well as the absence of reliably labeled training data. We introduce a method to deal with some of these features. The method is based on a thresholding approach (Sheng and Ling 2006) and aims at minimization of the mean square error of an estimator, which is derived from the classification. The method is designed to fit into the special requirements of the MAGIC data.

1 Introduction

Binary classification problems are quite common in scientific research. In very high energy (VHE) gamma-ray astronomy for example, the interest is in separating the gamma-ray signal from a hadronic background. The separation has to be done as exactly as possible since the number of gamma-ray events detected is needed for the calculation of energy spectra and light curves (Mazin 2007). There are some distinctive features characterizing the data we have to deal with.

T. Voigt (✉) · R. Fried
Faculty of Statistics, TU Dortmund University, Vogelpothsweg 87, 44227 Dortmund, Germany
e-mail: voigt@statistik.tu-dortmund.de; fried@statistik.tu-dortmund.de

M. Backes · W. Rhode
Physics Faculty, TU Dortmund University, Otto-Hahn-Straße 4, 44227 Dortmund, Germany
e-mail: backes@physik.tu-dortmund.de; rhode@physik.tu-dortmund.de

M. Spiliopoulou et al. (eds.), *Data Analysis, Machine Learning and Knowledge Discovery*, Studies in Classification, Data Analysis, and Knowledge Organization, DOI 10.1007/978-3-319-01595-8_13,

One feature is that there is a huge class imbalance in the data. It is known that hadron observations (negatives) are more than 100–1,000 times more common than gamma events (positives) (Weekes 2003; or Hinton and Hofman 2009). The exact ratio, however, is unknown. A second feature is that individual misclassification costs of gamma and hadron observations are unknown and not important in our context. We use classification as a preliminary step of an analysis, which aims at estimation of some quantity. The mean square error of the resulting estimator thus measures naturally also the expected loss resulting from our classification.

Throughout this paper we use random forests (Breiman 2001) as is usually done in the MAGIC experiment (Albert et al. 2008). One effective method of making these cost sensitive is the thresholding method (Sheng and Ling 2006). This method is not applicable as we do not know individual misclassification costs, but in the following we introduce a similar method based on the third feature of the data: In VHE gamma-ray astronomy one is not primarily interested in the best possible classification of any single event, but instead one wants to know the total number of gamma observations (positives) as this is the starting point for astrophysical interpretations. Statistically speaking this means estimation of the true number of positives based on a training sample. As said above the mean square error of this estimation measures naturally the expected loss of the classification, so we regard the mean square error (MSE) as overall misclassification risk in the thresholding method and choose the discrimination threshold which minimizes the MSE of the estimated number of positives in a data set. Additionally, the unknown class imbalance is taken into consideration by this method.

2 The MAGIC Experiment and Data

The MAGIC telescopes on the canary island of La Palma are two of the biggest Cherenkov telescopes in the world. Their purpose is to detect highly energetic gamma particles emitted by various astrophysical sources like Active Galactic Nuclei (AGNs). Gamma particles are of special interest to astrophysicists, because they are not scattered by magnetic fields, so that their point of origin can be reconstructed. When a gamma particle reaches Earth, it interferes with the atmosphere, inducing a so called air shower of secondary particles. The air shower then emits Cherenkov light in a cone, which can be seen by Cherenkov telescopes like the MAGIC telescopes. The somewhat elliptical shape of the shower is imaged in the telescopes' cameras. A major issue one has to solve in the MAGIC experiment is that not only gammas induce particle showers, but also many other particles, summarized as hadrons. Thus, gamma and hadron particles have to be separated through classification. Figure 1 shows camera images of the MAGIC I telescope of a gamma and hadron event. As can be seen, the gamma event has a more regular shape than the hadron. Note though that these images are almost ideal cases. Usually the difference between the two types of particles cannot be seen that easily. Figure 1 also shows the raw data we have for the analysis. It consists of one light intensity

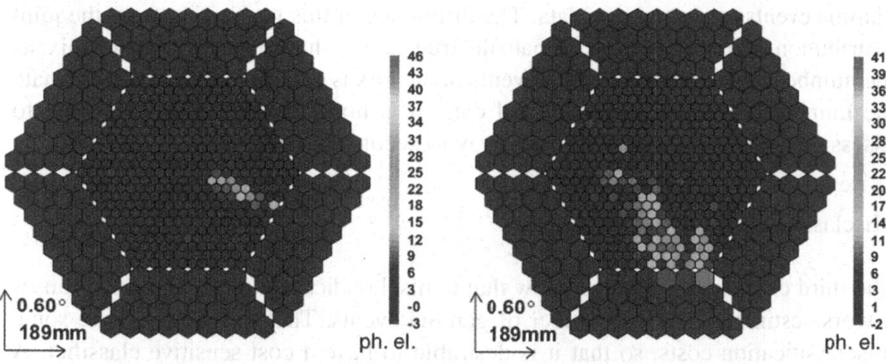

Fig. 1 Camera images of a gamma event (*left*) and a hadron event (*right*) in the MAGIC experiment

for each pixel in the camera. Additionally, but not shown here, a time information is given for each pixel.

One of the major goals of the MAGIC experiment is the Unfolding of Energy Spectra. Energy Spectra are basically histograms of the energy of observed gamma particles, that is an estimate of the unknown energy distribution of a source. From this distribution, characteristics of the source can be inferred. That means, what we are aiming for is to estimate the number of gamma observations in each of the histogram's energy bins as precisely as possible, to get a good estimation of the true energy distribution.

In order to achieve this goal one has to deal with some challenges.

Very Unfortunate Signal-Background Ratio

According to Weekes (2003) hadron events are around 100–1,000 times more common than gamma events. This leads to a very undesirable signal-to-background ratio for our classification. This high class imbalance makes classification of gammas and hadrons much more difficult than it could be with a more desirable ratio. Additionally, the true ratio is different for each source one is taking data from. So we cannot use any a priori knowledge about the ratio to make classification easier.

No Reliably Labeled Training Data

Another challenge one is facing in the analysis of MAGIC data is that we do not have access to training data to train the random forest. That means we cannot draw a sample from the joint distribution of gammas and hadrons with known labels. What we can do is to take real hadronic background data and mix it with simulated gamma events to get training data. The difference of this to drawing from the joint distribution is that we cannot estimate the true gamma-hadron-ratio from the mix, as the number of gamma and hadron events in the mix is chosen manually. To estimate the number of gamma events in real data, it is however necessary to be able to assess this ratio. So we have to find a way to accomplish this.

Misclassification Costs

The third challenge is that we know that a misclassification of observations causes a worse estimation of the number of gamma events. That means, there are some misclassification costs, so that it is desirable to have a cost-sensitive classifier. A random forest, which we use in the MAGIC analysis chain, can be made cost sensitive in various ways. One is the thresholding method by Sheng and Ling (2006). The idea of this method is to minimize the misclassification costs over the classification threshold in the random forest's output, that is the fraction of votes of the trees. However, to apply this method it is of course necessary to know the misclassification costs. So to make the classifier cost sensitive, we must first assess the misclassification costs.

3 Threshold Optimization

An example of how we try to achieve a good estimation of the energy spectrum is the optimization of the threshold in the outcome of the random forest.

Problem Setup

The problem we are facing is a binary classification problem. We have a random vector of input variables $\mathbf{X} = (X_1, \ldots, X_m)^T$ and a binary classification variable Y. $\mathbf{X}$ and Y have the joint distribution $P(\mathbf{X}, Y)$. We neither know this distribution, nor can we make any justifiable assumptions about it. Additionally, in our application it is not possible to draw a sample from this distribution. We are, however, able to draw samples from P(**X**) as well as the conditional distributions $P(\mathbf{X}|Y = 0)$ and $P(\mathbf{X}|Y = 1)$. Thus, we have independent realizations $(\mathbf{x}_1, 0), \ldots, (\mathbf{x}_{n_{0\cdot}}, 0)$ and $(\mathbf{x}_{n_{0\cdot}+1}, 1), \ldots, (\mathbf{x}_{n_{1\cdot}+n_{0\cdot}}, 1)$ from the respective distributions with sizes $n_{0\cdot}$ and $n_{1\cdot}$, respectively, and $n = n_{0\cdot} + n_{1\cdot}$.

Many classifiers can be interpreted as a function $f : \mathbb{R}^m \to [0, 1]$. In the MAGIC experiment we use random forests, but any classifier which can be regarded as such a function f can be used. For a final classification into 0 and 1 we need a threshold c, so that

$$g(\mathbf{x}; c) = \begin{cases} 0, & \text{if } f(\mathbf{x}) \leq c \\ 1, & \text{if } f(\mathbf{x}) > c \end{cases}.$$

Table 1 True and classified numbers of positives and negatives in a training sample (left), in a sample of actual data (middle) and in Off data (right)

		Classified		
		1	0	$\sum$
True	1	n_{11}	n_{10}	$n_{1\cdot}$
	0	n_{01}	n_{00}	$n_{0\cdot}$
	$\sum$	$n_{\cdot 1}$	$n_{\cdot 0}$	n

		Classified		
		1	0	$\sum$
True	1	N_{11}	N_{10}	$N_{1\cdot}$
	0	N_{01}	N_{00}	$N_{0\cdot}$
	$\sum$	$N_{\cdot 1}$	$N_{\cdot 0}$	N

		Classified		
		1	0	$\sum$
True	1	0	0	0
	0	N_{01}^{off}	N_{00}^{off}	$N_{0\cdot}^{off}$
	$\sum$	$N_{\cdot 1}^{off}$	$N_{\cdot 0}^{off}$	N^{off}

We consider f to be given and only vary c in this paper. There are several reasons why we consider f to be given. Among other reasons, we want to adapt the thresholding method by Sheng and Ling (2006) and we cannot change the MAGIC analysis chain too drastically, as all changes need approval of the MAGIC collaboration.

In addition to the training data, we have a sample of actual data to be classified, $\mathbf{x}_1^*, \ldots, \mathbf{x}_N^*$, for which the binary label is unknown. This data consists of N events with $N_{1\cdot}$ and $N_{0\cdot}$ defined analogously to $n_{1\cdot}$ and $n_{0\cdot}$, but unknown.

As we have stated above, we only have simulated gamma events as training data and therefore need additional information to assess the gamma-hadron-ratio in the real data. This additional information is given by Off data, which only consists of $N_{0\cdot}^{off}$ hadron events. $N_{0\cdot}^{off}$ can be assumed to have the same distribution as $N_{0\cdot}$, so the two sample sizes should be close to each other. In fact, a realization of $N_{0\cdot}^{off}$ is an unbiased estimate for $N_{0\cdot}$. From this Off data we are able to estimate the true gamma-hadron-ratio.

For a given threshold c we denote the numbers of observations in the training data after the classification as $n_{ij}, i, j \in \{1, 0\}$, where the first index indicates the true class and the second index the class as which the event was classified. We can display these numbers in a 2×2 table. With $N_{ij}, i, j \in \{1, 0\}$ and $N_{0j}, j \in \{1, 0\}$ defined analogously we get similar 2×2 tables for the actual data and the Off data, see Table 1.

It is obvious that we do not know the numbers in the first two rows of the table for the actual data as we do not know the true numbers of positives and negatives $N_{1\cdot}$ and $N_{0\cdot}$.

As we can see above, $N_{1\cdot}$ is considered to be a random variable and our goal is to estimate, or perhaps better predict, the unknown realization of $N_{1\cdot}$. The same applies to $N_{0\cdot}$. That is why we consider all the following distributions to be conditional on these values.

We additionally define the True Positive Rate (*TPR*), which is also known as Recall or (signal) Efficiency, and the False Positive Rate (*FPR*) as

$$TPR = \frac{n_{11}}{n_{1\cdot}} \tag{1}$$

and

$$FPR = \frac{n_{01}}{n_{0\cdot}}, \tag{2}$$

respectively. As we will see in the following section, these two values are important in the estimation of the number of gamma events.

Estimating the Number of Gamma-Events

To estimate the number of gamma events, we first have a look at the following estimator:

$$\tilde{N}_{1\cdot} = \frac{1}{p_{11}} \left(N_{\cdot 1} - N_{01}^{off} \right). \tag{3}$$

where p_{11} is the (unknown) probability of classifying a gamma correctly. This estimator could be used if we knew p_{11}. It takes the difference between $N_{\cdot 1}$ and N_{01}^{off} as an estimate for N_{11} and multiplies this with $\frac{1}{p_{11}}$ to compensate for the classification error in the signal events.

Since we want to estimate the number of positives as precisely as possible we want to assess the quality of the estimator $\tilde{N}_{1\cdot}$. A standard measure of the quality of an estimator is the mean square error (MSE). As in applications we usually have fixed samples in which we want to estimate $N_{1\cdot}$, we calculate the MSE conditionally on $N_{1\cdot}$, $N_{0\cdot}$ and $N_{0\cdot}^{off}$. Under the assumption that N_{i1}, N_{01}^{off} and n_{i1}, $i \in \{1, 0\}$ are independent and (conditionally) follow binomial distributions, the conditional MSE of $\tilde{N}_{1\cdot}$ can easily be calculated. It is:

$$\begin{aligned} \mathrm{MSE}\left(\tilde{N}_{1\cdot} | N_{1\cdot}, N_{0\cdot}, N_{0\cdot}^{off}\right) &= \frac{p_{01}^2}{p_{11}^2} \left(N_{0\cdot} - N_{0\cdot}^{off} \right)^2 \\ &\quad + N_{1\cdot} \left(\frac{1}{p_{11}} - 1 \right) + \frac{p_{01} - p_{01}^2}{p_{11}^2} \left(N_{0\cdot} + N_{0\cdot}^{off} \right) \end{aligned} \tag{4}$$

where p_{01} is the probability of classifying a hadron as gamma and p_{11} is the probability of classifying a gamma correctly. As we do not know these values we have to estimate them. Consistent estimators for these values are *TPR* and *FPR* [(1) and (2)]. Using *TPR* as an estimator for p_{11} in (3) we get

$$\hat{N}_{1\cdot} = \frac{n_{1\cdot}}{n_{11}} \left(N_{\cdot 1} - N_{01}^{off} \right) = \frac{1}{TPR} \left(N_{\cdot 1} - N_{01}^{off} \right). \tag{5}$$

By estimating p_{11} with *TPR* and p_{01} with *FPR* in (4) we get the estimate

$$\begin{aligned} \widehat{\mathrm{MSE}}\left(\hat{N}_{1\cdot} | N_{1\cdot}, N_{0\cdot}, N_{0\cdot}^{off}\right) &= \frac{FPR^2}{TPR^2} \left(N_{0\cdot} - N_{0\cdot}^{off} \right)^2 \\ &\quad + N_{1\cdot} \left(\frac{1}{TPR} - 1 \right) + \frac{FPR - FPR^2}{TPR^2} \left(N_{0\cdot} + N_{0\cdot}^{off} \right). \end{aligned} \tag{6}$$

As *TPR* and *FPR* are consistent estimators of p_{11} and p_{01} and the sample sizes $n_{1\cdot}$ and $n_{0\cdot}$ are usually high ($> 10^5$), using the estimates instead of the true probabilities should only lead to a marginal difference.

Algorithm

Equations (5) and (6) can be used in an iterative manner to find a discrimination threshold, although $N_{1\cdot}$ in (6) is unknown. To find a threshold we alternately estimate $N_{1\cdot}$ and calculate the threshold:

1. Set an initial value c for the threshold.
2. With this threshold estimate $N_{1\cdot}$ using equation (5).
3. Compute a new threshold through minimizing equation (6) over all thresholds using the estimates $\hat{N}_{1\cdot}$ for $N_{1\cdot}$ and $N - \hat{N}_{1\cdot}$ for $N_{0\cdot}$.
4. If a stopping criterion is fulfilled, compute a final estimate of $N_{1\cdot}$ and stop. Otherwise go back to step 2.

Because negative estimates $\hat{N}_{1\cdot}$ can lead to a negative estimate of the MSE, we set negative estimates to 0. As a stopping criterion, we require that the change in the cut from one iteration to the next is below 10^{-6}. First experiences with the algorithm show that the convergence is quite fast. The stopping criterion is usually reached in less than ten iterations.

In the following we refer to this algorithm as the MSEmin method. This method takes both into consideration: The problem of class imbalance and the minimization of the MSE, that is, the overall misclassification costs. In the next section we investigate the performance of this algorithm on simulated data and compare it to other possible approaches.

4 Application

It is now of interest, if the MSEmin method proposed above means an improvement over the currently used and other methods. As stated above, we want to estimate the number of gamma events depending on the energy range. We therefore use the methods on each of several energy bins individually by splitting the data sets according to the (estimated) energy of the observations and using the methods on each of these subsamples individually.

The method currently in use in the MAGIC experiment is to choose the threshold manually so that the *TPR* is "high, but not too high". Often *TPR* is set to values between 0.4 and 0.9 (e.g. Aleksić et al. 2010) and the threshold is chosen accordingly. For our comparison we look at values of 0.1–0.9 for *TPR*. We call these methods Recall01, …, Recall09.

An approach to avoid energy binning is to fit a binary regression model to the random forest output, with energy as the covariate. The fitted curve can then be regarded as discrimination threshold. In this paper we use a logistic regression model. We fit the model to the training data using a standard Maximum Likelihood approach to estimate the model coefficients.

As the proposed MSEmin method is quite general and not bound to optimizing a fixed threshold, we use it in an additional approach by combining it with logistic regression. Instead of minimizing the MSE over possible fixed thresholds, we search for optimal parameters of the logistic regression curve, so that the MSE becomes minimal. The procedure is the same as for the MSEmin method proposed in the algorithm above, only that we exchange the threshold c with the two parameters of the logistic regression, say β_0 and β_1. For initialization we use the ordinary ML-estimates of the two parameters.

We use all these methods on 500 test samples and check which method gives the best estimate for the number of gamma events. We focus here on the hardest classification task with a gamma-hadron-ratio of 1:1000. For the comparison we use the following data:

Test data: To represent actual data we simulate 500 samples for each gamma-hadron-ratio 1:100, 1:200, ..., 1:1000. The number of hadron-events in each sample is drawn from a Poisson distribution with mean 150,000. The number of gamma events is chosen to match the respective ratio.

Training data: We use 652,785 simulated gamma observations and 58,310 hadron observations to represent the training data from which *TPR* and *FPR* are calculated. Note that the ratio of gammas and hadrons in this data has no influence on the outcome, as only *TPR* and *FPR* are calculated from this data.

Off data: For each test sample we draw a sample of hadron observations to represent the Off data. The number of hadrons in each sample is drawn from a Poisson distribution with mean 150,000.

The result can be seen in Fig. 2. As we can see, all methods seem to give unbiased estimates. However, all Recall methods have comparably high variances when estimating the true number of gamma events, with the best one being Recall01. Our proposed method MSEmin leads to a smaller variance and therefore performs better than all of them. The results of the logistic regression approach is quite similar to the MSEmin method, but has a bit smaller errors. The best performance is given by the combination of the methods MSEmin and logistic regression.

5 Conclusions and Outlook

MAGIC data has some distinctive features, which make the analysis of the data difficult. We have illustrated that major challenges can be overcome when we focus on the overall aim of the analysis, which is the estimation of the number of signal events.

We introduced a method to choose an optimal classification threshold in the outcome of a classifier, which can be regarded as a function mapping to the interval

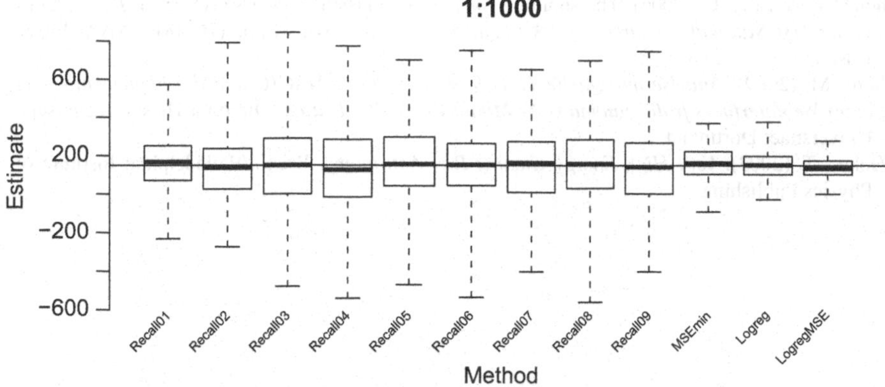

Fig. 2 Boxplots of the estimates in the 500 samples with a gamma-hadron-ratio of 1:1000. The *thick line* in the middle of each box represents the median of the estimates. Between the *upper* and *lower boundaries* of each box lie 50 % of the estimates. The whiskers range to the minimum and maximum of all the data. The true number is marked by the long *horizontal line*

[0,1]. In this paper we used random forests, but any classifier providing such a function can be used. The introduced method minimizes the MSE of the estimation of the number of signal events. In our experiments this method performs better than the method currently used. The method is also adaptable to combine it with other methods. The combination with a logistic regression approach gave even better results than the two methods on their own.

Acknowledgements This work has been supported by the DFG, Collaborative Research Center SFB 876. We thank the ITMC at TU Dortmund University for providing computer resources on LiDo.

References

Albert, J., et al. (2008). Implementation of the random forest method for the imaging atmospheric Cherenkov telescope MAGIC. *Nuclear Instruments and Methods in Physics Research A*, *588*, 424–432.

Aleksić, J., et al. (2010). MAGIC TeV gamma-ray observations of Markarian 421 during multiwavelength campaigns in 2006. *Astronomy and Astrophysics*, *519*, A32.

Aleksić, J., et al. (2012). Performance of the MAGIC stereo system obtained with Crab Nebula data. *Astroparticle Physics*, *35*, 435–448.

Breiman, L. (2001). Random forests. *Machine Learning*, *45*, 5.

Hillas, A. M. (1985). Cherenkov light images of EAS produced by primary gamma. In *Proceedings of the 19th International Cosmic Ray Conference ICRC* (Vol. 3, p. 445), San Diego.

Hinton, J. A., & Hofman, W. (2009). Teraelectronvolt astronomy. *Annual Review of Astronomy & Astrophysics*, *47*, 523–565.

Mazin, D. (2007). *A Study of Very High Energy-Ray Emission From AGNs and Constraints on the Extragalactic Background Light*, Ph.D. Thesis, Technische Universitaet Muenchen.

Sheng, V., & Ling, C. (2006). Thresholding for making classifiers cost sensitive. In *Proceedings of the 21st National Conference on Artificial Intelligence* (Vol. 1, pp. 476–481). AAAI Press, Boston.

Thom, M. (2009). *Anaylse der Quelle 1ES 1959 + 650 mit MAGIC und die Implementierung eines Webinterfaces fr die automatische Monte-Carlo-Produktion*, Diploma Thesis, Technische Universitaet Dortmund.

Weekes, T. (2003). *Very High Energy Gamma-Ray Astronomy*. Bristol/Philadelphia: Institute of Physics Publishing.

Part II
AREA Machine Learning and Knowledge Discovery: Clustering, Classifiers, Streams and Social Networks

Implementing Inductive Concept Learning For Cooperative Query Answering

Maheen Bakhtyar, Nam Dang, Katsumi Inoue, and Lena Wiese

Abstract Generalization operators have long been studied in the area of Conceptual Inductive Learning (Michalski, A theory and methodolgy of inductive learning. In: Machine learning: An artificial intelligence approach (pp. 111–161). TIOGA Publishing, 1983; De Raedt, About knowledge and inference in logical and relational learning. In: Advances in machine learning II (pp. 143–153). Springer, Berlin, 2010). We present an implementation of these learning operators in a prototype system for cooperative query answering. The implementation can however also be used as a usual concept learning mechanism for concepts described in first-order predicate logic. We sketch an extension of the generalization process by a ranking mechanism on answers for the case that some answers are not related to what user asked.

M. Bakhtyar (✉)
Asian Institute of Technology Bangkok, Bangkok, Thailand
e-mail: Maheen.Bakhtyar@ait.asia

L. Wiese
University of Göttingen, Göttingen, Germany
e-mail: lena.wiese@udo.edu

K. Inoue
National Institute of Informatics, Tokyo, Japan
e-mail: inoue@nii.ac.jp

N. Dang
Tokyo Institute of Technology, Tokyo, Japan
e-mail: namd@de.cs.titech.ac.jp

M. Spiliopoulou et al. (eds.), *Data Analysis, Machine Learning and Knowledge Discovery*, Studies in Classification, Data Analysis, and Knowledge Organization, DOI 10.1007/978-3-319-01595-8_14,

1 Introduction

Conceptual inductive learning is concerned with deriving a logical description of concepts (in a sense, a classification) for a given set of observations or examples; in induction, the resulting description is also called a hypothesis. Background knowledge can support the concept learning procedure. In his seminal paper on inductive learning, Michalski (1983) introduced and surveyed several learning operators that can be applied to a set of examples to obtain (that is, induce) a description of concepts; each concept subsumes (and hence describes) a subset of the examples. He further differentiates inductive learning into *concept acquisition* (where a set of examples must be classified into a predefined set of concepts) and *descriptive generalization* (where a set of observations must be classified into a newly generated and hence previously unknown set of concepts). In a similar vein, de Raedt (2010) emphasizes the importance of logic representations for learning processes as follows:

> To decide whether a hypothesis would classify an example as positive, we need a notion of coverage.[…] In terms of logic, the example e is a logical consequence of the rule h, which we shall write as $h \models e$. This notion of coverage forms the basis for the theory of inductive reasoning[…]
> Especially important in this context is the notion of generality. One pattern is *more general* than another one if all examples that are covered by the latter pattern are also covered by the former pattern.[…] The generality relation is useful for inductive learning, because it can be used (1) to prune the search space, and (2) to guide the search towards the more promising parts of the space.[…] Using logical description languages for learning provides us not only with a very expressive and understandable representation, but also with an excellent theoretical foundation for the field. This becomes clear when looking at the generality relation. It turns out that the generality relation coincides with logical entailment.[…]

In this paper we present the implementations of three logical generalization operators: Dropping Condition (*DC*), Anti-Instantiation (*AI*) and Goal Replacement (*GR*). The novelty of our approach lies in the fact that these operators are combined iteratively. In other words, several successive steps of generalization are applied and operators can be mixed. Our work is based on the soundness results of an optimized iteration that can be found in Inoue and Wiese (2011); the main result there is that it is sufficient to apply these three operators in a certain order: starting with *GR* applications, followed by *DC* applications and ending with *AI* applications (see Fig. 1). This order is employed in our system when applying the three operators iteratively in a tree-like structure.

To the best of our knowledge, the iteration of these three operators is also novel when learning concepts from a set of examples. In this paper we present generalization as an application for a cooperative query answer system CoopQA: it applies generalization operators to *failing* queries which unsatisfactorily result in empty answers; by applying generalization operators we obtain a set of logically more general queries which might have more answers (called *informative* answers) than the original query. The implementation can however also be used in a traditional concept learning setting—that is, learning concepts by iterative generalization: as

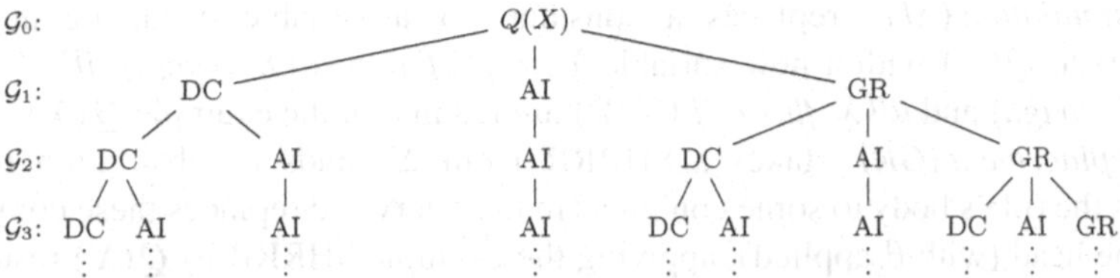

Fig. 1 Tree-shaped combination of DC, AI and GR

such we see the failing query as an initial description of a concept (which however does not cover all positive examples contained in a knowledge base); we then use our tree-shaped generalization until all positive examples can be derived as informative answers to the more general query (and hence we have obtained a more general description of the initial concept).

More formally, we concentrate on generalization of conjunctive queries that consist of a conjunction (written as $\wedge$) of literals l_i; a literal consists of a logical atom (a relation name with its parameters) and an optional negation symbol ($\neg$) in front. We write $Q(X) = l_1 \wedge \ldots \wedge l_n$ for a user's query where X is a free variable occurring in the l_i; if there is more than one free variable, we separate variables by commas. The free variables denote which values the user is looking for. Consider for example a hospital setting where a doctor asks for illnesses of patients. The query $Q(X) = ill(X, flu) \wedge ill(X, cough)$ asks for all the names X of patients that suffer from both flu and cough. A query $Q(X)$ is sent to a knowledge base Σ (a set of logical formulas) and then evaluated in Σ by a function *ans* that returns a set of answers (a set of formulas that are logically implied by Σ); as we focus on the generalization of queries, we assume the *ans* function and an appropriate notion of logical truth given. Note that (in contrast to the usual connotation of the term) we also allow negative literals to appear in conjunctive queries; we just require Definition 1 below to be fulfilled while leaving a specific choice of the $\models$ operator open. Similarly, we do not put any particular syntactic restriction on Σ. However, one of the generalization operators scans Σ for "single-headed range-restricted rules" (SHRRR) which consists of a body part left of an implication arrow ($\rightarrow$) and a head part right of the implication arrow. The body of a SHRRR consists of a disjunction of literals whereas the head consists only of one single literal: $l_{i_1} \wedge \ldots \wedge l_{i_m} \rightarrow l'$; range-restriction requires that all variables that appear in the head literal also appear in one of the body literals; again, we also allow negative literals in the body and in the head. As a simple example for a SHRRR consider $ill(X, flu) \rightarrow treat(X, medi)$ which describes that every patient suffering from flu is treated with a certain medicine.

CoopQA applies the following three operators to a conjunctive query (which—among others—can be found in the paper of Michalski 1983):

Dropping Condition (*DC*) removes one conjunct from a query; applying *DC* to the example $Q(X)$ results in $ill(X, flu)$ and $ill(X, cough)$.

Anti-Instantiation (*AI*) replaces a constant (or a variable occurring at least twice) in $Q(X)$ with a new variable Y; $ill(Y, flu) \wedge ill(X, cough)$, $ill(X, Y) \wedge ill(X, cough)$ and $ill(X, flu) \wedge ill(X, Y)$ are results for the example $Q(X)$.

Goal Replacement (*GR*) takes a SHRRR from Σ, finds a substitution θ that maps the rule's body to some conjuncts in the query and replaces these conjuncts by the head (with θ applied); applying the example SHRRR to $Q(X)$ results in $treat(X, medi) \wedge ill(X, cough)$.

These three operators all fulfill the following property of deductive generalization (which has already been used by Gaasterland et al. 1992) where ϕ is the input query and ψ is any possible output query:

Definition 1 (Deductive generalization wrt. knowledge base). Let Σ be a knowledge base, $\phi(\mathbf{X})$ be a formula with a tuple $\mathbf{X}$ of free variables, and $\psi(\mathbf{X}, \mathbf{Y})$ be a formula with an additional tuple $\mathbf{Y}$ of free variables disjoint from $\mathbf{X}$. The formula $\psi(\mathbf{X}, \mathbf{Y})$ is a *deductive generalization* of $\phi(\mathbf{X})$, if it holds in Σ that the less general ϕ implies the more general ψ where for the free variables $\mathbf{X}$ (the ones that occur in ϕ and possibly in ψ) the universal closure and for free variables $\mathbf{Y}$ (the ones that occur in ψ only) the existential closure is taken:

$$\Sigma \models \forall \mathbf{X} \exists \mathbf{Y} \left(\phi(\mathbf{X}) \rightarrow \psi(\mathbf{X}, \mathbf{Y})\right)$$

In the following sections, we briefly present the implementation of the CoopQA system (Sect. 2) and show a preliminary evaluation of the performance overhead of iterating generalization operators (Sect. 3).

2 Implementation Details

The focus of CoopQA lies on the efficient application of the underlying generalization operators. As CoopQA applies the generalization operators on the original query in a combined iterative fashion, the resulting queries may contain equivalent queries, which only differ in occurrences of variables or order of literals. CoopQA thus implements an equivalence checking mechanism to eliminate such duplicate queries. The most important step of query equivalence checking is finding substitutions between two queries. CoopQA tries to rearrange the literals in the two queries such that the substitutions can be obtained by mapping the variables in the two queries according to their positions. Instead of finding the substitution over the whole of the two queries, we segment the queries into *segments of pairwise equivalent literals* and find the correct ordering for each pair of corresponding segments. We now briefly sketch how each operator is implemented:

Dropping Condition (DC)*:* For a query of length n (i.e., n literals in the query), n generalized queries are generated by dropping one literal. As this involves replicating the $n-1$ remaining literals, run-time complexity of *DC* is $O(n^2)$.

Anti-Instantiation (AI)*:* If a query of length n contains M occurrences of constants and variables, at most M generalized queries (each of length n) are generated. Assume that the query was divided into r segments, then the literal affected by anti-instantiation has to be removed from its segment and the anti-instantiated literal with the new variable has to be placed into a (possibly different) segment; finding this segment by binary search requires $\log r$ time (which is less than n). Thus, run-time complexity of *AI* is $O(Mn)$.

Goal Replacement (GR)*:* *GR*, as described in Sect. 1 (and in more detail in Inoue and Wiese (2011)), requires finding parts of the query that are subsumed by the body of a given SHRRR. The definition of subsumption is that a logical formula $R(X)$ subsumes $Q(Y)$ if there is a substitution θ such that $R(X)\theta = Q(Y)$. Our implementation uses a generator that takes the query and the rule's body as input, and produces matchings of the body against the rule. We apply *relaxed* segmentation upon the two lists of literals (of the query and of the rule's body), which requires equivalent literals to only have the same predicate. For example, $q(X, Y)$ and $q(a, Z)$ are considered equivalent literals in this case. We then perform matching upon corresponding equivalent segments of the query against the segments of the rule's body to obtain a list of literals that are subsumed by the rule's body. Note that the segments in the query and the rule's body need not have the same size, as long as all the segments in the rule's body are found in the query. The matching literals are then replaced in the query by the rule's head, which already has the substitution applied.

The worst case scenario of *GR* is when both the query and the rule contain only one segment of (relaxed) equivalent literals. Given a query of size n, a rule with the body of size k ($n \geq k$), the number of possible matchings of k out of n literals from the query against the rule is $\binom{n}{k}$. For each combination of k literals, we have to perform permutation to find the correct ordering to determine the substitution against the rule body. Thus, in the worst case, the cost of finding the substitution for each combination is $k!$. Hence, complexity of finding all matchings of the query against the rule's body is $O(k!\binom{n}{k})$. In general, for a rule with a body of s segments, each of length k_i, and a query with s corresponding segments, each of length n_i, complexity is $O(\sum_{i=1}^{s}(k_i!\binom{n_i}{k_i}))$.

3 Performance Evaluation

We present some benchmarking and performance evaluation results of our implementation of the generalization operators. We focus on the cost of the tree-shaped generalization process.

Our benchmark suite generates a random knowledge base and query which consists large set of a combinations of illnesses and treatments from our medical example. We chose this synthetic data set because goal replacement has to capture the semantic dependencies in the data expressed by rules; so far we could not find any real-world benchmark with appropriate rules for goal replacement.

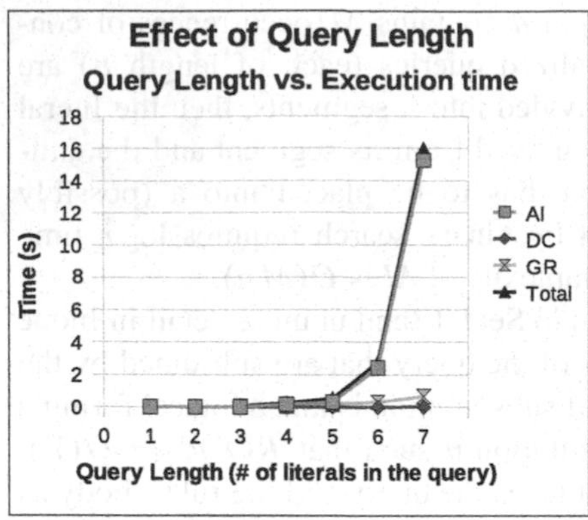

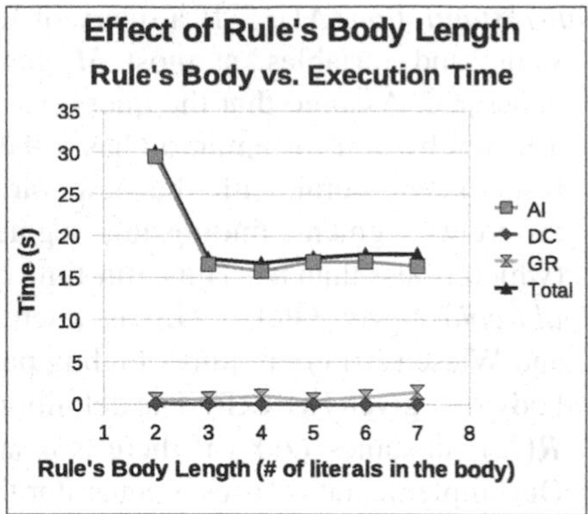

Fig. 2 Effects of query length and rule body length

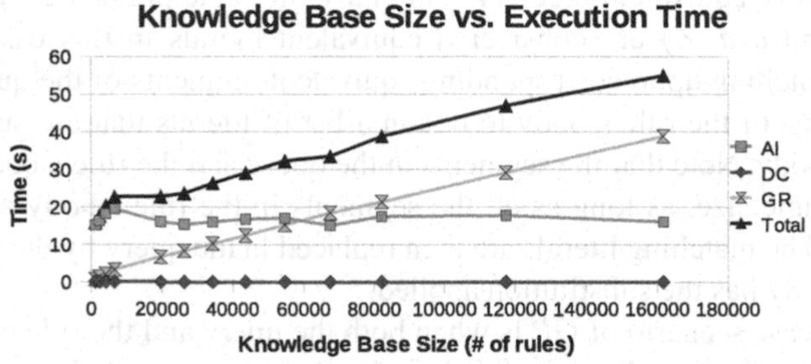

Fig. 3 Effects of number of rules

We analyze the effects on execution time by varying various parameters such as query length, number of SHRRRs in the knowledge base and the length of rule bodies. The runtime graphs show the time spent on each of the three operators alone as well as their summed total execution time. The tests were run on a PC under Linux 2.6 with a 64 bit Ubuntu 10.04 using Java 1.6 with OpenJDK Runtime Environment 6. The PC had 8 GB of main memory and a 3.0 GHz Quad Core processor.

We first analyze how query length affects execution time of generalization. Figure 2 shows the total time taken by each operator when running CoopQA for a certain query length (that is, number of literals). We observe an increase in the execution time of *AI* operator after query length 5 as potentially more *AI* operations are possible; in case of *DC* and *GR* operation, effect of query length on the execution time is negligible. Analyzing the effect of number of rules contained in the knowledge base shows that the total execution time increases with the number of rules as shown in Fig. 3. Closely investigating each operator reveals that there is no significant change in execution time in case of *AI*. In case of *DC* it is again negligible; however, we observe a linear increase in execution time in case of *GR* and that is because of increased matching and replacement. Lastly, there is no effect on *DC* and *GR* execution time when rule bodies are longer (see Fig. 2). Yet, time decreases for *AI*. This is due to the replaced part of the query: a long rule body makes the query shorter by replacing a longer part of the query with one literal; hence less *AI* operations are possible.

Profiling our code using VisualVM[1] shows that *AI* operation takes up 48.4 % of the total execution time; the main cost lies in performing equivalence checking to detect duplicated queries when a new one is generated (taking up to 33.6 % of total execution time). *GR*'s main bottleneck is to replace matched literals to generate queries, not the matching against rules itself.

4 Discussion and Conclusion

The CoopQA system uses three generalization operators that were widely used for inductive learning of concepts; it applies them in a cooperative query answering system in order to discover information for a user which might be related to his query intention. In contrast to other approaches using *DC*, *AI* and *GR*, tree-shaped generalization profits from an efficient combination of the three operators. The presented implementation shows favorable performance of the generalization operators. Future and ongoing work in the CoopQA system is covering the important issue of answer relevance which we discuss briefly: some answers might be generalized "too much" and are "too far away" from the user's query intention; a relevance ranking for answers can provide the user with the most useful answers while disregarding the irrelevant ones. In particular, a *threshold* value for answer ranks can be specified to return only the most relevant answers to the user and an aggregation of ranks reflecting the iteration of operators must be defined. Dropping Conditions and Goal Replacement are purely syntactic operators that do not introduce new variables. A relevance ranking for them can be achieved by assigning the answers to generalized queries a penalty for the dropped or replaced conditions. In contrast, anti-instantiation leads to the introduction of a new variable. Recall from Sect. 1 the example query $Q(X) = ill(X, flu) \wedge ill(X, cough)$ that asks for patients suffering from flu and cough at the same time. Applying AI on the constant cough leads to the generalized query $ill(X, flu) \wedge ill(X, Y)$ where the condition of cough is relaxed and any other disease would be matched to the new variable Y. These diseases might be very dissimilar to cough and hence totally irrelevant from the point of view of the original query $Q(X)$. A more intelligent version of the AI operator can hence rank the answer to the generalized query regarding their *similarity to the original query*. Here we have to differentiate between the case that a constant was anti-instantiated, and the case that a variable was anti-instantiated. More precisely, within a single application of the AI operator, we can assign each new (more general) answer ans_j the rank value $rank_j$, where the rank is calculated as follows:

[1] http://visualvm.java.net/

- if Y is the anti-instantiation of a constant c (like cough in our example), we obtain the similarity between the value of Y in answer ans_j (written as $val_j(Y)$) and the original constant c; that is, $rank_j = sim(val_j(Y), c)$.
- if Y is the anti-instantiation of a variable (like in the generalized query $ill(Y, flu) \wedge ill(X, cough)$ where different patients X and Y are allowed), we obtain the similarity between the value of Y in ans_j and the value of the variable (say, X) which is anti-instantiated by Y in the same answer; that is, $rank_j = sim(val_j(Y), val_j(X))$.

Let us assume we have predefined similarities $sim(bronchitis, cough) = 0.9$ and $sim(brokenLeg, cough) = 0.1$ for our example. An answer containing a patient with both flu and bronchitis would then be ranked high with 0.9; whereas an answer containing a patient with both flu and broken leg would be ranked low with 0.1. Such similarities can be based on a taxonomy of words (or an ontology); in our example we would need a medical taxonomy relating several diseases in a hierarchical manner. Several notions of distance in a taxonomy of words can be used (e.g., Shin et al. 2007) to define a similarity between each two words in the taxonomy (e.g., Wu and Palmer 1994). When the AI operator is applied repeatedly, similarities should be computed for each replaced constant or variable; these single similarities can then be combined into a rank for example by taking their *weighted sum*. To make the system more adaptive to user behavior, the taxonomy used for the similarities can be revised at runtime (with an approach as described in Nikitina et al. 2012).

References

De Raedt, L. (2010). About knowledge and inference in logical and relational learning. In *Advances in Machine Learning II* (pp. 143–153). Berlin: Springer.

Gaasterland, T., Godfrey, P., & Minker, J. (1992). Relaxation as a platform for cooperative answering. *Journal of Intelligent Information Systems, 1*(3/4), 293–321.

Inoue, K., & Wiese, L. (2011). Generalizing conjunctive queries for informative answers. In *9th International Conference on Flexible Query Answering Systems. Lecture notes in artificial intelligence* (Vol. 7022, pp. 1–12). New York: Springer.

Michalski, R. S. (1983). A theory and methodolgy of inductive learning. In *Machine learning: An artificial intelligence approach* (pp. 111–161). TIOGA Publishing.

Muslea, I. (2004). Machine learning for online query relaxation. In *Tenth ACM SIGKDD International Conference on Knowledge Discovery and Data Mining* (pp. 246–255). New York: ACM.

Nikitina, N., Rudolph, S., & Glimm, B. (2012). Interactive ontology revision. *Journal of Web Semantics, 12*, 118–130.

Shin, M. K., Huh, S.-Y., & Lee, W. (2007). Providing ranked cooperative query answers using the metricized knowledge abstraction hierarchy. *Expert Systems with Applications, 32*(2), 469–484.

Wu, Z., & Palmer, M. (1994). Verb semantics and Lexical selection. In *32nd Annual Meeting of the Association for Computational Linguistics* (pp. 133–138). Los Altos: Morgan Kaufmann Publishers.

Clustering Large Datasets Using Data Stream Clustering Techniques

Matthew Bolaños, John Forrest, and Michael Hahsler

Abstract Unsupervised identification of groups in large data sets is important for many machine learning and knowledge discovery applications. Conventional clustering approaches (k-means, hierarchical clustering, etc.) typically do not scale well for very large data sets. In recent years, data stream clustering algorithms have been proposed which can deal efficiently with potentially unbounded streams of data. This paper is the first to investigate the use of data stream clustering algorithms as light-weight alternatives to conventional algorithms on large non-streaming data. We will discuss important issue including order dependence and report the results of an initial study using several synthetic and real-world data sets.

1 Introduction

Clustering very large data sets is important for may applications ranging from finding groups of users with common interests in web usage data to organizing organisms given genetic sequence information. The data are often not only so large that they do not fit into main memory, but they even have to be stored in a distributed manner. Conventional clustering algorithms typically require repeated access to all the data which is very expensive in a scenario with very large data.

Data stream clustering has become an important field of research in recent years. A data stream is an ordered and potentially unbounded sequence of objects (e.g., data points representing sensor readings). Data stream algorithms have been

M. Bolaños · M. Hahsler (✉)
Southern Methodist University, Dallas, TX, USA
e-mail: mhahsler@lyle.smu.edu

J. Forrest
Microsoft, Redmond, WA, USA

M. Spiliopoulou et al. (eds.), *Data Analysis, Machine Learning and Knowledge Discovery*, Studies in Classification, Data Analysis, and Knowledge Organization, DOI 10.1007/978-3-319-01595-8_15,

developed in order to process large volumes of data in an efficient manner using a single pass over the data while having only minimal storage overhead requirements. Although these algorithms are designed for data streams, they obviously can also be used on non-streaming data.

In this paper we investigate how to use data stream clustering techniques on large, non-streaming data. The paper is organized as follows. We introduce the problems of clustering large data sets and data stream clustering in Sects. 2 and 3, respectively. Section 4 discusses issues of the application of data stream clustering algorithms to non-streaming data. We present results of first experiments in Sect. 5 and conclude with Sect. 6.

2 Clustering Large Data Sets

Clustering groups objects such that objects in a group are more similar to each other than to the objects in a different group (Kaufman and Rousseeuw 1990). Formally clustering can be defined as:

Definition 1 (Clustering). Partition a set of objects $O = \{o_1, o_2, \ldots, o_n\}$ into a set of clusters $C = \{C_1, C_2, \ldots, C_k, C_\epsilon\}$, where k is the number of clusters and C_ϵ contains all objects not assigned to a cluster.

We restrict clustering here to hard (non-fuzzy) clustering where $C_i \cap C_j = \emptyset$ for all $i, j \in \{1, 2, \ldots, k, \epsilon\}$ and $i \neq j$. Unassigned objects are often considered noise, however, many algorithms cannot leave objects unassigned (i.e., $C_\epsilon = \emptyset$). The ability to deal with noise becomes more important in very large data sets where manually removing noise before clustering is not practical. The number k is typically user-defined, but might also be determined by the clustering algorithm. In this paper, we assume that the objects are embedded in a d-dimensional metric space ($o \in \mathbb{R}^d$) where dissimilarity can be measured using Euclidean distance. We do not deal with the problem of finding the optimal number of clusters, but assume that a reasonable estimate for k is available.

The most popular conventional clustering methods are k-means type clustering, hierarchical clustering and density-based clustering. All these methods have in common that they do not scale well for very large data sets since they either need several passes over the data or they create data structures that do not scale linearly with the number of objects. We refer the reader to the popular book by (Jain and Dubes 1988) for details about the various clustering methods.

To cluster large data sets, researchers have developed parallel computing approaches, most notably using Google's MapReduce framework (e.g., for k-means see Zhao and He 2009). On the other hand, researchers started earlier to reduce the data size by sampling. For example, CLARA (Kaufman and Rousseeuw 1990) uses sampling and then applies Partitioning Around Medoids (PAM) on the samples and returns the best clustering. Another algorithm, BIRCH (Zhang et al. 1996), builds a height balanced tree (also known as a cluster feature or CF tree) in a single pass over the data. The tree stores information about subclusters in its leaf nodes.

During clustering, each data point is either added to an existing leaf node or a new node is created. Some reorganization is applied to keep the tree at a manageable size. After all data points are added to the CF tree, the user is presented with a list of subclusters. BIRCH was developed by the data mining community and resembles in many ways the techniques used in data stream clustering which we will discuss in the next section.

3 Data Stream Clustering

We first define data stream since data stream clustering operates on data streams.

Definition 2 (Data Stream). A data stream is an ordered and potentially unbounded sequence of objects $S = \langle \mathbf{o}_1, \mathbf{o}_2, \mathbf{o}_3, \ldots \rangle$.

Working with data streams imposes several restrictions on algorithms. It is impractical to permanently store all objects in the stream which implies that there is limited time to process each object and repeated access to all objects is not possible.

Over the past 10 years a number of data stream clustering algorithms have been developed. For simplicity, we will restrict the discussion in this paper to algorithms based on micro-clusters (see Gama 2010). Most data stream clustering algorithms use a two stage online/offline approach.

Definition 3 (Online Stage). Summarize the objects in stream S in real-time (i.e., in a single pass over the data) by a set of k' micro-clusters $\mathcal{M} = \{m_1, m_2, \ldots, m_{k'}\}$ where m_i with $i = \{1, 2, \ldots, k'\}$ represents a micro-cluster in a way such that the center, weight, and possibly additional statistics can be computed.

When the user requires a clustering, the offline stage reclusters the micro-clusters to form a final (macro) clustering.

Definition 4 (Offline Stage). Use the k' micro-clusters in $\mathcal{M}$ as pseudo-objects to produce a set C of $k \ll k'$ final clusters using clustering defined in Definition 1.

Note that while k often is specified by the user, k' is not fixed and may grow and shrink during the clustering process. Micro-clusters are typically represented as a center and each new object is assigned to its closest (in terms of a proximity measure) micro-cluster. Some algorithms also use a grid and micro-clusters represent non-empty grid-cells. If a new data point cannot be assigned to an existing micro-cluster, typically a new micro-cluster is created. The algorithm may also do some housekeeping (merging or deleting micro-clusters) to keep the number of micro-clusters at a manageable size.

Since the online component only passes over the data once and the offline component operates on a drastically reduced data set which typically fits into main memory, data stream clustering algorithms can be used efficiently on very large, disk-resident data. For a comprehensive treatment of data stream clustering algorithms including some single-pass versions of k-means and related issues, we refer the interested reader to the books by Aggarwal (2007) and Gama (2010).

4 Data Stream Clustering of Non-streaming Data

Applying a data stream clustering algorithm to non-streaming data is straightforward. To convert the set of objects $\mathcal{O}$ into a stream $\mathcal{S}$, we simply take one object at a time from $\mathcal{O}$ and hand it to the clustering algorithm. However, there are several important issues to consider.

A crucial aspect of data streams is that the objects are temporally ordered. Many data streams are considered to change over time, i.e., clusters move, disappear or new clusters may form. Therefore, data stream algorithms incorporate methods to put more weight on current data and forget outdated data. This is typically done by removing micro-clusters which were not updated for a while (e.g., in CluStream; Aggarwal et al. 2003) or using a time-dependent exponentially decaying weight for the influence of an object (most algorithms). For large, stationary data sets, where order has no temporal meaning and is often arbitrary, this approach would mean that we put more weight on data towards the end of the data set while loosing the information at the beginning. Some data stream clustering algorithms allow us to disable this feature, e.g., using a very large horizon parameter for CluStream forces the algorithm not to forget micro-clusters and instead merge similar clusters. For many other algorithms, the decay rate can be set to 0 or close to 0. For example in DenStream (Cao et al. 2006) a value close to 0 reduces the influence of the order in the data. However, setting it to 0 makes the algorithm unusable since removing small micro-clusters representing noise or outliers depends on the decay mechanism. It is very important to established if and how this type of order dependence can be removed or reduced before applying a data stream clustering algorithm to non-streaming data.

Since data stream clustering algorithms use a single pass over the data, the resulting clustering may still be order dependent. This happens for algorithms, where the location of the created micro-clusters is different if objects are added in a different order. However, this type of order dependency typically only effects micro-cluster placement slightly and our results below indicate that after reclustering, the final clustering is only effected minimally.

Another issue is that data stream clustering algorithms dispose of the objects after they are absorbed by a micro-cluster and since the data stream is expected to be unbounded, not even the cluster assignments of the objects are retained. The only information that is available after reclustering is the set of micro-clusters $\mathcal{M}$ and the mapping of micro-clusters onto final clusters $f_{\text{macro}} : \mathcal{M} \mapsto \mathcal{C}$. In order to infer the mapping of objects in $\mathcal{O}$ to cluster labels, we find for each object o the closest micro-cluster and then use f_{macro} to retrieve the cluster label in $\mathcal{C}$. Note that this approach works not only for reclustering that produces spherical clusters but also for reclustering that produces arbitrarily shaped clusters (e.g., with single-linkage hierarchical clustering or density based clustering).

Table 1 Data sets

Dataset	Number of objects	Dimensions	Clusters	Noise
Mixture of Gaussians	100,000	d^{a}	k^{a}	$n\ \%^{a}$
Covertype	581,012	10	7	Unknown
16S rRNA	406,997	64	110	Unknown

[a] Values correspond with the data sets' names

5 Comparing Different Clustering Methods

To perform our experiments we use *stream*,[1] a R-extension currently under development which provides an intuitive interface for experimenting with data streams and data stream algorithms. It includes the generation of synthetic data, reading of disk-resident data in a streaming fashion, and a growing set of data stream mining algorithms (including some from the MOA framework by Bifet et al. 2010). In this first study, we only evaluate sampling and the online component of three of the more popular clustering methods suitable for data streams.

- Reservoir sampling (Vitter 1985)
- BIRCH (Zhang et al. 1996)
- CluStream (Aggarwal et al. 2003)
- DenStream (Cao et al. 2006)

For evaluation, we use the data sets shown in Table 1. We use several mixture of Gaussians data sets with k clusters in a d-dimensional hypercube created with randomly generated centers and covariance matrices similar to the method suggested by Jain and Dubes (1988).[2] Some clusters typically overlap. For one set we add $n = 20\,\%$ noise in the form of objects uniformly distributed over the whole data space.

The Covertype data set[3] contains remote sensing data from the Roosevelt National Forest of northern Colorado for seven different forest cover types. We use for clustering the ten quantitative variables.

The 16S rRNA data set contains the 3-gram count for the more than 400,000 16S rRNA sequences currently available for bacteria in the Greengenes database.[4] 16S sequences are about 1,500 letters long and we obtain 64 different 3-gram counts for the 4 letters in the RNA alphabet. Since these sequences are mainly used for classification, we use the phylum, a phylogenetic rank right below kingdom, as ground truth.

---

[1] *stream* is available at http://R-Forge.R-Project.org/projects/clusterds/.

[2] Created with the default settings of function `DSD_Gaussian_Static()` in *stream*.

[3] Obtained from UCI Machine Learning Repository at http://archive.ics.uci.edu/ml/datasets/Covertype.

[4] Obtained from Greengenes at http://greengenes.lbl.gov/Download/Sequence_Data/Fasta_data_files/current_GREENGENES_gg16S_unaligned.fasta.gz.

All data sets come from a stationary distribution and the order in the data is arbitrary making them suitable for evaluating clustering of non-streaming data.

5.1 Evaluation Method

We cluster each data set with each data stream clustering method. We reduce decay and forgetting in the algorithms by using appropriate parameters (horizon $= 10^6$ for CluStream and $\lambda = 10^{-6}$ for DenStream). Then we tune each algorithm for each data set to generate approximately 1,000 micro-clusters to make the results better comparable. Finally, we recluster each data stream clustering algorithm's result using weighted k-means using the known number of clusters for k.

We chose data sets with available ground truth in order to be able to use an external evaluation measure. Studies showed that the corrected Rand index (Hubert and Arabie 1985) is an appropriate external measure to compare partitions for clustering static data (Milligan and Cooper 1986). It compares the partition produced via clustering with the partition given by the ground truth using the Rand index (a measure of agreement between partitions) corrected for expected random agreements. The index is in the interval $[-1, 1]$, where 0 indicates that the found agreements can be entirely explained by chance and the higher the index, the better the agreement. The index is also appropriate to compare the quality of different partitions given the ground truth (Jain and Dubes 1988) as is done in this paper.

5.2 Results

First, we take a look at how different algorithms place micro-clusters since this gives us a better idea of how well reclustering will work. In Fig. 1 we apply all four algorithms on a simple mixture of Gaussians data set with 10,000 data points and 5 % noise. We either set or tuned all algorithms to produce about 100 micro-clusters. Reservoir sampling (1 %) in Fig. 1 concentrates on the dense areas and only selects few noise points. However, we can see that some quite dense areas do not have a representative. Also, no weights are available for sampling. BIRCH places micro-clusters relatively evenly with heavier micro-clusters in denser areas. Since BIRCH tries to represent all objects, noise creates many very light micro-clusters. CluStream produces results very similar to BIRCH. DenStream tends to create a single heavy cluster for a dense area, but often micro-cluster compete for larger dense areas resulting in a cloud of many very light clusters. A big difference to the other methods is that DenStream is able to suppress noise very well.

Next, we look at the order dependence of each method. We use the same data set as above, but randomly reorder the data ten times, run the four algorithms on it and then recluster with weighted k-means and $k = 5$. We assign the 10,000 objects to the found clusters using the method discussed above in Sect. 4 and then compare the

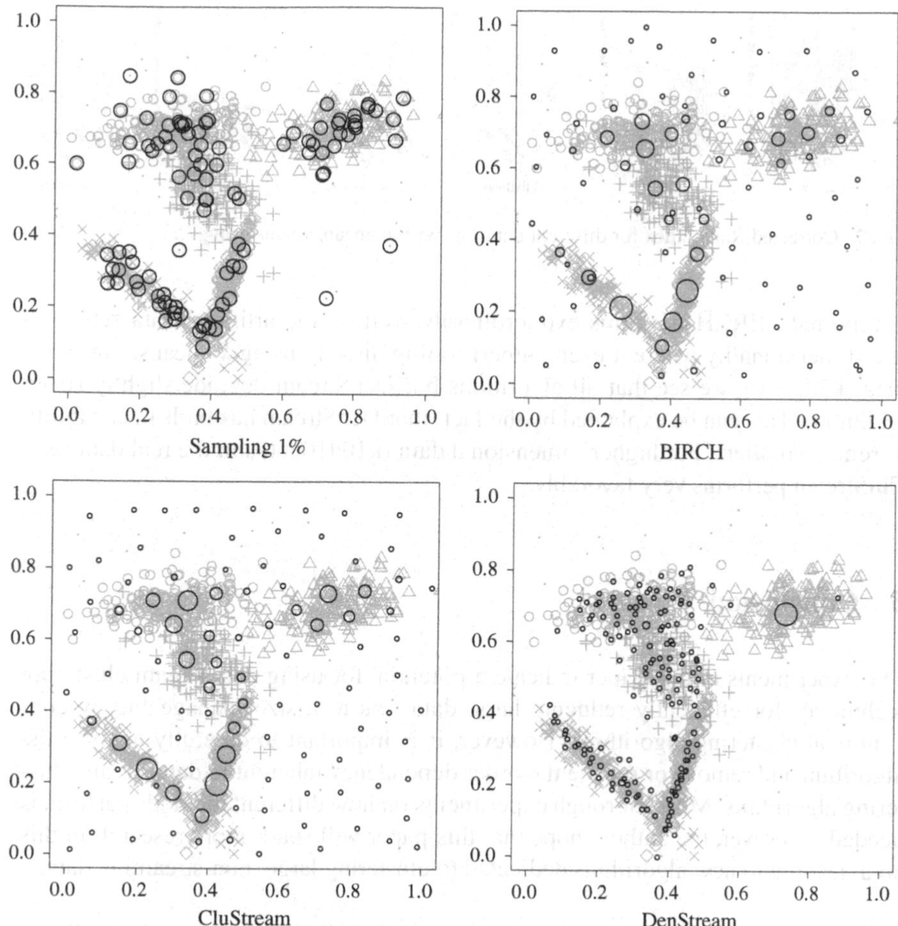

Fig. 1 About 100 micro-clusters placed by different algorithms on a mixture of $k = 5$ Gaussians in $d = 2$ (shown in *gray*) with 5 % noise. Micro-cluster weights are represented by *circle size*

assignments for the ten different orders (45 comparisons) using the corrected Rand index. The average corrected Rank index is relatively high with 0.74 (sampling), 0.85 (BIRCH), 0.81 (CluStream) and 0.79 (DenStream). Sampling has the lowest index, however, this is not caused by the order of the data since random sampling is order independent, but by the variation caused by choosing random subsets. The higher index for the other methods indicates that, using appropriate parameters, order dependence is below the variability of a 1 % sample.

Finally, we cluster and recluster the artificial and real data sets and report the corrected Rank index between the clustering and the known ground truth in Fig. 2. We replicate each experiment for the artificial data ten times and report the average corrected Rand index. For comparison, the result of k-means on the whole data set

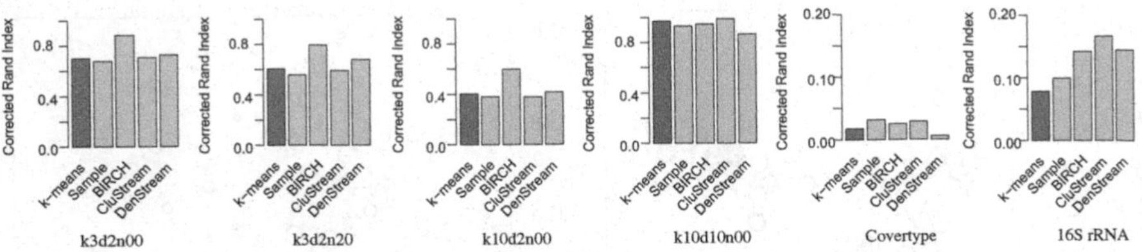

Fig. 2 Corrected Rand index for different data sets with k-means reclustering

is reported. BIRCH performs extraordinarily well on the artificial data sets with low dimensionality where it even outperforming directly using k-means. For noisy data (k3d2n20), we see that all algorithms but DenStream degrade slightly (from k3d2n00). This can be explained by the fact, that DenStream has built-in capability to remove outliers. For higher-dimensional data (k10d10n00 and the real data sets) CluStream performs very favorably.

6 Conclusion

The experiments in this paper indicate a potential for using data stream clustering techniques for efficiently reducing large data sets to a size manageable by conventional clustering algorithms. However, it is important to carefully analyze the algorithm and remove or reduce the order dependency inherent in data stream clustering algorithms. More thorough experiments on how different methods perform is needed. However, the authors hope that this paper will spark more research in this area, leading to new algorithms dedicated to clustering large, non-streaming data.

Acknowledgements This work is supported in part by the U.S. National Science Foundation as a research experience for undergraduates (REU) under contract number IIS-0948893 and by the National Institutes of Health under contract number R21HG005912.

References

Aggarwal, C. (2007). *Data streams: Models and algorithms. Advances in database systems* (Vol. 31). New York: Springer.

Aggarwal, C. C., Han, J., Wang, J., & Yu, P. S. (2003). A framework for clustering evolving data streams. In *Proceedings of the 29th International Conference on Very Large Data Bases (VLDB '03)* (Vol. 29, pp. 81–92). VLDB Endowment.

Bifet, A., Holmes, G., Kirkby, R., & Pfahringer, B. (2010). MOA: Massive online analysis. *Journal of Machine Learning Research, 99*, 1601–1604.

Cao, F., Ester, M., Qian, W., & Zhou, A. (2006). Density-based clustering over an evolving data stream with noise. In *Proceedings of the 2006 SIAM International Conference on Data Mining* (pp. 328–339). Philadelphia: SIAM.

Gama, J. (2010). *Knowledge discovery from data streams* (1st ed.). Boca Raton: Chapman & Hall/CRC.

Hubert, L., & Arabie, P. (1985). Comparing partitions. *Journal of Classification, 2*(1), 193–218.

Jain, A. K., & Dubes, R. C. (1988). *Algorithms for clustering data.* Upper Saddle River: Prentice-Hall.

Kaufman, L., & Rousseeuw, P. J. (1990). *Finding groups in data: An introduction to cluster analysis.* New York: Wiley.

Milligan, G. W., & Cooper, M. C. (1986). A study of the comparability of external criteria for hierarchical cluster analysis. *Multivariate Behavioral Research, 21*(4), 441–458.

Vitter, J. S. (1985). Random sampling with a reservoir. *ACM Transactions on Mathematical Software, 11*(1), 37–57.

Zhang, T., Ramakrishnan, R., & Livny, M. (1996). BIRCH: An efficient data clustering method for very large databases. In *Proceedings of the 1996 ACM SIGMOD International Conference on Management of Data* (pp. 103–114). New York: ACM.

Zhao, W., Ma, H., & He, Q. (2009) Parallel k-means clustering based on MapReduce. In *Proceedings of the 1st International Conference on Cloud Computing, CloudCom '09* (pp. 674–679). Berlin: Springer.

Feedback Prediction for Blogs

Krisztian Buza

Abstract The last decade lead to an unbelievable growth of the importance of social media. Due to the huge amounts of documents appearing in social media, there is an enormous need for the *automatic* analysis of such documents. In this work, we focus on the analysis of documents appearing in blogs. We present a proof-of-concept industrial application, developed in cooperation with Capgemini Magyarország Kft. The most interesting component of this software prototype allows to predict the number of feedbacks that a blog document is expected to receive. For the prediction, we used various predictions algorithms in our experiments. For these experiments, we crawled blog documents from the internet. As an additional contribution, we published our dataset in order to motivate research in this field of growing interest.

1 Introduction

The last decade lead to an unbelievable growth of the importance of social media. While in the early days of social media, blogs, tweets, facebook, youtube, social tagging systems, etc. served more-less just as an entertainment of a few enthusiastic users, nowadays news spreading over social media may govern the most important changes of our society, such as the revolutions in the Islamic world, or US president elections. Also advertisements and news about new products, services and companies are spreading quickly through the channels of social media. On the one hand, this might be a great possibility for promoting new products and services. On the other hand, however, according to sociological studies, negative opinions

K. Buza (✉)
Department of Computer Science and Information Theory, Budapest University of Technology and Economics, Budapest, Hungary
e-mail: buza@cs.bme.hu

M. Spiliopoulou et al. (eds.), *Data Analysis, Machine Learning and Knowledge Discovery*, Studies in Classification, Data Analysis, and Knowledge Organization, DOI 10.1007/978-3-319-01595-8_16,

spread much quicker than positive ones, therefore, if negative news appear in social media about a company, the company might have to react quickly, in order to avoid losses.

Due to the huge amounts of documents appearing in social media, analysis of all these documents by human experts is hopeless, and therefore there is an enormous need for the *automatic* analysis of such documents. For the analysis, however, we have to take some special properties of the application domain into account. In particular, the uncontrolled, dynamic and rapidly-changing content of social media documents: e.g. when a blog-entry appears, users may immediately comment this document.

We developed a software prototype in order to demonstrate how data mining techniques can address the aforementioned challenges. This prototype has the following major components: (1) the crawler, (2) information extractors, (3) data store and (4) analytic components. In this paper, we focus on the analytic components that allow to predict the number of feedbacks that a document is expected to receive in the next 24 h. For feedback prediction, we focused on the documents appearing in blogs and performed experiments with various predictions models. For these experiments we crawled Hungarian blog sites. As an additional contribution, we published our data.

2 Related Work

Data mining techniques for social media have been studied by many researchers, see e.g. Reuter et al. (2011) and Marinho et al. (2008). Our problem is inherently related to many web mining problems, such as opinion mining or topic tracking in blogs. For an excellent survey on opinion mining we refer to Pang and Lee (2008). Out of the works related to blogs we point out that Pinto (2008) applied topic tracking methods, while Mishne (2007) exploited special properties of blogs in order to improve retrieval.

Despite its relevance, there are just a few works on predicting the number of feedbacks that a blog-document is expected to receive. Most closely related to our work is the paper of Yano and Smith (2010) who used Naive Bayes, Linear and Elastic Regression and Topic-Poisson models to predict the number of feedbacks in political blogs. In contrast to them, we target various topics (do not focus on political blogs) and perform experiments with a larger variety of models including Neural Networks, RBF Networks, Regression Trees and Nearest Neighbor models.

3 Domain-Specific Concepts

In order to address the problem, first, we defined some domain-specific concepts that are introduced in this chapter. We say that a *source* produces *documents*. For example, on the site *torokgaborelemez.blog.hu*, new documents appear regularly, therefore, we say that torokgaborelemez.blog.hu is the source of these documents.

From the point of view of our work, the following parts of the documents are the most relevant ones: (1) *main text of the document:* the text that is written by the author of the document, this text describes the topic of the document, (2) *links to other documents:* pointers to semantically related documents, in our case, trackbacks are regarded as such links, (3) *feedbacks:* opinions of social media users about a document is very often expressed in form of feedbacks that the document receives. Feedbacks are usually short textual comments referring to the main text of the document and/or other feedbacks. Temporal aspects of all the above entities are relevant for our task. Therefore, we extract time-stamps for the above entities and store the data together with these timestamps.

4 Feedback Prediction

Feedback prediction is the scientifically most interesting component of the prototype, therefore we focus on feedback prediction. For the other components of the software prototype we refer to the presentation slides available at http://www.cs.bme.hu/~buza/pdfs/gfkl_buza_social_media.pdf.

4.1 Problem Formulation

Given some blog documents that appeared in the past, for which we already know when and how many feedbacks they received, the task is to predict how many feedbacks *recently* published blog-entries will receive in the next H hours. We regard the blog documents published in the last 72 h as recently published ones, we set $H = 24$ h.

4.2 Machine Learning for Feedback Prediction

We address the above prediction problem by machine learning, in particular by regression models. In our case, the instances are the recently published blog documents and the target is the number of feedbacks that the blog-entry will receive in the next H hours.

Most regression algorithms assume that the instances are vectors. Furthermore, it is assumed that the value of the target is known for some (sufficiently enough) instances, and based on this information, we want to predict the value of the target for those cases where it is unknown. First, using the cases where the target is known, a prediction model, *regressor*, is constructed. Then, the regressor is used to predict the value of the target for the instances with unknown valued target.

In our prototype we used neural networks (multilayer perceptrons in particular), RBF-networks, regression trees (REP-tree, M5P-tree), nearest neighbor models, multivariate linear regression and bagging out of the ensemble models. For more detailed descriptions of these models we refer to Witten and Franke (2005) and Tan et al. (2006).

In the light of the above discussion, in order to apply machine learning to the feedback prediction problem, we have to resolve two issues: (1) we have to transform the instances (blog documents) into vectors, and (2) we need some data for which the value of the target is already known (train data).

For the first issue, i.e., for turning the documents into vectors, we extract the following features from each document:

1. *Basic features*: Number of links and feedbacks in the previous 24 h relative to baseTime; number of links and feedbacks in the time interval from 48 h prior to baseTime to 24 h prior to baseTime; how the number of links and feedbacks increased/decreased in the past (the past is seen relative to baseTime); number of links and feedbacks in the first 24 h after the publication of the document, but before baseTime; aggregation of the above features by source,
2. *Textual features*: The most discriminative bag of words features,[1]
3. *Weekday features*: Binary indicator features that describe on which day of the week the main text of the document was published and for which day of the week the prediction has to be calculated,
4. *Parent features*: We consider a document d_P as a patent of document d, if d is a reply to d_P, i.e., there is a trackback link on d_P that points to d; parent features are the number of parents, minimum, maximum and average number of feedbacks that the parents received.

We solve the first issue as follows: we select some date and time in the past and simulate as if the current date and time would be the selected date and time. We call the selected date and time *baseTime*. As we actually know what happened after the baseTime, i.e., we know how many feedbacks the blog entries received in the next H hours after *baseTime*, we know the values of the target for these cases. While doing so, we only take blog pages into account that were published in the last 3 days relative to the baseTime, because older blog pages usually do not receive any more new feedbacks.

A similar approach allows us to quantitatively evaluate the prediction models: we choose a time interval, in which we select different times as baseTime, calculate the value of the target and use the resulting data to train the regressor. Then, we select a disjoint time interval in which we again take several baseTimes and calculate the

[1] In order to quantify how discriminative is a word w, we use the average and standard deviation of the number of feedbacks of documents that contain w, and the average and standard deviation of the number of feedbacks of documents that *do not* contain w. Then, we divide the difference of the number of average feedbacks with the sum of the both standard deviations. Then, we selected the 200 most discriminative words.

true values of the target. However, the true values of the target remain hidden for the prediction model, we use the prediction model to estimate the values of the targets for the second time interval. Then we can compare the true and the predicted values of the target.

5 Experiments

We examined various regression models for the blog feedback prediction problem, as well as the effect of different type of features. The experiments, in total, took several months of CPU time into account.

5.1 *Experimental Settings*

We crawled Hungarian blog sites: in total we downloaded 37,279 pages from roughly 1,200 sources. This collection corresponds approximately 6 GB of plain HTML document (i.e., without images). We preprocessed as described in Sect. 4.2. The preprocessed data had in total 280 features (without the target variable, i.e., number of feedbacks). In order to assist reproducibility of our results as well as to motivate research on the feedback prediction problem, we made the preprocessed data publicly available at http://www.cs.bme.hu/~buza/blogdata.zip.

In the experiments we aimed to simulate the real-world scenario in which we train the prediction model using the blog documents of the past in order to make predictions for the blog documents of the present, i.e., for the blog documents that have been published recently. Therefore, we used a temporal split of the train and test data: we used the blog documents from 2010 and 2011 as train data and the blog documents from February and March 2012 as test data. In both time intervals we considered each day as baseTime in the sense of Sect. 4.2.

For each day of the test data we consider ten blog pages that were *predicted* to have to largest number of feedbacks. We count how many out of these pages are among the 10 pages that received the largest number of feedbacks *in the reality*. We call this evaluation measure *Hits@10* and we average Hits@10 for all the days of the test data.

For the AUC, i.e., area under the receiver-operator curve, see Tan et al. (2006), we considered as positive the 10 blog pages receiving the highest number of feedbacks *in the reality*. Then, we ranked the pages according to their *predicted* number of feedbacks and calculated AUC. We call this evaluation measure *AUC@10*.

For the experiments we aimed at selecting a representative set of state-of-the-art regressors. Therefore, we used multilayer perceptrons (MLP), linear regressors, RBF-Networks, REP-Trees and M5P-Trees. These regressors are based on various theoretical background (see e.g. neural networks versus regression trees). We used the Weka-implementations of these regressors, see Witten and Franke (2005) for more details.

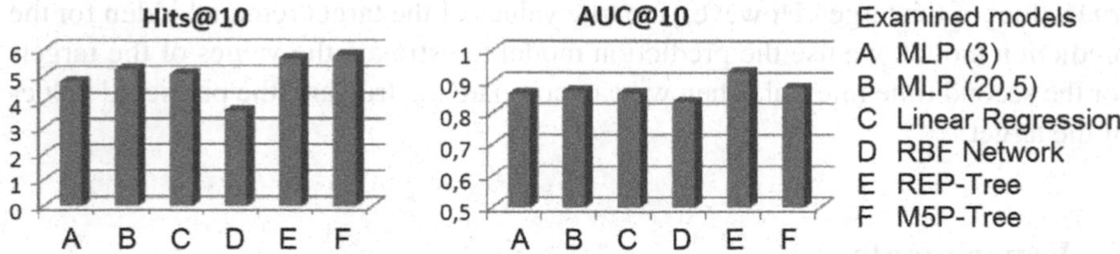

Fig. 1 The performance of the examined models

Table 1 The effect of different types of features and the effect of bagging

Model	Basic	Basic + weekday	Basic + parent	Basic + textual	Bagging
MLP (3)	5.533 ± 1.384	5.550 ± 1.384	5.612 ± 1.380	4.617 ± 1.474	5.467 ± 1.310
	0.886 ± 0.084	0.884 ± 0.071	0.894 ± 0.062	0.846 ± 0.084	0.890 ± 0.080
MLP (20,5)	5.450 ± 1.322	5.488 ± 1.323	5.383 ± 1.292	5.333 ± 1.386	5.633 ± 1.316
	0.900 ± 0.080	0.910 ± 0.056	0.914 ± 0.056	0.896 ± 0.069	0.903 ± 0.069
k-NN	5.433 ± 1.160	5.083 ± 1.345	5.400 ± 1.172	3.933 ± 1.223	5.450 ± 1.102
($k = 20$)	0.913 ± 0.051	0.897 ± 0.061	0.911 ± 0.052	0.850 ± 0.060	0.915 ± 0.051
RBF Net	4.200 ± 1.458	4.083 ± 1.320	3.414 ± 1.700	3.833 ± 1.428	4.750 ± 1.233
(clusters: 100)	0.860 ± 0.070	0.842 ± 0.069	0.846 ± 0.074	0.818 ± 0.074	0.891 ± 0.050
Linear	5.283 ± 1.392	5.217 ± 1.343	5.283 ± 1.392	5.083 ± 1.215	5.150 ± 1.327
Regression	0.876 ± 0.088	0.869 ± 0.097	0.875 ± 0.091	0.864 ± 0.096	0.881 ± 0.082
REP Tree	5.767 ± 1.359	5.583 ± 1.531	5.683 ± 1.420	5.783 ± 1.507	5.850 ± 1.302
	0.936 ± 0.038	0.931 ± 0.042	0.932 ± 0.043	0.902 ± 0.086	0.934 ± 0.039
M5P Tree	6.133 ± 1.322	6.200 ± 1.301	6,000 ± 1.342	6.067 ± 1.289	5.783 ± 1.305
	0.914 ± 0.073	0.907 ± 0.084	0.913 ± 0.081	0.914 ± 0.068	0.926 ± 0.048

The performance (Hits@10 and AUC@10) of the models for different feature sets

5.2 Results and Discussion

The performance of the examined models, for the case of using all the available features is shown in Fig. 1. For MLP, we used a feed-forward structure with (1) 3 hidden neurons and 1 hidden layer and (2) 20 and 5 hidden neurons in the first and second hidden layers. In both cases we set the number of training iteration of the Backpropagation Algorithm to 100, the learning rate to 0.1 and the momentum to 0.01. For the RBF-Network, we tried various number of clusters, but they did not have substantial impact on the results. We present results for 100 clusters.

The effect of different feature types and the effect of bagging is shown in Table 1. For bagging, we constructed 100 randomly selected subsets of the basic features and we constructed regressors for all of these 100 subsets of features. We considered the average of the predictions of these 100 regressors as the prediction of the bagging-based model.

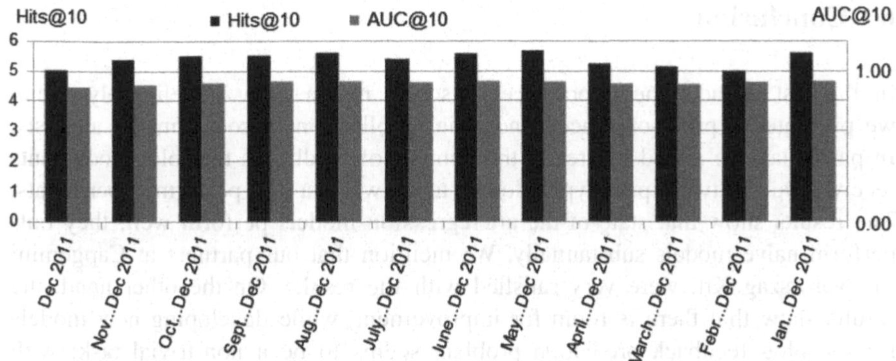

Fig. 2 The performance of the REP-tree classifier with basic features for various training intervals

The number of hits was around 5–6 for the examined models, which was much better than the prediction of a naive model, i.e., of a model that simply predicts the average number of feedbacks per source. This naive model achieved only 2–3 hits in our experiments. In general, relatively simple models, such as M5P Trees and REP Trees, seem to work very well both in terms of quality and runtime required for training of these models and for prediction using these models. Depending on the parameters of neural networks, the training may take relatively long time into account. From the quality point of view, while we observed neural networks to be competitive to the regression trees, the examined neural networks did not produce much better results than the mentioned regression trees.

Additionally to the presented results, we also experimented with support vector machines. We used the Weka-implementation of SVM, which had inacceptably long training times, even in case of simple (linear) kernel.

Out of the different types of features, the basic features (including aggregated features by source) seem to be the most predictive ones.

Bagging, see the last column of Table 1, improved the performance of MLPs and RBF-Network both in terms of Hits@10 and AUC@10, and the performance of REP-tree in terms of Hits@10. In the light of average and standard deviation, these improvement are, however, not significant.

We also examined how the length of the training interval affects the quality of prediction: both Hits@10 and AUC@10 of the REP-tree classifier are shown in Fig. 2 for various training intervals. As expected, recent training intervals, such as the last 1 or 2 months of 2011, seem to be informative enough for relatively good predictions. On the other hand, with using more and more historical data from larger time intervals, we did not observe a clear trend which may indicate that the user's behavior may (slightly) change and therefore historical data of a long time interval is not necessary more useful than recent data from a relatively short time interval.

6 Conclusion

In the last decade, the importance of social media grew unbelievably. Here, we presented a proof-of-concept industrial application of social media analysis. In particular, we aimed to predict the number of feedbacks that blog documents receive. Our software prototype allowed to crawl data and perform experiments. The results show that state-of-the art regression models perform well, they outperform naive models substantially. We mention that our partners at Capgemini Magyarország Kft. were very satisfied with the results. On the other hand, the results show that there is room for improvement, while developing new models for the blog feedback prediction problem seems to be a non-trivial task: with widely-used techniques, in particular ensemble methods, we only achieved marginal improvement. In order to motivate research in this area of growing interest, we made our data publicly available.

Acknowledgements We thank Capgemini Magyarország Kft. for the financial support of the project. The work reported in the paper has been developed in the framework of the project "Talent care and cultivation in the scientific workshops of BME" project. This project is supported by the grant TÁMOP-4.2.2.B-10/1–2010-0009.

References

Marinho, L. B., Buza, K., & Schmidt-Thieme, L. (2008). Folksonomy-based collabulary learning. *The Semantic Web—ISWC 2008. Lecture Notes in Computer Science* (Vol. 5318) (pp. 261–276). Heidelberg: Springer.

Mishne, G. (2007). Using blog properties to improve retrieval. In *International Conference on Weblogs and Social Media* (ICWSM'2007), Boulder, CO. http://www.icwsm.org/papers/3--Mishne.pdf.

Pang, B., & Lee, L. (2008). Opinion mining and sentiment analysis. *Journal Foundations and Trends in Information Retrieval, 2*, 1–135.

Pinto, J. P. G. S. (2008). Detection methods for blog trends. Report of Dissertation Master in Informatics and Computing Engineering. Faculdade de Engenharia da Universidade do Porto.

Reuter, T., Cimiano, P., Drumond, L., Buza, K., & Schmidt-Thieme, L. (2011). Scalable event-based clustering of social media via record linkage techniques. In *5th International AAAI Conference on Weblogs and Social Media*. Menlo Park: The AAAI Press.

Tan, P. N., Steinbach, M., & Kumar, V. (2006) *Introduction to data mining*. Boston: Pearson Addison Wesley.

Witten, I. H., & Franke, E. (2005). *Data mining. Practical machine learning tools and techniques* (2nd ed.). San Francisco: Elsevier.

Yano, T., & Smith, N. A. (2010). What's worthy of comment? Content and comment volume in political blogs. In *4th International AAAI Conference on Weblogs and Social Media* (pp. 359–362). Menlo Park: The AAAI Press.

Spectral Clustering: Interpretation and Gaussian Parameter

Sandrine Mouysset, Joseph Noailles, Daniel Ruiz, and Clovis Tauber

Abstract Spectral clustering consists in creating, from the spectral elements of a Gaussian affinity matrix, a low-dimensional space in which data are grouped into clusters. However, questions about the separability of clusters in the projection space and the choice of the Gaussian parameter remain open. By drawing back to some continuous formulation, we propose an interpretation of spectral clustering with Partial Differential Equations tools which provides clustering properties and defines bounds for the affinity parameter.

1 Introduction

Spectral clustering aims at selecting dominant eigenvectors of a parametrized Gaussian affinity matrix in order to build an embedding space in which the clustering is made. Many interpretations of this method were lead to explain why the clustering is made in the embedding space with graph theory with random walks (Meila and Shi 2001), matrix perturbation theory (Ng et al. 2002), Operators in Manifolds (Belkin and Niyogi 2003), physical models as inhomogeneous ferromagnetic Potts model (Blatt et al. 1996) or Diffusion Maps (Nadler et al. 2006). But all these analysis are investigated asymptotically for a large number of points and do not

S. Mouysset (✉)
University of Toulouse, IRIT-UPS, 118 route de Narbonne, 31062 Toulouse, France
e-mail: sandrine.mouysset@irit.fr

J. Noailles · D. Ruiz
University of Toulouse, IRIT-ENSEEIHT, 2 rue Camichel, 31071 Toulouse, France
e-mail: joseph.noailles@irit.fr; daniel.ruiz@irit.fr

C. Tauber
University of Tours, Hopital Bretonneau, 2 boulevard Tonnelle, 37044 Tours, France
e-mail: clovis.tauber@univ-tours.fr

M. Spiliopoulou et al. (eds.), *Data Analysis, Machine Learning and Knowledge Discovery*, Studies in Classification, Data Analysis, and Knowledge Organization, DOI 10.1007/978-3-319-01595-8_17,

explain why this method works for a finite data set. Moreover, another problem still arise: the affinity parameter influences the clustering results (Ng et al. 2002; Von Luxburg 2007). And the difficulty to define an adequate parameter seems to be slightly connected to the lack of some clustering property explaining how the grouping in this low-dimensional space correctly defines the partitioning in the original data.

In this paper, we propose a fully theoretical interpretation of spectral clustering whose first steps were introduced by Mouysset et al. (2010). From this, we define a new clustering property in the embedding space at each step of the study and new results showing the rule of the Gaussian affinity parameter. After recalling the spectral clustering method and the rule of the affinity parameter in Sect. 2.1, we propose a continuous version of the Spectral Clustering with Partial Differential Equations (PDE). To do so, we consider a sampling of connected components and, from this, we draw back to original shapes. This leads to formulate spectral clustering as an eigenvalue problem where data points correspond to nodes of some finite elements discretization and to consider the Gaussian affinity matrix A as a representation of heat kernel and the affinity parameter σ as the heat parameter t. Hence, the first step is to introduce an eigenvalue problem based on heat equation which is defined with a Dirichlet boundary problem. From this, in Sect. 2.2, we deduce an "almost" eigenvalue problem which can be associated to the Gaussian values. Thus identifying connected component appears to be linked to these eigenfunctions. Then, by introducing the Finite Elements approximation and mass lumping, we prove in Sect. 2.3 that this property is preserved with conditions on t when looking at eigenvectors given by spectral clustering algorithm. Finally, in Sect. 3, we study numerically the difference between eigenvectors from the spectral clustering algorithm and their associated discretized eigenfunctions from heat equation on a geometrical example, as a function of the affinity parameter t.

2 Interpretation

In the following, spectral clustering and its inherent problem are presented. Then we propose a continuous version of this method.

2.1 Spectral Clustering: Rule of Gaussian Parameter

Let consider a data set $\mathcal{P} = \{x_i\}_{i=1..N} \in \mathbb{R}^p$. Assume that the number of targeted clusters k is known. First, the spectral clustering consists in constructing the parametrized affinity matrix based on the Gaussian affinity measure between points of the data set $\mathcal{P}$. After a normalization step, by stacking the k largest eigenvectors, the spectral embedding in $\mathbb{R}^k$ is created. Each row of this matrix represents a data point x_i which is plotted in this embedding space and then grouped into clusters via

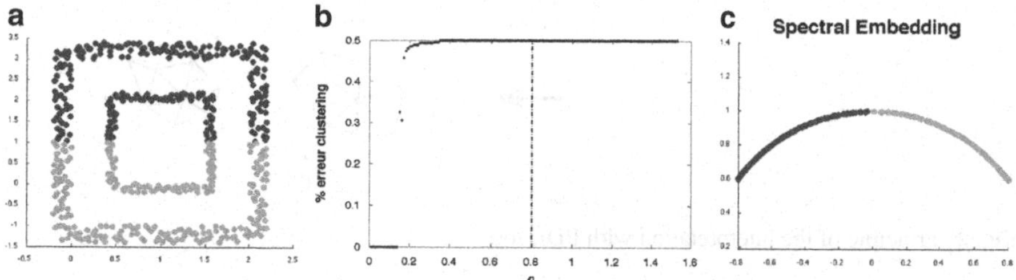

Fig. 1 Geometrical example: (**a**) clustering result for $\sigma = 0.8$, (**b**) percentage of clustering error function of σ, (**c**) spectral embedding space for $\sigma = 0.8$

the K-means method. Finally, thanks to an equivalence relation, the final partition of data set is directly defined from the clustering in the embedding space.

So this unsupervised method is mainly based on the Gaussian affinity measure, its parameter σ and its spectral elements. Moreover, it is known that the Gaussian parameter conditions the separability between the clusters in the spectral embedding space and should be well chosen (Von Luxburg 2007). The difficulty to fix this choice seems to be tightly connected to the lack of results explaining how the grouping in this low-dimensional space defines correctly the partitioning in the original data for a finite data set. Figure 1 summaries these previous remarks via a percentage of clustering which evaluated the percentage of mis-clustered points applied on a geometrical example of two concentric rectangles (Fig. 1a). For $\sigma = 0.8$, value which provides clustering errors (Fig. 1b), the two clusters defined with K-means are represented in the spectral embedding (Fig. 1c) by the respective black and grey colors. A piece of circle in which no separation by hyperplane is possible is described. Thus, in the original space, both rectangles are cut in two and define a bad clustering as show in Fig. 1a.

2.2 *Through an Interpretation with PDE Tools*

As spectral elements used in spectral clustering do not give explicitly this topological criteria for a discrete data set, we are drawing back to some continuous formulation wherein clusters will appear as disjoint subsets as shown in Fig. 2. In that way, we first have to define a clustering compatibility which establishes the link between continuous interpretation and the discrete case. So we consider an open set Ω subdivided by k disjoints connected components of Ω.

Definition 1 (Clustering Compatibility). Let Ω be a bounded open set in $\mathbb{R}^p$ made by Ω_i, $i \in 1, .., k$ disjoint connected components such that: $\Omega = \bigcup_{i=1}^{k} \Omega_i$. Let $\mathcal{P}$ be a set of points $\{x_i\}_{i=1}^{N}$ in the open set Ω. Let note $\mathcal{P}_j$, for $j = \{1, .., k\}$, the non empty set of points of $\mathcal{P}$ in the connected component Ω_j of Ω:$\mathcal{P}_j = \Omega_j \cap \mathcal{P}, \forall j \in \{1, .., k\}$. Let $C = \{C_1, .., C_{k'}\}$ be a partition of $\mathcal{P}$.

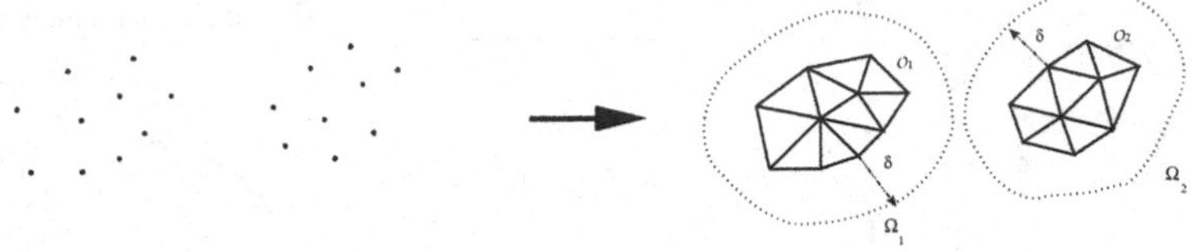

Fig. 2 Principle of the interpretation with PDE tools

Suppose that $k = k'$ then C is a compatible clustering if $\forall j = \{1, .., k'\}, \exists i \in \{1, .., k\}$, $C_j = \mathcal{P}_i$.

To make a parallel version in the $L^2(\Omega)$ space, data points which believe in a subset of Ω are equivalent to believe in the same connected component. In the following, we will formulate spectral clustering as an eigenvalue problem by assuming data points as nodes of some finite elements discretization and by considering Gaussian affinity matrix as a representation of heat kernel. But as the spectrum of heat operator in free space is essential, we will make a link with a problem defined on bounded domain in which the spectrum is finite. Then, due to the fact that we compare the discrete data defined by the elements of the affinity matrix with some L^2 functions which are the solutions of heat equation, we will introduce an explicit discretization with the Finite Element theory and the mass lumping to cancel all knowledge about the mesh. Then we will make a feedback of this analysis for the application of spectral clustering by defining clustering properties following the successive approximations. Finally, this study will lead to a functional rule of σ and a new formulation of a spectral clustering criterion.

2.2.1 Link Between Gaussian Affinity and Heat Kernel in $\mathbb{R}^p$

Let recall the Gaussian affinity element A_{ij} between two data points x_i and x_j is defined by $A_{ij} = \exp\left(-\left\|x_i - x_j\right\|^2/2\sigma^2\right)$. A direct link between the affinity A_{ij} and the heat kernel on $\mathbb{R}^*_+ \times \mathbb{R}^p$, defined by $K_H(t, x) = (4\pi t)^{-\frac{p}{2}} \exp\left(-\|x\|^2/4t\right)$ could be established as follows:

$$A_{ij} = (2\pi\sigma^2)^{\frac{p}{2}} K_H\left(\sigma^2/2, x_i - x_j\right), \ \forall i \neq j, \ \forall (i, j) \in \{1, .., N\}. \quad (1)$$

Equation (1) permits defining the affinity measure as a limit operator: the Gaussian affinity is interpreted as the heat kernel of a parabolic problem and its Gaussian parameter σ as a heat parameter t. Consider the following parabolic problem which is called heat equation, for $f \in L^2(\mathbb{R}^p)$:

$$(\mathcal{P}_{\mathbb{R}^p}) \begin{cases} \partial_t u - \Delta u = 0 \ \text{ for } \ (t, x) \in \mathbb{R}^+ \times \mathbb{R}^p, \\ u(x, 0) = f \ \text{ for } x \in \mathbb{R}^p. \end{cases}$$

Due to the fact that the spectrum of heat operator in free space, noted S_H, is essential and eigenfunctions are not localized in $\mathbb{R}^p$ without boundary conditions, we have to restrict the domain definition and make a link with a problem on a bounded domain Ω in which the eigenfunctions could be studied.

2.2.2 Clustering Property with Heat Equation

Let now introduce the initial value problem in $L^2(\Omega)$, for $f \in L^2(\Omega)$:

$$(\mathcal{P}_\Omega)\begin{cases} \partial_t u - \Delta u = 0 \text{ in } \mathbb{R}^+ \times \Omega, \\ u(t=0) = f, \text{ in } \Omega, \\ u = 0, \text{ on } \mathbb{R}^+ \times \partial\Omega. \end{cases}$$

Denote by K_D the Green's kernel of $(\mathcal{P}_\Omega)$. The solution operator in $H^2(\Omega) \cap H_0^1(\Omega)$ associated to this problem is defined, for $f \in L^2(\Omega)$, by:

$$S_D(t)f(x) = \int_\Omega K_D(t,x,y)f(y)dy,\ x \in \mathbb{R}^p.$$

Let consider $\{(\widetilde{v_{n,i}})_{n,i>0},\ i \in \{1,..,k\}\} \in H_0^1(\Omega)$ such that $(\widetilde{v_{n,i}})_{n,i>0}$ are the solutions of $\Delta \widetilde{v_{n,i}} = \lambda_{n,i}\widetilde{v_{n,i}}$ on Ω_i for $i \in \{1,..,k\}$ and $n > 0$ and extend $\widetilde{v_{n,i}} = 0$ on $\Omega \backslash \Omega_i$. These functions are eigenfunctions of $(\mathcal{P}_\Omega)$ and the union of these eigenfunctions is an Hilbert basis of $H_0^1(\Omega)$. Moreover, as $\Omega = \bigcup_{i=1}^k \Omega_i$, for all $i \in \{1,..,k\}$ and $n > 0$, the eigenfunctions, noted $\{(\widetilde{v_{n,i}})_{n,i>0},\ i \in \{1,..,k\}\}$, satisfied: $S_D(t)\widetilde{v_{n,i}} = e^{-\lambda_{n,i}t}\widetilde{v_{n,i}}$. So the eigenfunctions of S_D have a geometrical property: its support is included in only one connected component. Thus a clustering property in the spectral embedding space could be established.

Proposition 1 (Clustering Property). *For all point $x \in \Omega$ and $\epsilon > 0$, let note ρ_x^ϵ a regularized Dirac function centred in x: $\rho_x^\epsilon \in C^\infty(\Omega, [0,1])$, $\rho_x^\varepsilon(x) = 1$ and $supp(\rho_x^\varepsilon) \subset \mathcal{B}(x,\varepsilon)$. The eigenfunctions of S_D, noted $\widetilde{v_{n,i}}$, for $i \in \{1,..,k\}$ and $n > 0$ such that for all $x \in \Omega$ and all $i \in \{1,..,k\}$ and for all $t > 0$, the following result is satisfied:*

$$\left[\exists \varepsilon_0 > 0,\ \forall \varepsilon \in]0,\varepsilon_0[,\ \exists n > 0,\ (S_D(t)\rho_x^\varepsilon | \widetilde{v_{n,i}})_{L^2(\Omega)} \neq 0\right] \Longleftrightarrow x \in \Omega_i \tag{2}$$

where $(f|g)_{L^2(\Omega)} = \int_\Omega f(y)g(y)dy$, $\forall (f,g) \in L^2(\Omega)$ is the usual scalar product in L^2.

Proof. By contrapositive, let $i \in \{1,..,k\}$ and a point $x \in \Omega_j$ with any $j \neq i$. Let $d_x = d(x, \partial\Omega_j) > 0$ be the distance of x from the boundary of Ω_j. According to the hypothesis on Ω, we have $d_0 = d(\Omega_i, \Omega_j) > 0$. So for all $\varepsilon \in]0, \inf(d_x, d_0)[$, $\mathcal{B}(x,\varepsilon) \subset \Omega_j$. Then for all $t > 0$, $supp(S_D(t)\rho_x^\varepsilon) \subset \Omega_j$ and so, for $n > 0$,

$(S_D(t)\rho_x^{\varepsilon}|\widetilde{v_{n,i}})_{L^2(\Omega)} = 0$. So there does not any $\varepsilon_0 > 0$ which verifies the direct implication of (2). Reversely, let $x \in \Omega_i$ and $\varepsilon \in]0, \inf(d_x, d_O)[$, $\mathcal{B}(x,\varepsilon) \subset \Omega_i$. So the support of ρ_x^{ε} is in Ω_i. As the $(\widetilde{v_{n,i}})_{n>0}$ is an Hilbert basis of $L^2(\Omega_i)$ and that $\rho_x^{\varepsilon}(x) = 1 \neq 0$ then there exists $n > 0$ such that $(\rho_x^{\varepsilon}|\widetilde{v_{n,i}}) \neq 0$. In this case, $(S_D(t)\rho_x^{\varepsilon}|\widetilde{v_{n,i}})_{L^2(\Omega)} = e^{-\lambda_{n,i}t}(\rho_x^{\varepsilon}|\widetilde{v_{n,i}}) \neq 0$.

By considering an open subset O which approximates from the interior the open set Ω such that $Volume(\Omega \backslash O) \leq \epsilon$, for $\epsilon > 0$, both heat operators of $(\mathcal{P}_{\mathbb{R}^p})$ and $(\mathcal{P}_{\Omega})$ could be compared in O. Let δ be the distance from O to Ω as shown in Fig. 2. Due to the fact that the difference between the Green kernels K_H and K_D could be estimated in O and is function of the heat parameter t, the geometrical property could thus be preserved on the heat operator in free space restricted to O. Let $v_{n,i}$ be the eigenfunction $\widetilde{v_{n,i}}$ which support is restricted to O, for all $i \in \{1,..,k\}$ and $n > 0$. From this, we obtain, for $0 < t < \delta^2$:

$$S_H^O(t)v_{n,i} = \exp^{-\lambda_{n_i}t} v_{n,i} + \eta(t, v_{n,i}), \tag{3}$$

with $\|\eta(t, v_{n,i})\|_{L^2(O)} \to 0$ when $t \to 0, \delta \to 0$.
So we can prove that on O, the eigenfunctions for the solution operator for bounded heat equation are quasi-eigenfunctions for S_H^O plus a residual (Mouysset et al. 2010). The clustering property adapted to the restricted heat operator S_H^O remains introducing an hypothesis on the heat parameter t. Moreover, (2) is modified with non-null values by introducing a gap between scalar product with eigenfunctions such that for all $x > 0$ and all $i \in \{1,..,k\}$:

$$\left[\begin{array}{l} \exists\varepsilon_0 > 0,\ \exists\alpha > 0,\ \forall\varepsilon \in]0,\varepsilon_0[,\ \exists n > 0,\ \forall t > 0 \text{ small enough,} \\ v_{n,i} = \arg\max_{\{v_{m,j}, m\in\mathbb{N}, j\in[|1,k|]\}} \left|(S_H^O(t)\rho_x^{\epsilon}|v_{m,j})_{L^2(O)}\right| \\ \text{and } \left|(S_H^O(t)\rho_x^{\varepsilon}|v_{n,i})_{L^2(O)}\right| > \alpha \end{array}\right] \Longleftrightarrow x \in O_i. \tag{4}$$

These previous results prove that in infinite dimension, a clustering could be realized in the spectral embedding space because the eigenfunctions have a geometrical property. This study leads to the following question: do eigenvectors of the affinity matrix behave like eigenfunctions of $(\mathcal{P}_{\Omega})$?

2.3 Discretization with Finite Elements

From this, we will look for a similar behaviour onto eigenvectors of A by introducing a finite dimension representation matching with the initial data set $\mathcal{P}$ with help of the finite elements (Ciarlet 1978). So, we consider data points as finite dimensional approximation and elements of the affinity matrix built from data points as nodal values of S_H^O.

2.3.1 Approximation in Finite Dimension

Let τ_h be a triangulation on $\bar{O}$ such that: $h = \max_{K \in \tau_h} h_K$, h_K being a characteristic length of triangle K. Let consider a finite decomposition of the domain: $\bar{O} = \cup_{K \in \tau_h} K$ in which (K, P_K, Σ_K) satisfies Lagrange finite element assumptions for all $K \in \tau_h$. We define also the finite dimension approximation space: $V_h = \{w \in C^0(\bar{O}); \forall K \in \tau_h, w_{|K} \in P_K\}$ and denote Π_h the linear interpolation from $C^0(\bar{O})$ in V_h with the usual scalar product $(\cdot|\cdot)_{L^2(V_h)}$ (Ciarlet 1978). According to this notations, for $t > 0$, the Π_h-mapped operator S_H^O applied to each shape function ϕ_j is, for $h^{3p+2} < t^2$, for all $1 \leq j \leq N$:

$$(4\pi t)^{\frac{p}{2}} \Pi_h(S_H^O(t)\phi_j)(x) = \sum_{k=1}^{N} ((A + \mathbb{I}_N)M)_{kj}\, \phi_k(x) + O\left(\frac{h^{3p+2}}{t^2}\right), \tag{5}$$

where M stands for the mass matrix defined by: $M_{ij} = (\phi_i|\phi_j)_{L^2(V_h)}$. Equation (5) means that the affinity matrix defined in (1) in spectral algorithm is interpreted as the Π_h-projection of operator solution of $(\mathcal{P}_{\mathbb{R}^p})$ with M mass matrix from Finite Element theory (Mouysset et al. 2010).

So we could formulate finite elements approximation of continuous clustering result (3). From the eigenfunctions of S_D restricted to O, their projection in V_h, noted $W_{n,i}$, are defined by: $W_{n,i} = \Pi_h v_{n,i} \in V_h, \forall i \in \{1, .., k\}$. So, for $h^{\frac{3p+2}{2}} < t < \delta^2$, the following result could be established:

$$(4\pi t)^{\frac{-p}{2}} (A + \mathbb{I}_N) M\, W_{n,i} = e^{-\lambda_{n,i} t} W_{n,i} + \Psi(t, h), \tag{6}$$

where $\|\Psi(t,h)\|_{L^2(V_h)} \to 0$ and $\delta \to 0$. Equation (6) shows that the geometrical property is preserved in finite dimension on the eigenvectors of $(A + \mathbb{I}_N)M$. Moreover, a lower bound for the heat parameter was defined. But all this previous results include the mass matrix which is totally dependent of the finite elements. In order to cancel this dependence, mass lumping process is investigated.

2.3.2 Mass Lumping

The mass lumping method consists in using a quadrature formula whose integration points are the interpolation points of the finite element. So let I_k be the list of indices of points which are element of $K \in \tau_h$. Let consider the quadrature scheme exact for polynomials of degree ≤ 1:

$$\int_K \phi(x)dx \approx \sum_{k \in I_k} \frac{|K|}{3} \phi(x_{i_k}) \tag{7}$$

where $|K|$ is the area of the finite element K. So, with additional regularity condition on the mesh which bounds $|K|$, the mass lumping permits considering the mass matrix M as a homogeneous identity matrix. So (6) is modified so that, $\exists\, \alpha > 0$, such that:

$$\alpha\,(A + \mathbb{I}_N)\, W_{n,i} = e^{-\lambda_{n,i} t}\, W_{n,i} + \Psi'(t,h), \tag{8}$$

where $\|\Psi'(t,h)\|_{L^2(V_h)} \to 0$ and $\delta \to 0$. The approximation in finite dimension of the clustering property (4) is reformulated as follows, for all $x_r \in \mathcal{P}$, for all $i \in \{1, .., k\}$:

$$\begin{bmatrix} \exists \alpha > 0,\ \exists n > 0,\ \forall t > 0, t, h^2/t \text{ and } h^{(3p+1)}/t^2 \text{ small enough}, \\ W_{n,i} = \arg\max_{\{W_{m,j}, m\in\mathbb{N}, j\in[|1,k|]\}} \left| \left((A+\mathbb{I}_N)_{.r} \,|\, W_{m,j} \right)_{L^2(V_h)} \right| \\ \text{and } \left| \left((A+\mathbb{I}_N)_{.r} \,|\, W_{n,i} \right)_{L^2(V_h)} \right| > \alpha \end{bmatrix} \iff x_r \in O_i, \tag{9}$$

where $((A + \mathbb{I}_N))_{.r}$ is the rth column of the matrix $(A + \mathbb{I}_N)$, for all $r \in \{1, ..N\}$. This leads to the same clustering for a set of data points either we consider eigenfunctions in $L^2(\Omega)$ or Π_h-interpolated eigenfunction in the approximation space V_h. With an asymptotic condition on the heat parameter t (or Gaussian parameter σ), points which are elements of the same cluster have the maximum of their projection coefficient along the same eigenvector. So the clustering in spectral embedding space provides the clustering in data space.

3 Gaussian Parameter: A Geometrical Example

This previous theoretical interpretation proves that the Gaussian parameter should be chosen within a specific interval in order to improve the separability between clusters in the spectral embedding space. In order to experiment the parallel between continuous version and the approximate one, we consider a geometrical example with non convex shapes as shown in Fig. 3a. For each connected component (or each cluster) $i \in \{1, 2\}$, the discretized eigenfunction, noted $W_{1,i}$, associated to the first eigenvalue of each connected component and the eigenvectors, noted Y_i, which gives maximum projection coefficient with $W_{1,i}$ are respectively plotted in Fig. 3b, c and e, f. The correlation ω between $W_{1,i}$ and Y_i is represented as a function of the heat parameter t in Fig. 3d: $\omega = |(W_{1,i}|Y_i)|(\|W_{1,i}\|_2\|Y_i\|_2)^{-1}$. The vertical black dash dot lines indicate the lower and upper estimated bounds of the heat parameter. In this interval, the correlation between the continuous version and the eigenvectors of the Gaussian affinity matrix is maximum. So the clusters are well separated in the spectral embedding space.

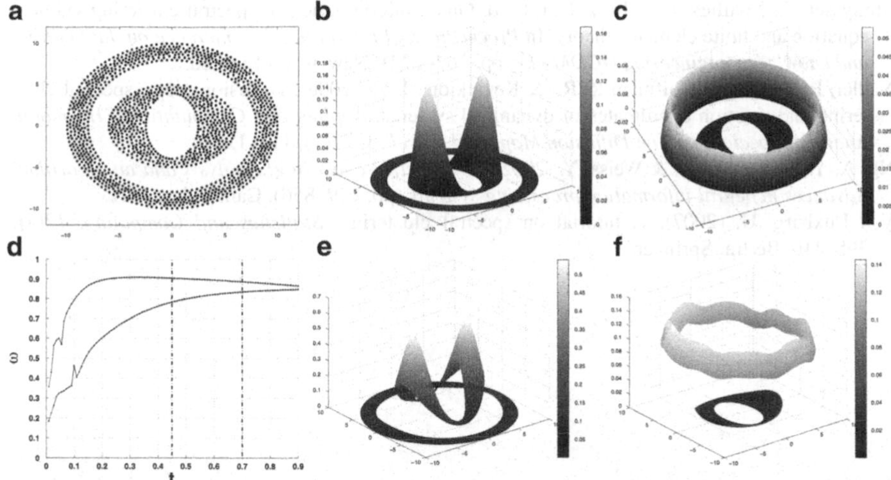

Fig. 3 (**a**) Data set ($N = 669$), (**b**) and (**c**) discretized eigenfunctions of S_D, (**d**) correlation ω between the continuous version and its discrete approximation function of t, (**e**) and (**f**) eigenvectors from A which provides the maximum projection coefficient with the eigenfunctions of S_D

4 Conclusion

In this paper, spectral clustering was formulated as an eigenvalue problem. From this interpretation, a clustering property on the eigenvectors and some conditions on the Gaussian parameter have been defined. This leads to understand how spectral clustering works and to show how clustering results could be affected with a bad choice of the affinity parameter. But we do not take into account the normalization step in the whole paper but its rule is crucial for ordering largest eigenvectors for each connected components to the first eigenvectors and should be studied.

References

Belkin, M., & Niyogi, P. (2003). Laplacian eigenmaps for dimensionality reduction and data representation. *Neural Computation, 15*(6), 1373–1396.

Blatt, M., Wiseman, S., & Domany, E. (1996). Superparamagnetic clustering of data. *Physical Review Letters, 76*(18), 3251–3254.

Ciarlet, P. G. (1978). *The finite element method for elliptic problems. Series studies in mathematics and its applications* (Vol. 4). Amsterdam: North-Holland.

Meila, M., & Shi, J. (2001). A random walks view of spectral segmentation. In *Proceedings of eighth international workshop on artificial intelligence and statistics (AISTATS) 2001*.

Mouysset, S., Noailles, J., & Ruiz, D. (2010). On an interpretation of spectral clustering via heat equation and finite elements theory. In *Proceedings of international conference on data mining and knowledge engineering (ICDMKE)* (pp. 267–272). Newswood Limited.

Nadler, B., Lafon, S., Coifman, R. R., & Kevrekidis, I. G. (2006). Diffusion maps, spectral clustering and reaction coordinates of dynamical systems. *Applied and Computational Harmonic Analysis: Special Issue on Diffusion Maps and Wavelets, 21*(1), 113–127.

Ng, A. Y., Jordan, M. I., & Weiss, Y. (2002). *On spectral clustering: Analysis and an algorithm. Advances in neural information processing systems* (pp. 849–856). Cambridge: MIT.

Von Luxburg, U. (2007). A tutorial on spectral clustering. *Statistics and Computing, 17*(4), 395–416. Berlin: Springer.

On the Problem of Error Propagation in Classifier Chains for Multi-label Classification

Robin Senge, Juan José del Coz, and Eyke Hüllermeier

Abstract So-called classifier chains have recently been proposed as an appealing method for tackling the multi-label classification task. In this paper, we analyze the influence of a potential pitfall of the learning process, namely the discrepancy between the feature spaces used in training and testing: while true class labels are used as supplementary attributes for training the binary models along the chain, the same models need to rely on estimations of these labels when making a prediction. We provide first experimental results suggesting that the attribute noise thus created can affect the overall prediction performance of a classifier chain.

1 Introduction

Multi-label classification (MLC) has attracted increasing attention in the machine learning community during the past few years (Tsoumakas and Katakis 2007).The goal in MLC is to induce a model that assigns a *subset* of labels to each example, rather than a single one as in multi-class classification. For instance, in a news website, a multi-label classifier can automatically attach several labels—usually called tags in this context—to every article; the tags can be helpful for searching related news or for briefly informing users about their content.

Current research on MLC is largely driven by the idea that optimal prediction performance can only be achieved by modeling and exploiting *statistical dependencies* between labels. Roughly speaking, if the relevance of one label may depend

R. Senge (✉) · E. Hüllermeier
Philipps-Universität Marburg, Marburg, Germany
e-mail: senge@mathematik.uni-marburg.de; eyke@mathematik.uni-marburg.de

J.J. del Coz
University of Oviedo, Gijón, Spain
e-mail: juanjo@aic.uniovi.es

M. Spiliopoulou et al. (eds.), *Data Analysis, Machine Learning and Knowledge Discovery*, Studies in Classification, Data Analysis, and Knowledge Organization, DOI 10.1007/978-3-319-01595-8_18,

on the relevance of others, then labels should be predicted *simultaneously* and not *separately*. This is the main argument against simple *decomposition techniques* such as binary relevance (BR) learning, which splits the original multi-label task into several independent binary classification problems, one for each label.

Until now, several methods for capturing label dependence have been proposed in the literature, including a method called *classifier chains* (CC) (Read et al. 2011). This method enjoys great popularity, even though it has been introduced only lately. As its name suggests, CC selects an order on the label set—a *chain* of labels—and trains a binary classifier for each label in this order. The difference with respect to BR is that the feature space used to induce each classifier is extended by the previous labels in the chain. These labels are treated as additional attributes, with the goal to model conditional dependence between a label and its predecessors. CC performs particularly well when being used in an ensemble framework, usually denoted as *ensemble of classifier chains* (ECC), which reduces the influence of the label order.

Our study aims at gaining a deeper understanding of CC's learning process. More specifically, we address a potential pitfall of this method: since information about preceding labels is only available for training, this information has to be replaced by estimations (coming from the corresponding classifiers) at prediction time. As a result, CC has to deal with a specific type of attribute noise: while a classifier is learned on "clean" training data, including the true values of preceding labels, it is applied on "noisy" test data, in which true labels are replaced by possibly incorrect predictions. Obviously, this type of noise may affect the performance of each classifier in the chain. More importantly, since each classifier relies on its predecessors, a single false prediction might be propagated and possibly even reinforced along the chain.

The rest of the paper is organized as follows. The next section introduces the setting of MLC, and Sect. 3 explains the classifier chains method. Section 4 is devoted to a deeper discussion of the aforementioned pitfalls of CC, along with some experiments for illustration purposes. The paper ends with a couple of concluding remarks in Sect. 5.

2 Multi-label Classification

Let $\mathcal{L} = \{\lambda_1, \lambda_2, \ldots, \lambda_m\}$ be a finite and non-empty set of class labels, and let $\mathcal{X}$ be an instance space. We consider an MLC task with a training set $S = \{(\mathbf{x}_1, \mathbf{y}_1), \ldots, (\mathbf{x}_n, \mathbf{y}_n)\}$, generated independently and identically according to a probability distribution $\mathbf{P}(\mathbf{X}, \mathbf{Y})$ on $\mathcal{X} \times \mathcal{Y}$. Here, $\mathcal{Y}$ is the set of possible label combinations, i.e., the power set of $\mathcal{L}$. To ease notation, we define $\mathbf{y}_i$ as a binary vector $\mathbf{y}_i = (y_{i,1}, y_{i,2}, \ldots, y_{i,m})$, in which $y_{i,j} = 1$ indicates the presence (relevance) and $y_{i,j} = 0$ the absence (irrelevance) of λ_j in the labeling of $\mathbf{x}_i$. Under this convention, the output space is given by $\mathcal{Y} = \{0, 1\}^m$. The goal in MLC is to induce from S a hypothesis $\mathbf{h} : \mathcal{X} \longrightarrow \mathcal{Y}$ that correctly predicts the subset of relevant labels for unlabeled query instances $\mathbf{x}$.

The most-straight forward and arguably simplest approach to tackle the MLC problem is *binary relevance* (BR). The BR method reduces a given multi-label problem with m labels to m *binary classification* problems. More precisely, m hypotheses $h_j : \mathcal{X} \longrightarrow \{0,1\}$, $j = 1,\ldots,m$, are induced, each of them being responsible for predicting the relevance of one label, using $\mathcal{X}$ as an input space. In this way, the labels are predicted independently of each other and no label dependencies are taken into account.

In spite of its simplicity and the strong assumption of label independence, it has been shown theoretically and empirically that BR performs quite strong in terms of decomposable loss functions (Dembczyński et al. 2010), including the well-known *Hamming loss* $L_H(\mathbf{y}, \mathbf{h}(\mathbf{x})) = \frac{1}{m}\sum_{i=1}^{m} [\![y_i \neq h_i(\mathbf{x})]\!]$. The Hamming loss averages the standard 0/1 classification error over the m labels and hence corresponds to the proportion of labels whose relevance is incorrectly predicted. Thus, if one of the labels is predicted incorrectly, this accounts for an error of $\frac{1}{m}$. Another extension of the standard 0/1 classification loss is the *subset 0/1 loss* $L_{ZO}(\mathbf{y}, \mathbf{h}(\mathbf{x})) = [\![\mathbf{y} \neq \mathbf{h}(\mathbf{x})]\!]$. Obviously, this measure is more drastic and already treats a mistake on a single label as a complete failure. The necessity to exploit label dependencies in order to minimize the generalization error in terms of the subset 0/1 loss has been shown in Dembczyński et al. (2010).

3 Classifier Chains

While following a similar setup as BR, classifier chains (CC) seek to capture label dependencies. CC learns m binary classifiers linked along a chain, where each classifier deals with the binary relevance problem associated with one label. In the training phase, the feature space of each classifier in the chain is extended with the actual label information of all previous labels in the chain. For instance, if the chain follows the order $\lambda_1 \rightarrow \lambda_2 \rightarrow \ldots \rightarrow \lambda_m$, then the classifier h_j responsible for predicting the relevance of λ_j is of the form

$$h_j : \mathcal{X} \times \{0,1\}^{j-1} \longrightarrow \{0,1\} . \tag{1}$$

The training data for this classifier consists of instances $(\mathbf{x}_i, y_{i,1}, \ldots, y_{i,j-1})$ labeled with $y_{i,j}$, that is, original training instances $\mathbf{x}_i$ supplemented by the relevance of the labels $\lambda_1, \ldots, \lambda_{j-1}$ preceding λ_j in the chain.

At prediction time, when a new instance $\mathbf{x}$ needs to be labeled, a label subset $\mathbf{y} = (y_1, \ldots, y_m)$ is produced by successively querying each classifier h_j. Note, however, that the inputs of these classifiers are not well-defined, since the supplementary attributes $y_{i,1}, \ldots, y_{i,j-1}$ are not available. These missing values are therefore replaced by their respective predictions: y_1 used by h_2 as an additional input is replaced by $\hat{y}_1 = h_1(\mathbf{x})$, y_2 used by h_3 as an additional input is replaced by $\hat{y}_2 = h_2(\mathbf{x}, \hat{y}_1)$, and so forth. Thus, the prediction $\mathbf{y}$ is of the form

$$\mathbf{y} = \big(h_1(\mathbf{x}),\, h_2(\mathbf{x}, h_1(\mathbf{x})),\, h_3(\mathbf{x}, h_1(\mathbf{x}), h_2(\mathbf{x}, h_1(\mathbf{x}))), \ldots\big)$$

Realizing that the order of labels in the chain may influence the performance of the classifier, and that an optimal order is hard to anticipate, the authors in Read et al. (2011) propose the use of an ensemble of CC classifiers. This approach combines the predictions of different random orders and, moreover, uses a different sample of the training data to train each member of the ensemble. *Ensembles of classifier chains* (ECC) have been shown to increase prediction performance over CC by effectively using a simple voting scheme to aggregate predicted relevance sets of the individual CCs: for each label λ_j, the proportion $\hat{w}_j$ of classifiers predicting $y_j = 1$ is calculated. Relevance of λ_j is then predicted by using a threshold t, that is, $\hat{y}_j = [\![\hat{w}_j \geq t]\!]$.

4 The Problem of Error Propagation in CC

The learning process of CC violates a key assumption of machine learning, namely that the training data is representative of the test data in the sense of being identically distributed. This assumption does not hold for the chained classifiers in CC: while using the *true* label data y_j as input attributes during the training phase, this information is replaced by *estimations* $\hat{y}_j$ at prediction time. Needless to say, y_j and $\hat{y}_j$ will normally not follow the same distribution.

From the point of view of the classifier h_j, which uses the labels $y_1, \ldots, y_{j-1}$ as additional attributes, this problem can be seen as a problem of *attribute noise*. More specifically, we are facing the "clean training data vs. noisy test data" case, which is one of four possible noise scenarios that have been studied quite extensively in Zhu and Wu (2004). For CC, this problem appears to be vital: Could it be that the additional label information, which is exactly what CC seeks to exploit in order to gain in performance (compared to BR), eventually turn out to be a source of impairment? Or, stated differently, could the additional label information perhaps be harmful rather than useful? This question is difficult to answer in general. In particular, there are several factors involved, notably the following:

- *The length of the chain*: The larger the number $j-1$ of preceding classifiers in the chain, the higher is the potential level of attribute noise for a classifier h_j. For example, if prediction errors occur independently of each other with probability ϵ, then the probability of a noise-free input is only $(1-\epsilon)^{j-1}$. More realistically, one may assume that the probability of a mistake is not constant but will increase with the level of attribute noise in the input. Then, due to the recursive structure of CC, the probability of a mistake will increase even more rapidly along the chain.
- *The order of the chain*: Since some labels might be inherently more difficult to predict than others, the order of the chain will play a role, too. In particular, it

would be advantageous to put simpler labels in the beginning and harder ones more toward the end of the chain.

- *The accuracy of the binary classifiers*: The level of attribute noise is in direct correspondence with the accuracy of the binary classifiers along the chain. More specifically, these classifiers determine the input distributions in the test phase. If they are perfect, then the training distribution equals the test distribution, and there is no problem. Otherwise, however, the distributions will differ.
- *The dependency among labels*: Perhaps most interestingly, a (strong enough) dependence between labels is a prerequisite for both, an improvement and a deterioration through chaining. In fact, CC cannot gain (compared to BR) in case of no label dependency. In that case, however, it is also unlikely to loose, because a classifier h_j will most likely[1] ignore the attributes $y_1, \ldots, y_{j-1}$. Otherwise, in case of pronounced label dependence, it will rely on these attributes, and whether or not this is advantageous will depend on the other factors above.

In the following, we present two experimental studies that are meant to illustrate the problem of error propagation in classifier chains.

4.1 Experiment with Real Data

Our intuition is that attribute noise in the test phase can produce a propagation of errors through the chain, thereby affecting the performance of the classifiers depending on their position in the chain. More specifically, we expect classifiers in the beginning of the chain to systematically perform better than classifiers toward the end. In order to verify this conjecture, we perform the following simple experiment: we train a CC classifier on 500 randomly generated label orders. Then, for each label order and each position, we compute the performance of the classifier on that position in terms of the relative increase of classification error compared to BR. Finally, these errors are averaged *position-wise* (not label-wise). For this experiment, we used three standard MLC benchmark data sets: emotions (593 examples, 72 attributes, 6 labels), scene (2,407, 294, 6), yeast-10 (2,417, 103, 10); the latter is a reduction of the original yeast data set to the ten most frequent labels and their instances.

The results in Fig. 1 clearly confirm our expectations. In two cases, CC starts to loose immediately, and the loss increases with the position. In the third case, CC is able to gain on the first positions but starts to loose again later on.

[1]The possibility to ignore parts of the input information does of course also depend on the type of classifier used.

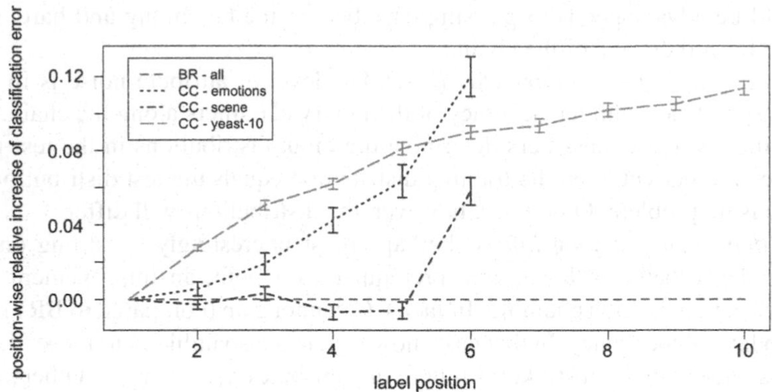

Fig. 1 Results of the first experiment: position-wise relative increase of classification error (mean plus standard error bars).

4.2 Experiment with Synthetic Data

In a second experiment, we used a synthetic setup that was proposed in Dembczynski et al. (2012) to analyze the influence of label dependence. The input space $\mathcal{X}$ is two-dimensional and the underlying decision boundary for each label is linear in these inputs. More precisely, the model for each label is defined as follows:

$$h_i(\mathbf{x}) = \begin{cases} 1 & a_{j1}x_1 + a_{j2}x_2 \geq 0 \\ 0 & \text{otherwise} \end{cases} \tag{2}$$

The input values are drawn randomly from the unit circle. The parameters a_{j1} and a_{j2} for the j-th label are set to $a_{j1} = 1 - \tau r_1, a_{j2} = \tau r_2$, with r_1 and r_2 randomly chosen from the unit interval. Additionally, random noise is introduced for each label by independently reversing a label with probability $\pi = 0.1$. Obviously, the level of label dependence can be controlled by the parameter $\tau \in [0, 1]$: the smaller τ, the stronger the dependence tends to be (see Fig. 2 for an illustration).

For different label cardinalities $m \in \{5, 10, 15, 20, 25\}$, we run ten repetitions of the following experiment: we created 10 different random model parameter sets (two for each label) and generated 10 different training sets, each consisting of 50 instances. For each training set, a model is learned and evaluated (in terms of Hamming and subset 0/1 loss) on an additional data set comprising 1,000 instances.

Figure 3 summarizes the results in terms of the average loss divided by the corresponding Bayes loss (which can be computed since the data generating process is known); thus, the optimum value is always 1. Comparing BR and CC, the big picture is quite similar to the previous experiment: the performance of CC tends to decrease with an increasing number of labels. In the case of less label dependence, this can already be seen for only five labels. The case of high label dependence

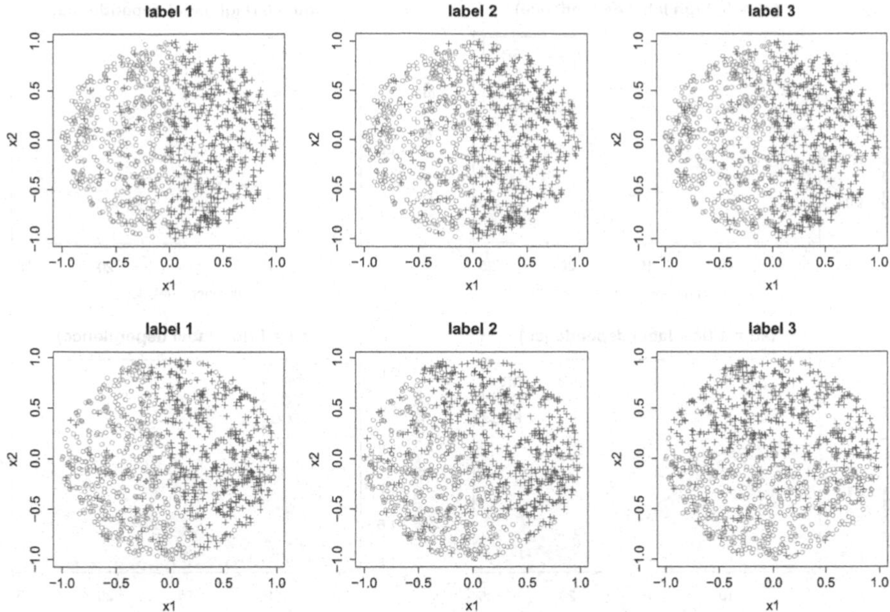

Fig. 2 Example of synthetic data: the *top* three labels are generated using $\tau = 0$, the three at the *bottom* with $\tau = 1$.

is more interesting: while CC seems to gain from exploiting the dependency for a small to moderate number of labels, it cannot extend this gain to more than 15 labels.

5 Conclusion

This paper has thrown a critical look at the classifier chains method for multi-label classification, which has been adopted quite quickly by the MLC community and is now commonly used as a baseline when it comes to comparing methods for exploiting label dependency. Notwithstanding the appeal of the method and the plausibility of its basic idea, we have argued that, at second sight, the chaining of classifiers begs an important flaw: a binary classifier that has learned to rely on the values of previous labels in the chain might be misled when these values are replaced by possibly erroneous estimations at prediction time. The classification errors produced because of this attribute noise may subsequently be propagated or even reinforced along the entire chain. Roughly speaking, what looks as a gift at training time may turn out to become a handicap in testing.

Our results clearly show that this problem is relevant, and that it may strongly impair the performance of the CC method. There are several lines of future work. First, it is of course desirable to complement this study by meaningful theoretical

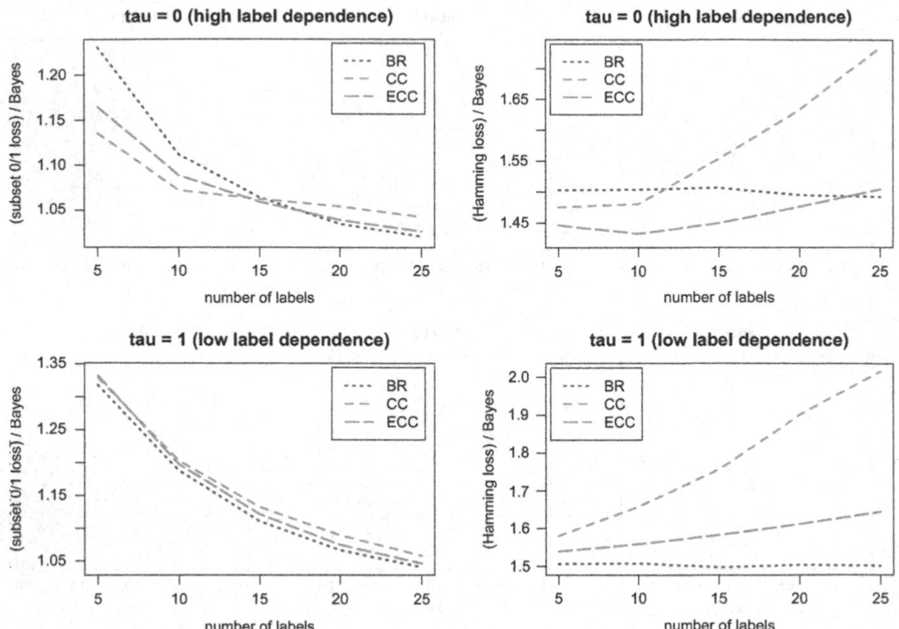

Fig. 3 Results of the second experiment for $\tau = 0$ (*top*—high label dependence) and $\tau = 1$ (*bottom*—low label dependence).

results supporting our claims. Second, it would be interesting to investigate to what extent the problem of attribute noise also applies to the probabilistic variant of classifier chains introduced in Dembczyński et al. (2010).

Acknowledgements This research has been supported by the Germany Research Foundation (DFG) and the Spanish Ministerio de Ciencia e Innovación (MICINN) under grant TIN2011-23558.

References

Dembczyński, K., Cheng, W., & Hüllermeier, E. (2010). Bayes optimal multilabel classification via probabilistic classifier chains. In *International Conference on Machine Learning* (pp. 279–286).

Dembczynski, K., Waegeman, W., Cheng, W., & Hüllermeier, E. (2012). On label dependence and loss minimization in multi-label classification. *Machine Learning, 88*(1–2), 5–45.

Read, J., Pfahringer, B., Holmes, G., & Frank, E. (2011). Classifier chains for multi-label classification. *Machine Learning, 85*(3), 333–359.

Tsoumakas, G., & Katakis, I. (2007). Multi label classification: An overview. *International Journal of Data Warehouse and Mining, 3*(3), 1–13.

Zhu, X., & Wu, X. (2004). Class noise vs. attribute noise: A quantitative study of their impacts. *Artificial Intelligence Review, 22*(3), 177–210.

Statistical Comparison of Classifiers for Multi-objective Feature Selection in Instrument Recognition

Igor Vatolkin, Bernd Bischl, Günter Rudolph, and Claus Weihs

Abstract Many published articles in automatic music classification deal with the development and experimental comparison of algorithms—however the final statements are often based on figures and simple statistics in tables and only a few related studies apply proper statistical testing for a reliable discussion of results and measurements of the propositions' significance. Therefore we provide two simple examples for a reasonable application of statistical tests for our previous study recognizing instruments in polyphonic audio. This task is solved by multi-objective feature selection starting from a large number of up-to-date audio descriptors and optimization of classification error and number of selected features at the same time by an evolutionary algorithm. The performance of several classifiers and their impact on the pareto front are analyzed by means of statistical tests.

1 Introduction

A large share of interdisciplinary research as music information retrieval (MIR) (Downie 2003) corresponds to experimental studies with comparison and evaluation of established and new algorithms. However, it can be observed that in many cases the suggestions or improvements of a novel technique are not properly evaluated: e.g. only one evaluation metric like accuracy is estimated, the holdout set is not completely independent, or the final assumptions are not underlined by any statistical tests which provide a solid estimation of the investigations reliability.

I. Vatolkin (✉) · G. Rudolph
TU Dortmund, Chair of Algorithm Engineering, Dortmund, Germany
e-mail: igor.vatolkin@tu-dortmund.de; guenter.rudolph@tu-dortmund.de

B. Bischl · C. Weihs
TU Dortmund, Chair of Computational Statistics, Dortmund, Germany
e-mail: bischl@statistik.tu-dortmund.de; weihs@statistik.tu-dortmund.de

M. Spiliopoulou et al. (eds.), *Data Analysis, Machine Learning and Knowledge Discovery*, Studies in Classification, Data Analysis, and Knowledge Organization, DOI 10.1007/978-3-319-01595-8_19,

Especially the lack of statistical testing holds also for the most of our own previous studies in music classification. Therefore we decided to check again the results of our study for instrument recognition in polyphonic recordings (Vatolkin et al. 2012) and to apply exemplary tests on two experimental results. The target of this paper is not to provide a comprehensive introduction into statistical testing—but to encourage the MIR community to use statistical tests for a better and more reliable evaluation of algorithms and results.

In the following subsections we introduce shortly the instrument recognition problem and refer to the relevant works about statistical tests for algorithm comparisons in MIR. Then we describe our study and discuss the application of two different tests for multi-objective classifier comparison. Finally, we conclude with recommendations for further research.

1.1 MIR and Instrument Recognition

Almost all MIR tasks deal directly or indirectly with classification: identification of music harmony and structure, genre recognition, music recommendation etc. One of these subtasks is instrument identification, allowing for further promising applications: music recommendation, organization of music collections or understanding of instrument role in a certain musical style. The largest challenge for successful instrument recognition in audio is that it is usually polyphonic: several simultaneously playing sources with different overtone distribution, noisy components and frequency progress over time make this task very complicated if the number of instruments is too large. Another problematic issue is that many different instrument playing possibilities (for example open or fretted strings) hinder the creation of well generalizable classification models which distinguish not only between different instruments but are also not influenced by these playing techniques. One of the recent comprehensive works related to instrument recognition in polyphonic audio is Fuhrmann (2012). An overview of the previous works mainly for recognition of singular instrument samples is provided by Eronen (2001).

1.2 Statistical Tests in Music Classification

Statistical hypothesis testing is a formal methodology for making judgments about stochastically generated data. In this article we will mainly consider two sample-location tests. In detail this means: we have observed numerical observations $v_1, \ldots, v_n$ and $w_1, \ldots, w_n$ and want to compare these two w.r.t. a specific "location parameter", e.g. their mean or median value. In a one-sample test we would compare the location parameter of only one population to a constant value, while "paired" means that we are actually interested in the location of $v_i - w_i$, because both observations have been measured at the same object and/or belong together.

A common example of pairing in machine learning is that v_i and w_i are performance values of predictive models and have both been observed during resampling in iteration i on the same training and test sets. Depending on whether the statistics of interest is approximately normally distributed, the two most popular tests for this scenario are the paired t-test and the Wilcoxon signed-rank test (Hollander and Wolfe 1973).

During the last years a vast number of conference and journal papers has been published in the MIR research area, but only a rather small fraction of them apply statistical tests. From a collection of 162 MIR-related publications we studied (from rather short conference papers to dissertations and master theses) only about 15 % apply or directly mention statistical tests. Furthermore, in almost all of these works hypothesis tests were employed only in very few cases instead of a systematic analysis for algorithm comparison.

To name a couple of examples for further reading, Gillick and Cox (1989) argue about the importance of applying statistical tests to speech recognition, in particular they mention McNemar and matched-pairs test. The k-fold cross-validated t-test for comparison of temporal autoregressive feature aggregation techniques in music genre classification is applied in Meng et al. (2007) and demonstrates the significantly improved performance of the proposed methods. Mckay (2010) uses the Wilcoxon signed-rank test for the comparison of features from different sources (symbolic, audio and cultural) for genre classification. Another evaluation of audio feature subsets for instrument recognition by statistical testing was performed by Bischl et al. (2010). Noland and Sandler (2009) mention the application of the z-test for correlation measurements in key estimation based on chord progression.

2 Instrument Identification in Intervals and Chords

Here we provide a short description of our study, for details please refer to Vatolkin et al. (2012). The binary classification task was to detect piano, guitars, wind or strings in the mixtures of 2 up to 4 samples playing at the same time. The complete set included 3,000 intervals (2 tone mixtures) and 3,000 chords (3 and 4 tone mixtures). 2,000 intervals and 2,000 chords were used for model training and optimization based on cross-validation and the remaining mixtures were used as an independent holdout set for validation.

A 1,148-dimensional audio feature vector was preprocessed and provided as input for four classifiers: decision tree C4.5, random forest (RF), naive Bayes (NB) and support vector machine (SVM). Since using a too large feature set comes with the additional costs of increased prediction time and storage space (both very relevant in MIR, e.g. see Blume et al. 2011) and the trade-off between the size of the feature set and the prediction performance of the model is difficult to specify a priori we decided to perform multi-objective feature selection by means of an evolutionary algorithm (EA) w.r.t. to classification error E^2 and the proportion of selected features f_r. Three different initial feature rates $i_{FR} \in \{0.5; 0.2; 0.05\}$

(probabilities that each feature was selected for model building at the beginning of the optimization process) and three different crossover operators for EA were tested as optimization parameters. The results confirmed our suggestion that the feature selection is an important step providing successful and generalizable models.

The formal definition of multi-objective feature selection (MO-FS) can be described as:

$$\theta^* = \arg\min_{\theta} \left[m_1 \left(Y ; \Phi(X, \theta) \right), \ldots, m_O \left(Y ; \Phi(X, \theta) \right) \right], \tag{1}$$

where X is the full feature set, Y the corresponding labels, θ the indices of the selected features, $\Phi(X, \theta)$ the selected feature subset and $m_1, \ldots, m_O$ are O objectives to minimize.

The output of a MO-FS algorithm is a solution set, each of them corresponding to the subset of initial features. The *non-dominated front* of solutions consisted of the feature subsets with the best compromises between E^2 and f_r. Such front can be evaluated by a corresponding hypervolume:

$$S(\mathbf{x}_1, \ldots, \mathbf{x}_N) = \bigcup_i vol(\mathbf{x}_i), \tag{2}$$

where $vol(\mathbf{x}_i)$ relates to the hypercube volume spanned between the solution $\mathbf{x}_i$ and the reference point which should be set to the worst possible solution responding to all metrics ([1;1] in our case).

3 Application of Tests

In Vatolkin et al. (2012) we provided nine experimental results examining the overall performance of our method and comparing different classifiers and settings of EA parameters. These observations were in most cases clearly underlined by the corresponding experimental statistics and figures—however no significance measurements were done in a proper way. For the following subsections we selected two results and considered appropriate statistical tests for the reliability measurements (for instrument detection in chords).

3.1 All Classifiers are Important

The first result was that if all selected feature sets after FS were compared, it was hardly possible to claim that some of the classification methods were irrelevant: the non-dominated fronts of solutions contained solutions from all classification methods. This statement is illustrated by Fig. 1. Here we plotted all final solutions

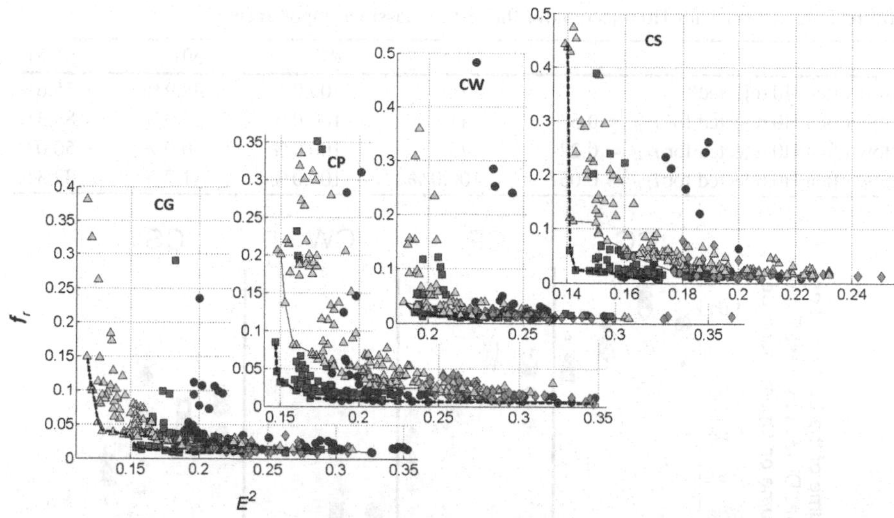

Fig. 1 Solutions after optimization from 10 statistical runs for each classification method optimizing mean classification error E^2 and feature rate f_r ($i_{FR} = 0.5$). *Circles*: C4.5; *rectangles*: RF; *diamonds*: NB; *triangles*: SVM. From left to right: CG: identification of guitar in chords; CP: piano; CW: wind; CS: strings

from ten statistical repetitions and marked the non-dominated fronts by thick dashed lines. It can be stated that often certain classifiers occupy specific regions of the front: RF and SVM provide often the smallest E^2 values but require larger feature sets whereas C4.5 and NB perform worse corresponding to E^2 but may build models from extremely small feature sets.

For the measurement of statistical significance of the observation, that all classifiers are reasonable for non-dominated solution fronts, we need at first a null hypothesis. Therefore, H0 can be formulated as follows: given a classifier $\mathcal{A}$, the hypervolumes for all solutions fronts $\mathcal{S}_{all}$ and fronts built of solutions without this classifier $\mathcal{S}_{all/\mathcal{A}}$ have the same distribution across r statistical repetitions of the experiment. Since: (1) the number of statistical repetitions was rather low ($r = 10$ because of large computing time and overall experiment number); (2) no assumption of the normal distribution and (3) the clear relationship between $\mathcal{S}_{all}$ and $\mathcal{S}_{all/\mathcal{A}}$, we selected the Wilcoxon signed rank test for paired observations. We run the test for 9 optimizer parameter settings (3 i_{FR} values $\times$ 3 crossover operators) separately. The frequency of H0 rejections for each classifier averaged across all combinations of optimization parameters is given in the first row of Table 1. It means, that the removal of RF solutions from the non-dominated front leads to decrease of hypervolume in all cases. The "least important" NB still contributes to the hypervolumes in 38.9 % of all experiments. Another interesting observation is the dependency of the classifier performance on the feature set size. We observed already in Vatolkin et al. (2012), that SVM performs better starting with large feature sets whereas C4.5 suffers from too large amount of features despite of an

Table 1 Frequencies for H0 rejection for the test of classifier importance

	C4.5	RF	NB	SVM
How often H0 rejected?	72.2 %	100.0 %	38.9 %	55.6 %
How often H0 rejected for $i_{FR} = 0.5$?	41.7 %	100.0 %	25.0 %	83.3 %
How often H0 rejected for $i_{FR} = 0.2$?	75.0 %	100.0 %	50.0 %	50.0 %
How often H0 rejected for $i_{FR} = 0.05$?	100.0 %	100.0 %	41.7 %	33.3 %

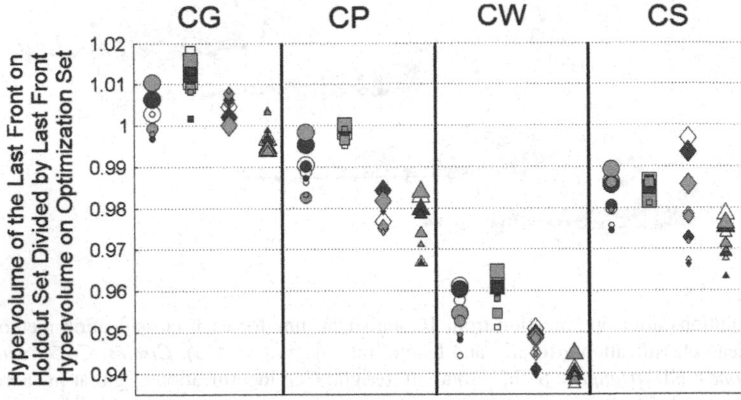

Fig. 2 Mean hypervolumes of the holdout set divided by mean hypervolumes of the optimization set. CG: recognition of guitar in chords; CP: piano; CS: strings; CW: wind. *Circles*: C4.5; *rectangles*: RF; *diamonds*: NB; *triangles*: SVM. *Large signs*: $i_{FR} = 0.5$; *medium signs*: $i_{FR} = 0.2$; *small signs*: $i_{FR} = 0.05$. *Different shades* correspond to different crossover settings

integrated pruning technique. Now we can underline this by statistical test results: for experiments started with initial feature rate of 0.5 the removal of SVM solutions leads in 83.3 % of the cases to hypervolume decrease. For $i_{FR} = 0.05$ this holds only for 33.3 % of the runs. For C4.5 the situation is exactly opposite: C4.5 solutions were required even in all runs with $i_{FR} = 0.05$ for the fronts with largest hypervolumes. For NB no such clear behavior can be observed, but it seems to perform worse with larger feature sets.

3.2 Generalization Ability

The second important observation is that the classifiers provided models with different generalization ability, i.e. performance on an independent data set. Figure 2 lists hypervolumes of the last populations on the holdout set (1,000 chords) divided by hypervolumes on the optimization set (2,000 chords). A value above 1 means that the models perform better for holdout set than for optimization set. From the figure it can be clearly seen that SVM models are almost all less generalizable than RF models; in general C4.5 and RF provide the most robust models. For statistical

Table 2 Frequencies for H0 rejection for the test of model generalization ability

Classifier $\mathcal{A}$	Classifier $\mathcal{B}$	How often H0 rejected?	How often $\widetilde{h_{\mathcal{A}}} > \widetilde{h_{\mathcal{B}}}$?
RF	SVM	88.9 %	100.0 %
RF	NB	66.7 %	86.1 %
RF	C4.5	22.2 %	91.7 %
C4.5	SVM	61.1 %	88.9 %
C4.5	NB	27.8 %	69.4 %
NB	SVM	22.2 %	88.9 %

analysis of model generalization ability between two classifiers $\mathcal{A}, \mathcal{B}$ we compare the distributions of $\widetilde{h_A}$ and $\widetilde{h_B}$, where $h_C(r_i) = S_{holdout}(C, r_i)/S_{opt}(C, r_i)$ is the rate of holdout hypervolume divided by optimization hypervolume for classifier C and run r_i and $\widetilde{h_C}$ is the mean value across ten statistical repetitions. The H0 hypothesis is that the $\widetilde{h_{\mathcal{A}}}$ and $\widetilde{h_{\mathcal{B}}}$ distributions are equal, meaning that there is no significant difference between the model generalization abilities for classifiers $\mathcal{A}$ and $\mathcal{B}$. In Table 2, the first table row can be interpreted as follows: the mean $\widetilde{h_C}$ across all optimizer parameters and statistical repetitions for RF was in 100 % cases larger than for SVM (last column). H0 was rejected in 88.9 % cases—although this is below 100 %, we can indeed state that RF tends to create significantly more generalizable models than SVM. The further lines provide less clear results, however we can state, that RF provides rather more generalizable models than NB and C4.5 than SVM. This behaviour can be also observed from Fig. 2—but it does not illustrate all concrete values from the statistical repetitions and provides no statistical significance testing.

4 Final Remarks

Another important issue for statistical test design is that the hypotheses must be created before the data analysis—otherwise they may hold only for the concrete data set. The first hypothesis (all classifiers are important) was already influenced by our multi-objective feature selection study in Vatolkin et al. (2011)—and the second one (different model generalization performances) was considered after the creation of Fig. 2. The final and only accurate way to underline the significance of this statement—which was here not possible because of the large optimization times for all experiments—is to rerun the complete study for another 3,000 chords and to apply the test again.

Concluding our short excursion with two examples of statistical test application in music instrument recognition, we strongly recommend the following three steps to be carefully planned for design of any new study comparing performance of classification algorithms (in MIR as well as in other domains): (a) design of an independent holdout set neither involved in training of classification models nor any optimization and parameter tuning (see Fiebrink and Fujinaga 2006 especially for feature selection in MIR and our previous publications from the reference list). (b) Consideration of multi-objective optimization or at least evaluation comparing the methods: if an algorithm performs better than another one with response to e.g. accuracy, it may be on the other side slower, fail on highly imbalanced sets or provide less generalizable models (see Vatolkin et al. 2011 for different evaluation scenarios). (c) Application of statistical tests for reliable comparison of methods and significance measurements as discussed in this work.

References

Bischl, B., Eichhoff, M., & Weihs, C. (2010). Selecting groups of audio features by statistical tests and the group lasso. In *Proceedings of the 9th ITG Fachtagung Sprachkommunikation*. Berlin: VDE Verlag.

Blume, H., Bischl, B., Botteck, M., Igel, C., Martin, R., & Rötter, G. (2011). Huge music archives on mobile devices. *IEEE Signal Processing Magazine, 28*(4), 24–39.

Downie, S. (2003). Music information retrieval. *Annual Review of Information Science and Technology, 37*(1), 295–340.

Eronen, A. (2001). *Automatic Musical Instrument Recognition* (Master's thesis). Department of Information Technology, Tampere University of Technology.

Fiebrink, R., & Fujinaga, I. (2006). Feature selection pitfalls and music classification. In *Proceedings of the 7th International Conference on Music Information Retrieval (ISMIR)* (pp. 340–341). University of Victoria.

Fuhrmann, S. (2012). *Automatic Musical Instrument Recognition from Polyphonic Music Audio Signals* (Ph.D. thesis). Department of Information and Communication Technologies, Universitat Pompeu Fabra, Barcelona.

Gillick, L., & Cox, S. (1989). Some statistical issues in the comparison of speech recognition algorithms. In *Proceedings of the IEEE Conference on Acoustics, Speech and Signal Processing (ICASSP)* (pp. 532–535). New York: IEEE.

Hollander, M., & Wolfe, D. A. (1973). *Nonparametric statistical methods*. New York: Wiley.

Mckay, C. (2010). *Automatic Music Classification with jMIR* (Ph.D. thesis). Department of Music Research, Schulich School of Music, McGill University, Montreal.

Meng, A., Ahrendt, P., Larsen, J., & Hansen, L. K. (2007). Temporal feature integration for music genre classification. *IEEE Transactions on Audio, Speech and Language Processing, 15*(5), 1654–1664.

Noland, K., & Sandler, M. (2009). Influences of signal processing, tone profiles, and chord progressions on a model for estimating the musical key from audio. *Computer Music Journal, 33*(1), 42–56.

Vatolkin, I., Preuß, & Rudolph, G. (2011). Multi-objective feature selection in music genre and style recognition tasks. In *Proceedings of the 2011 Genetic and Evolutionary Computation Conference (GECCO)* (pp. 411–418). New York: ACM.

Vatolkin, I., Preuß, Rudolph, G., Eichhoff, M., & Weihs, C. (2012). Multi-objective evolutionary feature selection for instrument recognition in polyphonic audio mixtures. *Soft Computing, 16*(12), 2027–2047.

Part III
AREA Data Analysis and Classification in Marketing

The Dangers of Using Intention as a Surrogate for Retention in Brand Positioning Decision Support Systems

Michel Ballings and Dirk Van den Poel

Abstract The purpose of this paper is to explore the dangers of using intention as a surrogate for retention in a decision support system (DSS) for brand positioning. An empirical study is conducted, using structural equations modeling and both data from the internal transactional database and a survey. The study is aimed at evaluating whether the DSS recommends different product benefits for brand positioning when intention is used as opposed to retention as a criterion variable. The results show that different product benefits are recommended contingent upon the criterion variable (intention vs. retention). The findings also indicate that the strength of the structural relationships is inflated when intention is used. This study is limited in that it investigates only one industry; the newspaper industry. This research provides guidance for brand managers in selecting the most appropriate benefit for brand positioning and advices against the use of intention as opposed to retention in DSSs. To the best of our knowledge this study is the first to challenge and refute the commonly held belief that intention is a valid surrogate for retention in a DSS for brand positioning.

1 Introduction

Given an ongoing evolution from transaction-based marketing to relationship-based marketing (Grönroos 1997), that is primarily driven by the assertion that selling an additional product to an existing customer is several times less expensive than selling the product to a new customer (Rosenberg and Czepiel 1984), it has been argued that building enduring bonds with customers is a profitable strategy

M. Ballings (✉) · D. Van den Poel
Department of Marketing, Ghent University, Ghent, Belgium
e-mail: Michel.Ballings@UGent.be; Dirk.VandenPoel@UGent.be

M. Spiliopoulou et al. (eds.), *Data Analysis, Machine Learning and Knowledge Discovery*, Studies in Classification, Data Analysis, and Knowledge Organization, DOI 10.1007/978-3-319-01595-8_20,

(Reichheld 1996). Hence customer retention has gained a central role in marketing strategies (Buckinx et al. 2007).

One way of reducing churn, or driving retention, is establishing brand associations in consumers' minds (brand-positioning; Keller and Lehmann 2006) that have been shown to be positively linked to retention (Reynolds et al. 2001). In order to find these associations, and subsequently use them as building blocks in their brand positioning, companies resort to decision support systems (DSSs). A successful DSS consists in a model that enables marketing professionals to determine the brand-positioning that would reinforce the relationship between a provider and a customer. Because this requires linking attitudinal variables to behavior, and extant literature has shown that this link is hierarchical in nature (Shim and Eastlick 1998), a structural equations model is in order (e.g., brand positions influence attitudes such as product satisfaction and brand commitment that in turn influence retention). Although retention should be the criterion variable in these models, research has focused on survey measures of behavior leaving observed behavior lingering in the background (e.g., Homer and Kahle (1988). This can severely bias inferences (see Bolton 1998; Morwitz 1997 for details).

Although gains are made along several fronts (e.g. Milfont et al. 2010), the link between high level bases for brand positioning and observed behavior still remains unexplored. In this study we build on the studies Homer and Kahle (1988), Shim and Eastlick (1998), and Milfont et al. (2010) and assess the impact on the final brand-positioning recommendation of using observed behavior (retention) as opposed to declared behavior (intention) as the criterion variable. More specifically, the question is whether different values or benefits are recommended for brand positioning contingent upon whether the dependent variable is declared or observed.

2 Conceptual Model

Schwartz' value inventory is a comprehensive framework consisting of ten basic human values (see Schwartz and Boehnke 2004). It is widely used in industry for brand positioning. Literature indicates that value groupings depend on the context (see Schwartz and Boehnke 2004; Homer and Kahle 1988). From our analysis, we see a three factor solution emerging: (1) power and achievement, (2) hedonism and self-direction and (3) tradition, conformity and security. The values "Stimulation", "Universalism" and "Benevolence" display cross loadings on two factors and that why is we eliminate them from the analysis. In this study we will call the three emerging factors (1) self-enhancement, (2) openness to change and (3) conservation and we will level our hypotheses at this structure. Allen and Meyer's (1990) three-component model of commitment consists of normative commitment (social obligation based bond- ought to), affective commitment (emotional desire based bond- want to) and calculative or continuance commitment (rational cost based or need based bond- need to). Meyer et al. (2002) find, in an organizational context, that all three forms of commitment are negatively related to turnover intentions. Whereas commitment concerns the brand, involvement concerns the product class

(Zaichkowsky 1994; Ros et al. 1999). A person that identifies with either the product (involvement), or the brand (commitment) will be less likely to churn (Sjöberg and Sverke 2000). Research suggests that affective commitment might also affect normative commitment (see Ros et al. 1999; Meyer et al. 2002). There is ample evidence of a positive relationship between satisfaction and behavioral intentions (Cronin and Taylor 1992) and of commitment's mediating role of the relationship between satisfaction and intentions (Paul et al. 2009). Meta-analytic research points out that overall job satisfaction positively correlates with affective commitment and normative commitment (Meyer et al. 2002). Adversely, a small negative relationship is reported between satisfaction and continuance commitment (Meyer et al. 2002). Hence, we formulate the hypotheses as follows:

- H1: Affective commitment has a positive impact on renewal intentions.
- H2: Normative commitment has a positive impact on renewal intentions.
- H3: Calculative commitment has a positive impact on renewal intentions.
- H4: Involvement has a positive impact on renewal intentions.
- H5: Affective commitment has a positive influence on normative commitment.
- H6: Satisfaction has a positive impact on affective commitment.
- H7: Satisfaction has a positive impact on normative commitment.
- H8: Satisfaction has a small negative impact on calculative commitment.

Several authors found support for the relationship between values on the one hand and attitudes (Homer and Kahle 1988) and declared behavior (Shim and Eastlick 1998) on the other hand. Overall satisfaction then, is an evaluation of outcomes relative to a certain expected internal or external standard. Consequently, when thinking about possible relationships with satisfaction we have to keep substitutes (e.g. the internet) in mind. Because the purpose of this study is to discover which benefits display the strongest link to intention (retention), and subsequently satisfaction and involvement we will not hypothesize about which link is strongest.

- H9: Self-Enhancement is related to satisfaction.
- H10: Openness-to-change is related to satisfaction.
- H11: Conservation is related to satisfaction.
- H12: Self-Enhancement is related to involvement.
- H13: Openness to change is related to involvement.
- H14: Conservation is related to involvement.

3 Empirical Study

3.1 Sample

The customers of two Belgian newspaper brands were invited to participate in a study. Both brands can be considered similar, except their geographical targeting at the province/state level. One of the two brands consisted of different editions. 25,897 emails were sent out, inviting customers to click on a link to go to an online questionnaire (an incentive in the form of a prize was offered). 2,142 of them reached the end of the questionnaire and were subsequently used in the analysis.

3.2 Measures

Except for observed retention all constructs used items measured with a 7-point Likert scale. To acquire an accurate translation, all measures were translated to Dutch and back-translated to English by two independent translation agencies. We used the Short Schwartz's Value Survey (Lindeman and Verkasalo 2005) and the following introductory question: "Imagine a typical moment at which you are reading [newspaper brand]. In general, how do you feel at that moment? Indicate to which degree the following concepts are consistent with that feeling". All measurement scales are based on extant literature: renewal intentions (Lindeman and Verkasalo 2005), normative commitment (Ros et al. 1999), calculative commitment (Gounaris 2005), affective commitment (Gustafsson et al. 2005), and satisfaction (Fornell et al. 1996). Benefits, satisfaction, commitment and intention are all measured at time t, to predict retention at time $t + 1$. The time window appears in Fig. 1.

Consistent with the approach of Gustafsson et al. (2005), we computed retention from the account data as the number of days a customer is retained. The average retention is 141.70 days (standard deviation = 34.37).

3.3 Analysis and Results

In order to test our hypotheses we used the procedure of structural equation modeling proposed by Anderson and Gerbing (1988). This procedure was applied in AMOS, version 18. Because our data is multivariate non-normal we cannot use a normal theory estimator such as ML or GLS. Hence we estimated the model parameters using weighted least squares which is an asymptotic distribution free method (Browne 1984), making no assumptions of normality.

The final model is displayed in Fig. 2. All direct paths in the final model were significant (at least at the $p < 0.05$ level) for the intention model. To compare it with the retention model we left all relationships in the model.

Cronbach's Alpha for the (latent) variables ranges from 0.721 to 0.954 with an average of 0.871. The standard deviations range from 0.963 to 1.582 for the Likert scales indicating a substantial amount of variance to be explained. The correlations range from 0.198 to 0.952 with a mean of 0.563. The mean of retention is 141.7 and the standard deviation is 34.37.

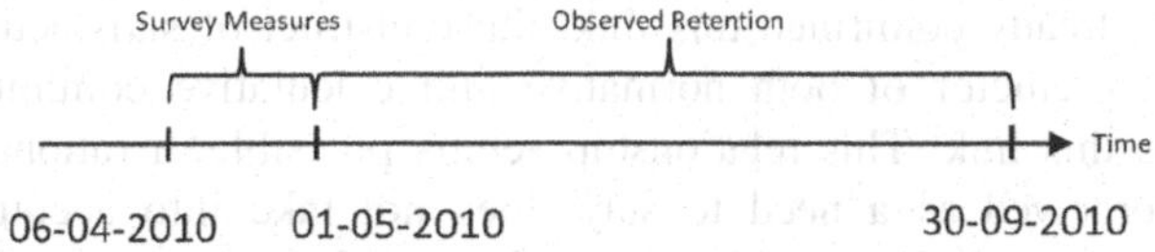

Fig. 1 Time window

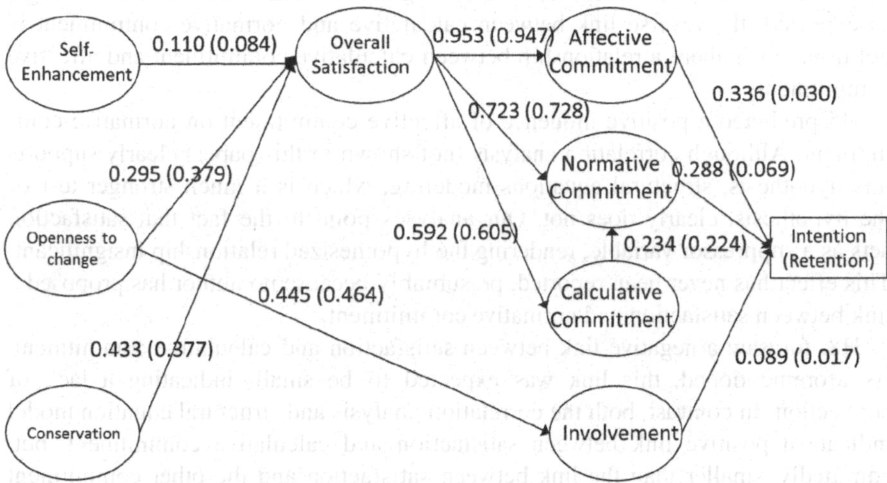

Fig. 2 Estimated model

The squared multiple correlations (SMC) reveal that a considerable amount of variance is explained by the model: intention (retention) = 0.396 (0.010), overall satisfaction = 0.627 (0.637), calculative commitment = 0.350 (0.366), normative commitment = 0.777 (0.778), affective commitment = 0.908 (0.896), involvement = 0.198 (0.215). Consistent with extant literature (Hennig-Thurau and Klee (1997); Newman and Werbe (1973)) only a very low amount of variance is explained in retention. In terms of model fit, both models are very much alike: intention (retention) : Chi-square (385):1081.424 (1046.501), Chi-square/DF: 2.809 (2.718), AGFI: 0.824 (0.835), GFI: 0.854 (0.863), PGFI: 0.707 (0.715), CFI: 0.782 (0.791), PNFI: 0.621 (0.627), RMSEA: 0.029 (0.028).

As the results show (Fig. 2), all hypotheses are supported, except hypothesis 3, 5, 8, 12 and 14. In what follows we'll provide possible explanations. H3 predicted a positive influence of calculative commitment on renewal intentions. As Meyer et al. (2002) point out, calculative commitment is the weakest commitment predictor of intentions. Consequently we deem it is plausible that the relationship becomes insignificant when modeled together with affective and normative commitment. Related to this, our analyses indicated that calculative commitment has a positive influence on normative commitment. Although correlation analysis (not shown in this paper) already confirmed this link, the construct of satisfaction, modeled as a common predictor of both normative and calculative commitment seems to substantiate this link. This relationship seems plausible; a rational cost-based motivation, perceived as a need to stay does not take into account the other side's interests and can be considered as taking unfair advantage of the partner. This can induce feelings of guilt and indebtedness toward the partner and a sense of obligation to return a favor. As such, calculative commitment can influence normative commitment. This line of reasoning is consistent with the other findings; as expected, the reverse link between calculative and normative commitment is not true, nor is there a relationship between calculative commitment and affective commitment.

H5 predicted a positive influence of affective commitment on normative commitment. Although correlation analysis (not shown in this paper) clearly supports this hypothesis, structural equations modeling, which is a much stronger test of the hypothesis, clearly does not. Our analyses point to the fact that satisfaction acts as a suppressor variable, rendering the hypothesized relationship insignificant. This effect has never been reported, presumably because no author has proposed a link between satisfaction and normative commitment.

H8, foresaw a negative link between satisfaction and calculative commitment. As aforementioned, this link was expected to be small, indicating a lack of connection. In contrast, both the correlation analysis and structural equation model indicate a positive link between satisfaction and calculative commitment, but, admittedly, smaller than the link between satisfaction and the other commitment constructs. A positive link seems plausible, because satisfied customers' lack of involvement with how to end the relationship, and the associated costs, may drive this positive relationship. Moreover, a positive connection between satisfaction and lack of worthwhile alternatives also seems justified in that satisfaction measures, among others, the brand's performance relative to competitors. Finally tests of H12 and 14 indicated a lack of relationship between self-enhancement and involvement and conservation and involvement. This hiatus is probably due to the strong link between openness-to-change and involvement, accounting for the majority of the correlation.

Finally, having discussed the validity of our model, we arrive at the main question of this study: the standardized total effects of the values/benefits on intention (retention) is as follows: Self Enhancement 0.063, $p < 0.30$ (0.007, $p < 0.20$), Openness to change 0.207, $p < 0.40$ (0.041, $p < 0.05$) and Conservation 0.245, $p < 0.05$ (0.033, $p < 0.05$). In the following section we discuss these findings.

4 Discussion

The purpose of the decision support system in this study is to make a recommendation about which value should serve as the primary benefit for brand positioning. When using intention as a dependent variable the recommendation is Conservation with 0.245 (which is 18.7 % higher than the value second in position, i.e., Openness-to-change 0.207), but when using retention it is the inverse, namely Openness-to-change with 0.041 (which is 24.24 % higher than the second in position, i.e., Conservation with 0.033). Using intention as a proxy for retention will not only overestimate the strength of the relationship between benefits and retention but will provide erroneous recommendations that are based on a 18.7 % difference. Given the fact that the practice of brand positioning involves substantial communications budget at the one hand and drives sales on the other hand this could be a serious threat to brands. If the wrong brand position is chosen, the wrong associations and expectations may be created in the consumer's mind, which may inflict permanent damage to a brand's performance in the market. Although survey measures are easier to obtain, we advise against their use as a surrogate for observed retention in decision support systems for brand positioning.

5 Future Research and Limitations

A direction for further research is modeling other dependent variables, such as acquisition, up-sell or cross-sell. Although brands should display a certain level of consistency to come across trustworthy, core values and subsequently positions are dynamic and evolve over time (McCracken 2005). It is highly important to monitor this evolution. Managers could just as well choose to position their brand based on customer acquisition if it is first launched, reposition it based on cross-sell when it is in the maturity stage and reposition it based on retention when the brand is in the decline phase. For example, if a car maker notices that its brand positioned on the value power is entering the decline phase, it might be time to reposition the brand on values that drive retention (e.g., green and efficiency). This study is limited in that it is restricted to one industry (the newspaper business). Another direction for future research is replication in other industries.

References

Allen, N. J., & Meyer, J. P. (1990). The measurement and antecedents of affective, continuance, and normative commitment to the organization. *Journal of Occupational Psychology, 63*(1), 1–18.

Anderson, J. C., & Gerbing, D. W. (1988). Structural equation modeling in practice: A review and recommended two-step approach. *Psychological Bulletin, 103*(3), 411–423.

Bolton, R. N. (1998). A dynamic model of the duration of the customer's relationship with a continuous service provider: The role of satisfaction. *Marketing Science, 17*(1), 45–65.

Browne, M. W. (1984). Asymptotically distribution-free methods for the analysis of covariance structures. *British Journal of Mathematics and Statistical Psychology, 37*, 62–83.

Buckinx, W., Verstraeten, G., & Van den Poel, D. (2007). Predicting customer loyalty using the internal transactional database. *Expert Systems with Applications, 32*, 125–134.

Cronin, J. J. Jr., & Taylor, S. A. (1992). Measuring service quality: A reexamination and extension. *Journal of Marketing, 56*(3), 55–68.

Fornell, C., Johnson, M. D., Anderson, E. W., Cha, J., & Bryant, B. E. (1996). The American customer satisfaction index: Nature, purpose, and findings. *Journal of Marketing, 60*(4), 7–18.

Gounaris, S. P. (2005). Trust and commitment influences on customer retention: Insights from business-to-business services. *Journal of Business Research, 58*(2), 126–140.

Grönroos, C. (1997). From marketing mix to relationship marketing towards a paradigm shift in marketing. *Management Decision, 35*(4), 839–843.

Gustafsson, A., Johnson, M. D., & Roos, I. (2005). The effects of customer satisfaction, relationship commitment dimensions, and triggers on customer retention. *Journal of Marketing, 69*(4), 210–218.

Hennig-Thurau, T., & Klee, A. (1997). The impact of customer satisfaction and relationship quality on customer retention: A critical reassessment and model development. *Psychology & Marketing, 14*(8), 737–764.

Homer, P. M., & Kahle, L. R. (1988). A structural equation test of the value-attitude-behavior hierarchy. *Journal of Personality and Social Psychology, 54*(4), 638–646.

Keller, K. L., & Lehmann, D. R. (2006). Brands and branding: Research findings and future priorities. *Marketing Science, 25*(6), 740–759.

Lindeman, M., & Verkasalo, M. (2005). Measuring values with the short Schwartz's value survey. *Journal of Personality Assessment, 85*(2), 170–178.

McCracken, G. (2005). *Culture and consumption II: Markets, meaning, and brand management*. Bloomington: Indiana University Press.

Meyer, J. P., Stanley, D. J., Herscovitch, L., & Topolnytsky, L. (2002). Affective, continuance, and normative commitment to the organization: A meta-analysis of antecedents, correlates, and consequences. *Journal of Vocational Behavior, 61*(1), 20–52.

Milfont, T. L., Duckitt, J., & Wagner, C. (2010). A cross-cultural test of the value-attitude-behavior hierarchy. *Journal of Applied Social Psychology, 40*(11), 2791–2813.

Morwitz, V. G. (1997). Why consumers don't always accurately predict their own future behavior. *Marketing Letters, 8*(1), 57–70.

Newman, J. W., & Werbe, R. A. (1973). Multivariate analysis of brand loyalty for major household appliances. *Journal of Marketing Research, 10*, 404–409.

Paul, M., Hennig-Thurau, T., Gremler, D. D., Gwinner, K. P., & Wiertz, C. (2009). Toward a theory of repeat purchase drivers for consumer services. *Journal of the Academy of Marketing Science, 37*(2), 215–237.

Reichheld, F. F. (1996). *The loyalty effect*. Cambridge: Harvard Business School Press.

Reynolds, T. J., Dethloff, C., & Westberg, S. J. (2001). Advancements in laddering. In T. J. Reynolds & J. C. Olson (Eds.), *Understanding consumer decision making, the means-end approach to marketing and advertising strategy* (pp. 92–118). Mahwah: Lawrence Erlbaum Associates.

Ros, M., Schwartz, S. H., & Surkiss, S. (1999). Basic individual values, work values, and the meaning of work. *Applied Psychology: An International Review, 48*(1), 49–71.

Rosenberg, L. J., & Czepiel, J. A. (1984). A marketing approach to customer retention. *Journal of Consumer Marketing, 1*, 45–51.

Schwartz, S., & Boehnke, K. (2004). Evaluating the structure of human values with confirmatory factor analysis. *Journal of Research in Personality, 38*(3), 230–255.

Shim, S., & Eastlick, M. A. (1998). The hierarchical influence of personal values on mall shopping attitude and behavior. *Journal of Retailing, 74*(1), 139–160.

Sjöberg, A., & Sverke, M. (2000). The interactive effect of job involvement and organizational commitment on job turnover revisited: A note on the mediating role of turnover intention. *Scandinavian Journal of Psychology, 41*(3), 247–252.

Zaichkowsky, J. L. (1994). The personal involvement inventory – Reduction, revision, and application to advertising. *Journal of Advertising, 23*(4), 59–70.

Multinomial SVM Item Recommender for Repeat-Buying Scenarios

Christina Lichtenthäler and Lars Schmidt-Thieme

Abstract Most of the common recommender systems deal with the task of generating recommendations for assortments in which a product is usually bought only once, like books or DVDs. However, there are plenty of online shops selling consumer goods like drugstore products, where the customer purchases the same product repeatedly. We call such scenarios repeat-buying scenarios (Böhm et al., Studies in classification, data analysis, and knowledge organization, 2001). For our approach we utilized the results of information geometry (Amari and Nagaoka, Methods of information geometry. Translation of mathematical monographs, vol 191, American Mathematical Society, Providence, 2000) and transformed customer data taken from a repeat-buying scenario into a multinomial space. Using the multinomial diffusion kernel from Lafferty and Lebanon (J Mach Learn Res 6:129–163, 2005) we developed the multinomial SVM (Support Vector Machine) item recommender system MN-SVM-IR to calculate personalized item recommendations for a repeat-buying scenario. We evaluated our SVM item recommender system in a tenfold-cross-validation against the state of the art recommender BPR-MF (Bayesian Personalized Ranking Matrix Factorization) developed by Rendle et al. (BPR: Bayesian personalized ranking from implicit feedback, 2009). The evaluation was performed on a real world dataset taken from a larger German online drugstore. It shows that the MN-SVM-IR outperforms the BPR-MF.

C. Lichtenthäler (✉)
Institute for Advanced Study, Technische Universität München, Boltzmannstr. 3, 85748 Garching, Germany
e-mail: christina.lichtenthaeler@in.tum.de

L. Schmidt-Thieme
Information Systems and Machine Learning Lab, University of Hildesheim, Marienburger Platz 22, 31141 Hildesheim, Germany
e-mail: schmidt-thieme@ismll.de

M. Spiliopoulou et al. (eds.), *Data Analysis, Machine Learning and Knowledge Discovery*, Studies in Classification, Data Analysis, and Knowledge Organization, DOI 10.1007/978-3-319-01595-8_21,

1 Introduction

Gathering customer purchase data by recording online purchases or via customer cards like Payback (www.payback.de) to determine personalized product recommendations is getting more and more important. Common recommender systems deal with the task of generating recommendations for assortments in which a product is usually bought only once (one-buying scenario) like books or DVDs. However, there are plenty of online shops selling consumer goods like drugstore products where the customer purchases the same product repeatedly. We call such scenarios repeat-buying scenarios (Böhm et al. 2001). Especially customer cards record purchase data in such repeat-buying scenarios. For this reason we developed a personalized recommender system considering the special issues of a repeat-buying scenario. The purchase history in repeat-buying scenarios differs to the purchase history of one-buying scenarios. It contains not only the information whether a product was bought or how it was rated, it also contains how often the product was bought over the recording period. For example one time a customer bought a shampoo, a conditioner and two toilet paper packages. Next time he bought a shampoo and a conditioner again. The purchase history of the customer contains therefore two shampoos, two conditioners, two toilet paper packages. This purchase history data is very similar to the word count representation (bag of words) used in text classification tasks. Counting product purchases is therefore similar to counting words in a document. The similarity of the data representations leads to the use of similar approaches. In text classification tasks the application of information geometry concepts produces outstanding results (see Zhang et al. 2005; Lafferty and Lebanon 2005). Due to the similarity of the data representation and the outstanding results of Lafferty and Lebanon (2005) we decided to use information geometry approaches. Therefore we derived our recommender system from the method Lafferty and Lebanon (2005) presented for text classification.

In the remainder of this paper we first state how our approach fits the context of related research (Sect. 2). Afterwards we introduce information geometry and present our embedding of user profiles into a multinomial space, which is used in our SVM based item recommender system described in Sect. 5 and evaluated in Sect. 6. Finally, we conclude and provide an outlook for future work (Sect. 7).

2 Related Work

There is a vast literature presenting recommender systems for one-buying scenarios like collaborative filtering (Deshpande and Karypis 2004), matrix factorization (Hu et al. 2008), etc. and only a few publications concerning the special issues of repeat-buying scenarios (Böhm et al. 2001). As one state-of-the-art recommender system we want to mention the BPR-MF (Bayesian Personalized Ranking Matrix Factorization) recommender system developed by Rendle et al. (2009). Similar to our proposed system BPR-MF uses the information if a product was bought by a user as implicit feedback. As aforementioned Lafferty and Lebanon (2005) as well as Zhang et al. 2005 present text classification methods using information geometry approaches. They embedded a text document into a multinomial space and utilized a multinomial kernel function to train an SVM (Support Vector Machine) for text classification. We derived our recommender from these approaches.

3 Notation and Problem Formulation

For the sake of clarity, we introduce some notions in the following describing a repeat-buying scenario. Let U be the set of customers (users) and I the set of items. For a fixed recording period T the purchases of all customers I are recorded. $I_u \subset I$ denotes the set of items the user u purchased in the recording time T. Implicit feedback is given by the purchase frequency $\delta(u,i) : U \times I \rightarrow \mathbb{N}$, which determines how often product i was bought by user u in the recording period T. The user profile $x_u \in X \subset \mathbb{N}^{|I|}$ of user u comprises the purchase frequency of each item $i \in I$:

$$x_u = (\delta(u,i_1),\ldots,\delta(u,i_n)), \{i_1,\ldots,i_n\} = I, n = |I| \,. \quad (1)$$

The aim of our recommender system is to predict the interest of an user in an item which he/she has not bought yet. For example a user who bought regularly diapers is eventually interested in other baby products. More specific, we want to learn a model $\mathbb{M}$ based on the given user profiles $\delta(u,i)$ that predicts the future interest $\iota(u,j)$ of a user u in an item $j \notin I_u$ as good as possible. To solve this problem we utilized approaches from information geometry (Amari and Nagaoka 2000), which provide very good results in text classification problems (Lebanon 2005; Zhang et al. 2005).

4 Information Geometry

In the following section we present a brief introduction to the concepts of information geometry, which we will use later on in our recommender system. For further details and proofs of the following concepts see Amari and Nagaoka (2000), Lafferty and Lebanon (2005) and Zhang et al. (2005).

4.1 Multinomial Manifold

The parameter space $\Theta \subset \mathbb{R}^n$ of a statistical distribution $p(x;\theta), \theta \in \Theta$, where x is a random variable, is referred to as a statistical manifold. If the manifold Θ is differentiable, then $(\Theta, \mathcal{J}_\theta)$ is a Riemannian manifold, where $\mathcal{J}_\theta$ is the Fisher information metric. The metric is used to compute distances within the manifold.

We now consider the multinomial distribution $p^{\mathrm{MN}}(x,\theta)$, where $\theta = (\theta_1,\ldots,\theta_n)$ with $\theta \geq 0$ and $\sum_{j=1}^{n} \theta_j = 1$. The parameter space Θ^{MN} is defined as follows:

$$\Theta^{\mathrm{MN}} := \left\{ \theta \in \mathbb{R}^n : \sum_{j=1}^{n} \theta_j = 1, \forall j \theta_j \geq 0 \right\} . \tag{2}$$

Obviously $\Theta^{\mathrm{MN}} = \mathbb{P}^{n-1}$ is the n-1 simplex, which is a differentiable manifold.

4.2 Geodesic Distance

In the following we describe how to compute distances within the multinomial manifold denoted as geodesic distance d. Using the aforementioned Fisher information metric to calculate the geodesic distance leads to an error-prone and complex calculation. For this reason we transform the simplex into a manifold for which the geodesic distance can be calculated in a simpler way.

It is a well known fact that the $n-1$ simplex $\mathbb{P}^{n-1}$ is isometric to the positive portion of the $n-1$ sphere $\mathbb{S}_{+}^{n-1}$ of radius 2,

$$\mathbb{S}_{+}^{n-1} := \left\{ x \in \mathbb{R}^n : \sum_{j=1}^{n} x_j^2 = 2, \forall j x_j \geq 0 \right\} , \tag{3}$$

through the diffeomorphism $F : \mathbb{P}^{n-1} \rightarrow \mathbb{S}_{+}^{n-1}$ follows:

$$F(\theta) = (2\sqrt{\theta_1},\ldots,2\sqrt{\theta_n}) . \tag{4}$$

The geodesic distance within the $n-1$ sphere $\mathbb{S}_{+}^{n-1}$ is given by $d_{\mathbb{S}^{n-1}}(x,x') = \arccos\langle x,x'\rangle$. Due to the isometry of $\mathbb{S}^{n-1}$ and $\mathbb{P}^{n-1}$ it holds that $d_{\mathbb{P}^{n-1}}(\theta,\theta') = d_{\mathbb{S}_{+}^{n-1}}(F(\theta),F(\theta'))$. From this it follows that

$$d_{\mathrm{MN}} = d_{\mathbb{P}^{n-1}}(\theta,\theta') = d_{\mathbb{S}_{+}^{n-1}}(F(\theta),F(\theta')) = \arccos\left(\sum_{j=1}^{n} \sqrt{\theta_j \times \theta_j'}\right) . \tag{5}$$

4.3 Embedding

In order to use the geometry of a statistical manifold we have to embed our data into the manifold. Our approach to embed a user profile $x_u \in X$ into a multinomial space is inspired by the embedding of text documents as described in Lebanon (2005) and Zhang et al. (2005). We assume that a user profile x_u is a sample of a unique multinomial distribution $p_{\text{MN}}(x_u, \theta)$. To determine the parameters of the unique distribution, we use a maximum likelihood estimation MLE $\hat{p}_{\text{ML}} : X \rightarrow \Theta^{\text{ML}}$ given by

$$\hat{p}_{\text{ML}}(x_u) = \left(\frac{\delta(u, i_1)}{\sum_{j=1}^{n} \delta(u, i_1)}, \ldots, \frac{\delta(u, i_n)}{\sum_{j=1}^{n} \delta(u, i_j)} \right) = \hat{\theta}_u, \tag{6}$$

which embeds the user profile x_u into the multinomial space.

4.4 Multinomial Kernel

The following two kernel functions are using the geodesic distance of the multinomial manifold instead of the Euclidean distance. For further information and proofs see Lafferty and Lebanon (2005). By replacing the Euclidean distance with the geodesic distance in the negative distance kernel K_{ND} you get the negative geodesic distance kernel

$$K_{\text{NGD}}(\theta, \theta') = -\arccos\left(\sum_{i=1}^{n} \sqrt{\theta_i \times \theta_i'} \right), \theta \times \theta' \in \Theta^{\text{MN}} \subset \mathbb{R}^n. \tag{7}$$

(see Zhang and Chen 2005). Just as well you get the multinomial diffusion kernel

$$K_t^{\text{MN}}(\theta, \theta') = (4\pi t)^{\frac{n}{2}} - \arccos\left(\sum_{i=1}^{n} \sqrt{\theta_i \times \theta_i'} \right), \theta \times \theta' \in \Theta^{\text{MN}} \subset \mathbb{R}^n \tag{8}$$

by replacing the Euclidean distance of the Euclidean diffusion kernel (see Lafferty and Lebanon 2005).

5 Multinomial SVM Item Recommender System MN-SVM-IR

The item recommender system we will present in the following utilizes the multinomial geometry by using a Support Vector Machine SVM with a multinomial kernel function (see Sect. 4.4) to compute recommendations of the future interest $\iota(u, i^*)$ of user u in item $i^* \in I$. For further details regarding Support Vector Machines and kernel functions see Cristianini and Shawe-Taylor (2000). In the following we present the *MN-SVM-IR - Method*:

1. First of all you have to split the set of users U into two parts. One part $U_T \subset U$ is used for training and one part $U_R \subset U$ for which the recommendation should be calculated. It holds $U_T \cap U_R = \emptyset$
2. Create the training data set:
   - The positive labeled training data contains all user profiles x_u of the users who have bought product $i*$:
   $$U_+ := \{u \in U_T : i^* \in I_u\} \subset U_T \tag{9}$$
   For the positive training data the item i^* will be removed. It holds:
   $$x_u \setminus i^* := (\delta(u, i_1), \dots, \delta(u, i_n)), i_1, \dots, i_n \in I \setminus \{i^*\}, n = |I| - 1 \tag{10}$$
   - The negative labeled training data contains all user profiles x_u of the users who have not bought product $i*$:
   $$U_- := \{u \in U_T : i^* \notin I_u\} \subset U_T \tag{11}$$
   It holds $U_+ \cup U_- = U_T$
   - Embed the user profiles $x_u \setminus i*, u \in U_+$ and $x_u, u \in U_-$ into the multinomial space using (6).
   - The training data contains all embedded user profiles $\hat{\theta}_u \setminus i*, u \in U_+$ and $\hat{\theta}_u, u \in U_-$ with the corresponding label.
   $$\text{T} := \{(\hat{\theta}_u \setminus i^*, +1) : u \in U_+\} \cup \{(\hat{\theta}_u, -1) : u \in U_-\} \tag{12}$$
3. Train the SVM with the training data T utilizing one of the multinomial kernel functions described in Sect. 4.4 to get the decision function f.
4. To determine the future interest $\iota(u, i*)$ of a user $u \in U_R$ in item i^* we first predict the class with the decision function f
   $$\text{sign}(f(\hat{\theta}_u)) = \begin{cases} +1 & \text{user } u \text{ is interested in } i^* \\ -1 & \text{user } u \text{ is not interested in } i^* \end{cases} \tag{13}$$
   Additionally we determine how much the user is interested in item I^* by calculating the distance $d_{i*}(\hat{\theta}_u, \mathbf{w})$ of the vector $\hat{\theta}_u$ to the decision boundary of the SVM in the multinomial space. The higher the distance, the higher is the interest of user u in item $i*$. It holds:
   $$\iota(u, i*) = (f(\hat{\theta}_u) \times d_{i*}(\hat{\theta}_u, \mathbf{w})) \, . \tag{14}$$

With the method at hand you get an one-vs-all classifier for item $i*$. In order to get a recommender system to determine recommendations for all items $i \in I$ you have to repeat the described method for all items and users.

6 Evaluation

We evaluated our item recommender system against the state-of-the-art recommender system BPR-MF developed by Rendle et al. (2009). As evaluation measure we used the Area under the ROC Curve AUC (Bradley 1997).

6.1 Dataset

We performed our experiments on a real world dataset taken from a larger German online drugstore. The dataset contained the purchases of 316,385 users on $7,057 = |I|$ items recorded over the recording period T of 61 weeks. In the experiment we used only data of users who purchased a minimum of ten items. Thus the number of users had been reduced to $64,311 = |U|$ the number of items remained the same.

6.2 Evaluation Method

To evaluate the two different methods regarding AUC we used a combination of tenfold-cross-validation and leave-one-out validation scheme because of the differing learning and prediction strategies of the two methods. First we randomly selected six different items $i*$ with three different purchase frequencies, low (0–5,000 purchases), middle (5,001–10,000 purchases) and high (10,000 and higher purchases) and performed the following procedure for each item $i*$. We randomly selected test datasets T of size $|T| = 1,000$ and $|T| = 2,000$. Following the tenfold-cross-validation scheme we divided the test set T into ten parts. Nine parts where used for training and one part for testing. In the testing part the item $i*$ had been removed from the user profiles x_u. The MN-SVM-IR was trained on the nine training parts and separately tested on the user profiles x_u of the test part whereas the BPR-MF was trained on all parts and tested on the removed items $i*$. To compare the methods we determined the AUC value based on the calculated predictions for the item $i*$.

6.3 Results

We compared the MN-SVM-IR using the multinomial diffusion kernel (DK) and the negative geodesic distance kernel (NGD) with BPR-MF regarding the AUC. Table 1 shows the AUC prediction quality determined in the experiments. The results show that MN-SVM-IR outperforms the BPR-MF in each experiment. Furthermore you can see that the diffusion kernel obtains better AUC values for middle and high

Table 1 AUC prediction quality for the MN-SVM-IR vs BPR-MF experiment (bold numers indicating the best AUC value)

	size T	BPR-MF	MN-SVM-IR (DK)	MN-SVM-IR (NGD)
Low purchase frequency	1,000	0.79376	0.79376	**0.81044**
		0.77244	0.77244*	**0.78984***
	2,000	0.79376	0.79376	**0,82778**
		0.79459	0.80003	**0.80206**
Middle purchase frequency	1,000	0.82704	**0.85072**	0.83316
		0.79892	**0.83540**	0.83396
	2,000	0.85082	**0.85092***	0.83316*
		0.79790	**0.84797***	0.84769*
High purchase frequency	1,000	0.77764	0.81632*	**0.81708***
		0.78792	0.83176*	**0.83301***
	1,000	0.75577	0.80382*	**0.80621***
		0.81326	**0.86464***	0.86455*

* Result is statistically significant ($p < 0.05$)

frequent items whereas the negative geodesic distance kernel obtains better results for the low frequent items. To determine the statistical significance of our results we used a paired t-test (1). For high frequent items statistical significance could be shown ($p < 0.05$) for all AUC values and also for two of the middle frequency test sets.

7 Conclusion

The main contribution of the work at hand is, at first the embedding of a repeat-buying user profile x_u into a multinomial space Θ^{MN} and second the development of an SVM based item recommender system MN-SVM-IR utilizing the multinomial geometry. Our evaluation on a real world data set shows that our MN-SVM-IR outperforms the state-of-the-art recommender system BPR-MF with statistical significance regarding the AUC.

In the future one could investigate different approaches to embed the user profiles into the multinomial space. Here a tf-idf similar embedding could be considered. Furthermore the development of additional kernel functions would be useful to extend the presented recommender system.

References

Amari, S., & Nagaoka, H. (2000). *Methods of information geometry. Translation of mathematical monographs* (Vol. 191). Providence: American Mathematical Society.

Böhm, W., Geyer-Schulz, A., Hahsler, M., Jahn, M. (2001). Repeat-buying theory and its application for recommender services. In O. Opitz (Ed.), *Studies in classification, data analysis, and knowledge organization*. Heidelberg: Springer.

Bradley, A. P. (1997). The use of the area under the ROC curve in the evaluation of machine learning algorithms. *Pattern Recognition, 30*, 1145–1159.

Cristianini, N., & Shawe-Taylor, J. (2000). *An introduction to support vector machines and other kernel-based learning methods*. Cambridge: Cambridge University Press.

Deshpande, M., & Karypis, G. (2004). Item-based top-n recommendation algorithms. *ACM Transactions on Information Systems, 22*(1), 143–177.

Hu, Y., Koren, Y., & Volinsky, C. (2008). Collaborative filtering for implicit feedback datasets. In *IEEE International Conference on Data Mining (ICDM 2008)*.

Lafferty, J., & Lebanon, G. (2005). Diffusion kernels on statistical manifolds. *Journal of Machine Learning Research, 6*, 129–163.

Lebanon, G. (2005). Information geometry, the embedding principle, and document classification. In *Proceedings of the 2nd International Symposium on Information Geometry and its Applications*.

Rendle, S., Freudenthaler, C., Gantner, Z., & Schmidt-Thieme, L. (2009). BPR: Bayesian personalized ranking from implicit feedback. In *Proceedings of the 25th Conference on Uncertainty in Artificial Intelligence*.

Zhang, D., Chen, X., & Lee, W. S. (2005). Text classification with kernels on the multinomial manifold. In *Proceedings of the 28th Annual International ACM SIGIR Conference on Research and Development in information Retrieval*.

Predicting Changes in Market Segments Based on Customer Behavior

Anneke Minke and Klaus Ambrosi

Abstract In modern marketing, knowing the development of different market segments is crucial. However, simply measuring the occurred changes is not sufficient when planning future marketing campaigns. Predictive models are needed to show trends and to forecast abrupt changes such as the elimination of segments, the splitting of a segment, or the like. For predicting changes, continuously collected data are needed. Behavioral data are suitable for spotting trends in customer segments as they can easily be recorded. For detecting changes in a market structure, fuzzy-clustering is used since gradual changes in cluster memberships can implicate future abrupt changes. In this paper, we introduce different measurements for the analysis of gradual changes that comprise the currentness of data and can be used in order to predict abrupt changes.

1 Introduction

The planning of future marketing campaigns highly depends on the developments of different market segments as it is not reasonable to plan strategies while disregarding future market needs. Therefore, an early knowledge of trends in a known market structure is crucial and hence, predictive models are needed rather than just descriptive ones. For this kind of analysis, internet market places bear the advantage that data can be collected easily and continuously due to, e.g., automatic transaction logs.

The basis for this research were the data of an internet market place of a German drug store. The transaction data had been collected over approximately

A. Minke (✉) · K. Ambrosi
Institut für Betriebswirtschaft und Wirtschaftsinformatik, Universität Hildesheim, Hildesheim, Germany
e-mail: minke@bwl.uni-hildesheim.de; ambrosi@bwl.uni-hildesheim.de

M. Spiliopoulou et al. (eds.), *Data Analysis, Machine Learning and Knowledge Discovery*, Studies in Classification, Data Analysis, and Knowledge Organization,
DOI 10.1007/978-3-319-01595-8_22,

2 years and was analyzed on a monthly basis. For the first months, five general segments where detected based on different behavioral data concerning recency of the last purchase, frequency and monetary ratio of purchases, the basket sizes with regard to the number of objects, and the variety of products. Naturally, when comparing data of later months to the first results, several changes in the customer segments were revealed. But the challenge is not detecting the changes retrospectively but predicting and reacting to changes proactively. As the prediction includes a descriptive analysis, it is a higher-level analysis in the area of change mining (c.f. Böttcher et al. 2008). In this paper, different measures are identified which show gradual changes in a cluster structure and which can implicate future structural changes in market segments based on cluster analysis. It provides the basis for future research for predicting abrupt changes such as the formation of new segments, the elimination of segments, and the merging or splitting of segments.

In Sect. 2, the fundamentals of this change mining problem are given. Afterwards, the analysis and evaluation of gradual changes in a known structure are described in Sect. 3 as they are the basis for predicting future abrupt changes, and different measures to display gradual changes are introduced. In order to show the relationship between the introduced measures and different gradual changes, experimental results based on artificial data sets are shown in Sect. 4. In Sect. 5, conclusions and an outlook on further research tasks are given.

2 Fundamentals

The objective of market segmentation based on behavioral data is finding groups of customers showing similar characteristics, i.e. detecting homogeneous segments with regard to the actions of the customers, and the handling of economically promising segments. The basis for predicting abrupt changes in these market segments, i.e. changes with regard to the number of segments, is the evaluation of gradual changes within separate segments. For example, a continuous decrease in the number of customers assigned to a particular segment, indicating a shrinking of this segment, can indicate a future segment elimination. Therefore, the analysis of gradual changes is crucial in this context. In order to analyze these changes, an adequate method is needed. In the field of market segmentation, clustering algorithms are commonly employed for automatically detecting homogeneous groups, so called clusters, based on numerical attributes. But to be able to show gradual changes in a cluster structure, each cluster has to be evaluated separately without taking into account the other clusters. Hence, possibilistic membership degrees are needed rather than probabilistic degrees or crisp cluster assignments because possibilistic fuzzy cluster analysis focusses on the typicality of objects rather than their probability of belonging to a particular cluster. In order to avoid the problem of cluster coincidence in possibilistic clustering (c.f. Krishnapuram and Keller 1996) while still regarding each cluster separately, we employ possibilistic

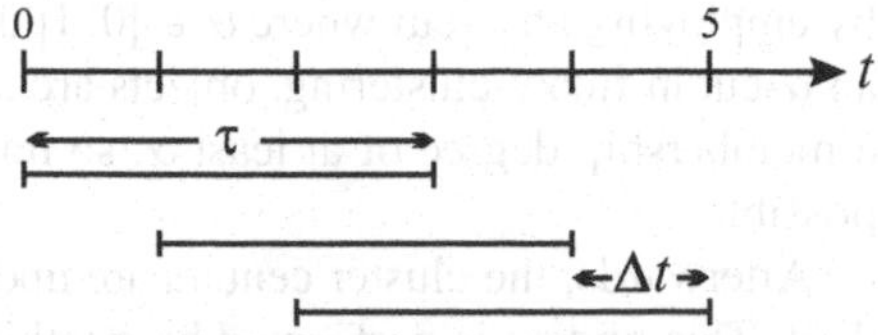

Fig. 1 Example for sliding time window

fuzzy clustering with a penalty term based on cluster homogeneity which we introduced in Minke et al. (2009).

For change detection, sliding time windows with a window-length of τ and an analyzing frequency of Δt are used as it is essential to distinguish between slight random changes in the data structure and significant changes; in the example shown in Fig. 1, $\tau = 3$ and $\Delta t = 1$, i.e. the analysis is performed periodically.

General Procedure

The general procedure for detecting changes in a market segmentation and, more generally, a cluster structure consists of three main steps:

1. *Detection of significant changes in the cluster structure*
   This is the descriptive part of the analysis including an estimation of the current cluster number and the updated cluster prototypes.
2. *Trend detection*
   By evaluating the gradual changes in the structure over several periods in time, future abrupt changes can be predicted. This is the predictive part of the analysis.
3. *Reclustering of current objects*
   The objects of the current time window are reclustered when needed, i.e. if there are any significant changes. The reclustering is initialized with the estimated cluster prototypes and cluster number.

Based on the general procedure, three problems emerge: the extraction of a time-dependent cluster drift, the detection of gradual changes in the structure of a single cluster which is the main topic of this research, and—based on the first two—the prognosis of future abrupt changes in the general cluster structure.

3 Analysis of Gradual Changes

The first step of the general procedure, i.e. the detection of significant changes in the cluster structure, includes the analysis of gradual changes of individual clusters. In order to analyze the differences between a known cluster structure and the current objects that are added in a new time period, the new objects are assigned to the existing clusters based on their membership degrees. This assignment is executed by employing an α-cut where $\alpha \in [0, 1]$ denotes the absorption threshold. By using an α-cut in fuzzy-clustering, objects are assigned to all clusters to which they have a membership degree of at least α, so multiple assignments as well as outliers are possible.

Afterwards, the cluster centers are updated incrementally to evaluate the cluster drift. This update is performed by combining each existing cluster center with the center of the new objects absorbed by this cluster, e.g. by applying the incremental update approach introduced in Crespo and Weber (2005). The extraction of the cluster drift is done preliminarily to the analysis of structural changes of a cluster because otherwise, false positive changes might be detected.

Subsequently, the gradual changes in the structure of a cluster are examined by evaluating different local measures and—in case of ellipsoid clusters—incrementally updating the covariance matrices in order to detect additional changes.

3.1 Local Measures

As there are already several measures that can show different structural characteristics of a cluster, known measures can be applied in the process of analyzing gradual changes. First of all, the absolute cardinality n_i^α of the α-cut can be used:

$$n_i^\alpha = | [\mathbf{u}_i]_\alpha | \tag{1}$$

where $\mathbf{u}_i = (u_{ij})$ is the vector of membership degrees u_{ij} of objects j to cluster i and $[\mathbf{u}_i]_\alpha$, $\alpha \in [0, 1]$, is the α-cut of $\mathbf{u}_i$. A change in the absolute cardinality can show an increase or decrease in the size of a segment.

Additionally, the density and the fuzzy cardinality of an α-cut can be evaluated, the local density being calculated as the fuzzy cardinality with $\alpha = 0.5$:

$$\text{card}\left([\mathbf{u}_i]_\alpha\right) = \sum_{j \in [\mathbf{u}_i]_\alpha} u_{ij} \tag{2}$$

The fuzzy cardinality takes into account the distribution of the objects absorbed by a cluster, revealing differences in a distribution and density shifts.

Thirdly, the index of compactness κ_i^α according to Bensaid et al. (1996) can be applied:

$$\kappa_i^\alpha = \frac{\sum_{j \in [\mathbf{u}_i]_\alpha} u_{ij}^m d_{A_i}^2\left(\mathbf{v}_i, \mathbf{x}_j\right)}{\text{card}\left([\mathbf{u}_i]_\alpha\right)} \tag{3}$$

where m is the fuzzifier and $d_{A_i}^2\left(\mathbf{v}_i, \mathbf{x}_j\right)$ is the squared Mahalanobis distance between cluster center $\mathbf{v}_i$ and object vector $\mathbf{x}_j$. A_i is a positive semi-definite norm matrix based on the fuzzy covariance matrix Σ_i with $A_i = \det(\Sigma_i)^{\frac{1}{P}} \Sigma_i^{-1}$ where P denotes the dimensionality. In case of spherical clusters, Σ_i and A_i are given by the identity matrix I. The smaller the index of compactness, the smaller the variance and the more compact the cluster.

Furthermore, local measures for evaluating structural changes can be derived from global validity measures. E.g., the local partition coefficient (LPC) can help to show differences in the membership degrees and in the distribution within a cluster as it focuses on rather crisp cluster assignments:

$$LPC\left([\mathbf{u}_i]_\alpha\right) = \frac{1}{n_i^\alpha} \sum_{j \in [\mathbf{u}_i]_\alpha} u_{ij}^2 \tag{4}$$

Additionally, the local partition entropy can provide further information when evaluating the density distribution in a cluster.

3.2 Covariance Matrices

In case of ellipsoid clusters, the development of the covariance matrices can be analyzed as well. Although fuzzy clustering techniques are applied to the overall problem, the crisp covariance matrices of the α-cut, Σ_i^α, are needed when evaluating changes between different points in time. The fuzzy covariance matrices are calculated based on membership degrees. However, these degrees might be outdated due to their estimation based on the previously measured cluster prototypes and, therefore, might lead to erroneous results.

The update of the crisp covariance matrices is performed incrementally, similarly to the update of the cluster centers, generating a weighted combination of the existing matrices and the matrices of the newly absorbed objects. Based on the estimated matrices, further characteristics of a cluster can be evaluated. For example, the Fuzzy Hyper-Volume given in (5) is valuable when describing the extension of a cluster in general and can be used to detect changes in the cluster volume.

$$FHV\left([\mathbf{u}_i]_\alpha\right) = \sqrt[P]{\det\left(\Sigma_i^\alpha\right)} \tag{5}$$

As the name states, it is usually applied to fuzzy covariance matrices, but it works for the hard case as well.

Furthermore, the eigenvalues λ_i. and eigenvectors $\mathbf{e}_{ip}$, $p = 1, \ldots, P$, contain further information regarding the extension as well as the cluster alignment. Hence, in addition to revealing changes in the cluster volume, they can help detecting a rotation of a cluster, indicating a different development of individual dimensions.

4 Experimental Results

In order to show the relationship between the introduced measures and different structural changes, several sequences of artificial data sets were created because the changes in the real-life data set mentioned in Sect. 1 were caused by a combination of different structural changes. To be able to assign different measures to particular cluster changes, simple examples are needed. Therefore, each of the first set of the created sequences only includes one of the following changes; in later experiments, different changes were combined.

- *Number of objects*:
  A change indicates a growing or shrinking of a cluster and thus, a future cluster elimination or, if the development of outlier clusters is monitored, the formation of a new cluster.
- *Cluster volume*:
  A difference in the cluster volume can show similar changes as the number of objects as well as changes with respect to the similarity of the objects absorbed.
- *Cluster alignment*:
  Especially in case of analyzing customer segments, a change in the cluster alignment can reveal changes in the customer behavior since different dimensions might show a different development.
- *Cluster density*:
  A change in the density represents the most complex situation because it often includes changes with respect to the cluster volume and the cluster alignment and can be caused by different reasons. Thus, a detailed analysis will not be shown here.

For each change, several sequences of normally-distributed data were created and analyzed. Apart from general changes regarding structure and separation of individual clusters, the created data sets differed in object number per period, number of clusters and outliers, dimensionality, and parameters concerning the sliding time windows. Table 1 provides an overview of the influence of each particular change on the measures introduced in Sects. 3.1 and 3.2; the notation will be explained in the following subsections where each change will be regarded separately. The results shown in Table 1 were stable through all tests.

4.1 Number of Objects

When the number of objects of a cluster is changed, the absolute and the fuzzy cardinality show a similar development, i.e. a decrease ($\downarrow$) has a negative influence ($-$) on the fuzzy cardinality and vice versa, which is quite intuitional. For the index of compactness and the local partition coefficient, there are no significant changes ($\sim$). For the measures regarding the cluster extension, there are slight variations,

Table 1 Influence of gradual changes on different measures

Measure	Number of objects	Cluster volume	Cluster alignment
n_i^α	$\downarrow$: $-$ / $\uparrow$: $+$	$\downarrow$: (+)/$\uparrow$: ($-$)	($-$)
card ($[\mathbf{u}_i]_\alpha$)	$\downarrow$: $-$ / $\uparrow$: $+$	$\downarrow$:+ / $\uparrow$: $-$	$\sim$
κ_i^α	$\sim$	$\downarrow$: $-$ / $\uparrow$: (+)	$\sim$
LPC ($[\mathbf{u}_i]_\alpha$)	$\sim$	$\downarrow$: + / $\uparrow$: ($-$)	$\sim$
FHV ($[\mathbf{u}_i]_\alpha$)	+,$-$	$\downarrow$: $-$ / $\uparrow$: +	$\sim$
$\lambda_{i\cdot}$	+,$-$	$\downarrow$: $-$ / $\uparrow$: +	+,$-$
$\lvert\mathbf{e}_{ip}^{t-1}\cdot\mathbf{e}_{ip}^{t}\rvert$	≈ 1	≈ 1	< 1

but they are rather random and independent of the direction of the occurred change; they only depend on where exactly objects are added or eliminated. The inner product of the eigenvectors of different time frames is approximately one as the cluster alignments do not change significantly.

4.2 Cluster Volume

When the volume of a cluster is constantly decreased, i.e. the objects are getting more similar, several measures are affected. The volume decrease results in a slight increase of the absolute as well as the fuzzy cardinality. The index of compactness is reduced, revealing that the cluster is getting more compact. Similarly, the local partition coefficient increases as the membership degrees of the objects absorbed by the cluster grow.

However, it turns out that in case of a volume increase, the effects are not as clear due to the observation area limited by α. If the volume increases, objects leave the observation area as their membership degrees are estimated based on the old cluster prototype and, hence, the old cluster volume. This fact yields results similar to a small decrease in the number of objects. The changes of the index of compactness and the local partition coefficient are rather small (see parentheses in Table 1). However, in case of ellipsoid clusters, the evaluation of changes in the covariance matrix can uncover the volume change when examining the measures regarding the cluster extension. Again, the inner products of the eigenvectors of the different time windows are close to one, showing an unchanged cluster alignment.

4.3 Cluster Alignment

The cluster alignment can only change in case of ellipsoid clusters due to unlike developments in different dimensions. Since the inner structure of the clusters is not changed but the alignment, local measures as introduced in Sect. 3.1 are not of help, i.e. there are no significant changes. The characteristics of the covariance matrix, especially the eigenvectors, contain the needed information. Due to the rotation, there are considerable changes in the eigenvectors as the inner products of the vectors of subsequent time windows are smaller than one. Some slight changes in the eigenvalues emerge as well, again due to the limited observation area.

4.4 Combination of Different Changes

The cluster alignment is analyzed independently as it influences the inner product of the eigenvectors only. However, when changes in number of objects and volume occur simultaneously, an additional analysis becomes elementary. Such co-occurrence can be uncovered by combining the correspondent columns in Table 1. There are no universally valid influences on the measures because the changes in the measures depend on the proportion of the different changes; the changes might even compensate each other. In empirical studies based on artificial data sets, it was possible to show the changes when first assigning the change with the strongest overlap. Afterwards, when eliminating the influences on the measures caused by the dominant change, the second change became visible.

5 Outlook

In this paper, different measures for detecting gradual changes in a cluster structure were introduced and evaluated. These are needed when predicting future abrupt changes regarding the cluster number and—in the context of marketing—the number of market segments as abrupt changes usually follow previous gradual changes. Experimental results based on artificial data sets show the relationship between simple structural changes within a cluster and the introduced measures. These relationships form the basis for describing simple changes and can be considered when evaluating more complex changes and predicting future needs. However, the experiments are limited to artificially generated data. The transfer of the findings to the real-life data set remains to be researched so a prediction can be achieved when analyzing changing market structures based on internet market places.

References

Bensaid, A. M., Hall, L. O., Bezdek, J. C., Clark, L. P., Silbiger, M. L., Arrington, J. A., et al. (1996). Validity-guided (re)clustering with applications to image segmentation. *IEEE Transactions on Fuzzy Systems, 4*(2), 112–123.

Böttcher, M., Höppner, F., & Spiliopoulou, M. (2008). On exploiting the power of time in data mining. *ACM SIGKDD Explorations Newsletter, 10*(2), 3–11.

Crespo, F., & Weber, R. (2005). A methodology for dynamic data mining based on fuzzy clustering. *Fuzzy Sets and Systems, 150*, 267–284.

Krishnapuram, R., & Keller, J. (1996). The possibilistic C-means algorithm: Insights and recommendations. *IEEE Transactions on Fuzzy Systems, 4*, 385–393.

Minke, A., Ambrosi, K., & Hahne, F. (2009). Approach for dynamic problems in clustering. In I. N. Athanasiadis, P. A. Mitkas, A. E. Rizzoli, & J. M. Gómez (Eds.), *Proceedings of the 4th International Symposium on Information Technologies in Environmental Engineering (ITEE'09)* (pp. 373–386). Berlin: Springer.

Symbolic Cluster Ensemble based on Co-Association Matrix versus Noisy Variables and Outliers

Marcin Pełka

Abstract Interval-valued data arise in practical situations such as recording monthly interval temperatures at meteorological stations, daily interval stock prices, etc. Ensemble approach based on aggregating information provided by different models has been proved to be a very useful tool in the context of the supervised learning. The main goal of this approach is to increase the accuracy and stability of the final classification. Recently the same techniques have been applied for cluster analysis, where by combining a set of different clusterings, a better solution can be received. Ensemble clustering techniques might be not a new problem, but their application to the symbolic data case is a quite new area. The article presents a proposal of application of the co-association based approach in cluster analysis when dealing symbolic data with noisy variables and outliers. In the empirical part simulation experiment results are compared based on artificial data (containing noisy variables and/or outliers). Besides that ensemble clustering results of real data set is shown (segmentation example). In both cases ensemble clustering results are compared with results obtained from a single clustering method.

1 Introduction

Generally speaking clustering methods seek to organize some set of items (objects) into clusters in such way that objects from the same cluster are more similar to each other than to objects from other (different) clusters. Clustering methods have been applied with a success in many different areas—such as taxonomy, image processing, data mining, etc. In general clustering techniques can be divided

M. Pełka (✉)
Department of Econometrics and Computer Science, Wrocław University of Economics, Wroclaw, Poland
e-mail: marcin.pelka@ue.wroc.pl

M. Spiliopoulou et al. (eds.), *Data Analysis, Machine Learning and Knowledge Discovery*, Studies in Classification, Data Analysis, and Knowledge Organization, DOI 10.1007/978-3-319-01595-8_23,

into two main groups of methods—hierarchical (agglomerative or divisive) and partitioning (see for example Gordon 1999; Jain et al. 1999).

In cluster analysis objects are usually described by single-valued variables (single values or categories). Such approach allows to represent them as a vector of quantitative or qualitative measurements. In such approach each column represents a variable. However this kind of data representation is too restrictive and does not allow to represent more complex data. To take into account the uncertainty and/or variability of the data variables must assume sets of categories or intervals. Such kind of data have been mainly studied in symbolic data analysis (SDA; see for example Bock and Diday 2000; Billard and Diday 2006).

Ensemble techniques that aggregate information provided by many different (diverse) base clusterings has been originally designed to deal supervised (discrimination or regression) learning problems. The main goal of this approach is to increase the accuracy and stability of the final classification. Recently ensemble learning has been applied with a success to clustering (unsupervised) tasks for classical data. Nevertheless this approach can be quite easily adapted to symbolic data situation.

This paper presents results ensemble clustering for symbolic interval-valued data, when dealing artificial data with noisy variables and/or outliers and real data set, with application of co-association matrix (co-occurrence matrix).

2 Symbolic Data

There are six main symbolic variable types (Bock and Diday 2000, p. 2; Billard and Diday 2006):

1. Single quantitative value,
2. Categorical value,
3. Quantitative variable of interval type (interval-valued variables),
4. Set of values or categories (multivalued variable),
5. Set of values or categories with weights (multivalued variable with weights),
6. Modal interval-valued variable.

Regardless of their type symbolic variables also can be (Bock and Diday 2000, p. 2) taxonomic—which presents prior known structure, hierarchically dependent—rules which decide if a variable is applicable or not have been defined and logically dependent—logical rules that affect variable's values have been defined.

Generally speaking there are two main types of symbolic objects (see for example Pelka 2010, pp. 342–343). First order objects—single product, respondent, company, etc. (single individuals) described by symbolic variables. These objects are symbolic due to their nature. Second order objects (aggregate objects, super individuals)—more or less homogeneous classes, groups of individuals described by symbolic variables. In the empirical part second order objects are applied in the real data set.

3 Ensemble Learning for Symbolic Data

There are two ensemble learning approaches that can be applied for symbolic data case. First of them is the *clustering algorithm for multiple distance matrices* proposed by de Carvalho et al. (2012). This approach is based on different distance matrices. Those matrices are treated as different points of view. Different distance matrices can be obtained by applying different distance measures, subsets of variables, subsets of objects. Second are based on *consensus functions in clustering ensembles* (see de Carvalho et al. 2012; Fred and Jain 2005; Fred 2001; Stehl and Gosh 2002).

There are four main ways to obtain different base partitions for co-association based functions (Fred and Jain 2005, pp. 835, 842; Hornik 2005)—combining results of different clustering algorithms, producing different partitions by resampling the data, applying different subsets of variables, applying the same clustering algorithm with different values of parameters or initializations.

There are five main types of consensus functions in clustering ensembles that can be applied for symbolic data case (Fred and Jain 2005; Fred 2001; Ghaemi et al. 2009):

1. *Hypergraph partitioning* in which clusters are represented as hyperedges on a graph, their vertices correspond to the objects to be clustered and each hyperedge describes a set of objects belonging to the same cluster. The problem of consensus clustering is reduced to finding the minimum-cut of a hypergraph.
2. *Voting approach*. The main idea is to permute cluster labels such that best agreement between the labels of two partitions is obtained. The partitions from the ensemble must be relabeled according to a fixed reference partition. The voting approach attempts to solve the correspondence problem.
3. *Mutual information* approach. Objective function—the mutual information measure (MI) between the empirical probability distribution of labels in the consensus partition and the labels in the ensemble.
4. *Finite mixture models*. The main assumption is that labels are modeled as random variables. These variables are drawn form a probability distribution. That distribution is described as a mixture of multinominal component densities. The objective of consensus clustering is formulated as a maximum likelihood estimation.
5. *Co-association based functions*. The consensus function operates on the co-association (co-occurrence) matrix which is build from initial N base partitions. Objects belonging to the same cluster ("natural cluster") are likely to be co-located in the same cluster in different data partitions. The co-association matrix is used to find final partition of the set E by applying some clustering algorithm—like single-link, complete-link, etc.

The elements of co-association matrix are defined as follows (Fred and Jain 2005, p. 844; Ghaemi et al. 2009, p. 640):

$$C(i, j) = \frac{n_{ij}}{N}, \tag{1}$$

where: i, j—pattern numbers, n_{ij}—number of times patterns i and j are assigned to the same cluster among N partitions, N—total number of partitions.

4 Simulation Studies

To compare the performance of ensemble clustering versus noisy variables and outliers with a comparison to single clustering method, five artificially generated interval-valued symbolic data sets were prepared (see also Fig. 1).

1. Data set 1—Shapes—the Gaussian, square, triangle and wave in two dimensions.
2. Data set 2—Spherical—three elongated clusters in two dimensions. The observations are independently drawn from bivariate normal distribution with means $(0, 0)$, $(1.5, 7)$, $(3, 14)$ and covariance matrix $\sum(\sigma_{jj} = 1, \sigma_{jl} = -0.9)$.
3. Data set 3—Cassini—the inputs of the Cassini problem are uniformly distributed on a two-dimensional space within three structures. The two external structures (classes) are banana-shaped structures and in between them, the middle structure (class) is a circle.
4. Data set 4—Spirals.
5. Data set 5—Smiley—the smiley consists of two Gaussian eyes, a trapezoid nose and a parabola mouth (with vertical Gaussian noise).

In order to obtain symbolic data from `mlbench` package data sets each point (z_1, z_2) of the data is considered as the "seed" of the rectangle. Each rectangle is therefore a vector of two intervals defined by: $[z_1 - \gamma_1/2, z_1 + \gamma_1/2]$, $[z_2 - \gamma_2/2, z_2 + \gamma_2/2]$. The parameters γ_1 and γ_2 are drawn randomly from the interval $[0; 1]$. In `clusterSim` package symbolic interval data the data is generated for each model twice into sets A and B and minimal (maximal) value of $\{x_{ij}^A, x_{ij}^B\}$ is treated as the beginning (the end) of an interval. The noisy variables are simulated independently from the uniform distribution. It is required that the variations of noisy variables in the generated data are similar to non-noisy variables (see Milligan and Cooper 1985; Qiu and Joe 2006; Walesiak and Dudek 2011; Leisch and Dimitriadou 2010).

For each data set 20 simulation runs were made and average adjusted Rand index (MR; Hubert and Arabie 1985) was calculated.

In ensemble approach different clusterings were merged—following methods that are available in R software have been applied: pam, hierarchical agglomerative (Ward, single, complete, average, McQuitty, median, centroid) and hierarchical divisive (diana). The number of clusters was drawn randomly from the interval $(2; 30)$. Then k-means method was applied to obtain final clustering from the co-association matrix.

In single clustering following methods that are available in R software have been applied: pam, SClust, hierarchical agglomerative (Ward, single, complete, average, McQuitty, median, centroid) and hierarchical divisive (diana).

Rousseeuws Silhouette cluster quality index (Rousseeuw 1987) was applied in both cases to find the final number of clusters.

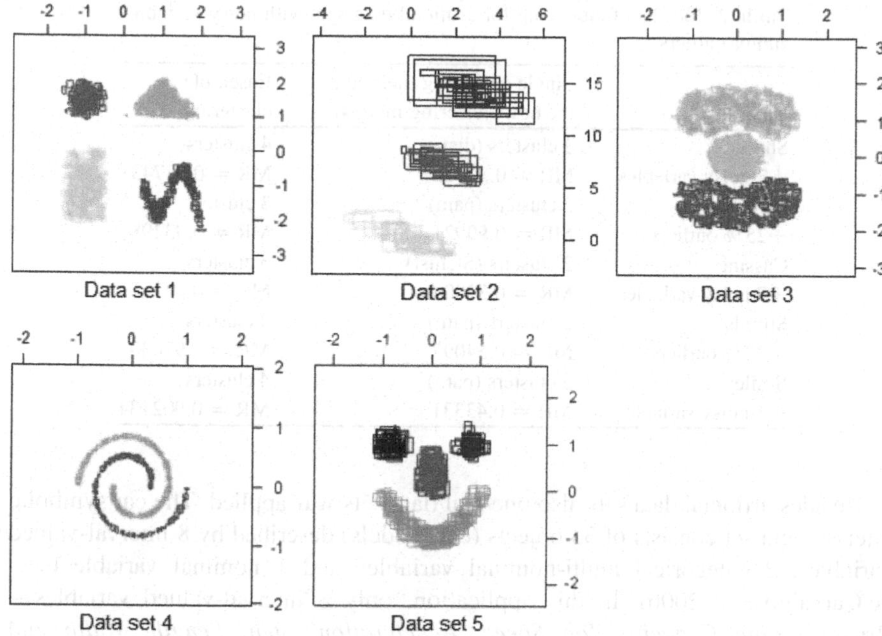

Fig. 1 Artificial data sets

Table 1 Results of clustering for artificial data sets without noisy variables and/or outliers

Data set	Single clustering method & the best clustering method	Ensemble clustering
1	2 clusters (SClust)	4 clusters
Shapes	MR = 0.34123	MR = 0.97561
2	3 clusters (pam)	3 clusters
Spherical	MR = 1	MR = 1
3	2 clusters (pam)	3 clusters
Cassini	sMR = 0.53112	MR = 1
4	2 clusters (pam)	2 clusters
Spirals	MR = 0.78972	MR = 1
5	3 clusters (pam)	4 clusters
Smiley	MR = 0.65132	MR = 1

In both cases—ensemble clustering and single clustering—Ichino and Yaguchi unnormalized distance measure was applied (see Bock and Diday 2000, pp. 153–183, for details on distance measurement for symbolic data). Table 1 presents and compares results of ensemble and single clustering methods for artificial data sets (without noisy variables and outliers). Table 2 presents and compares results of ensemble and single clustering for artificial data sets with noisy variables and/or outliers.

Table 2 Results of clustering for artificial data sets with noisy variables and/or outliers

Data set	Single clustering method & the best clustering method	Ensemble clustering
Shapes	2 clusters (diana)	4 clusters
+2 noisy variables	MR = 0.32010	MR = 0.87743
Spherical	3 clusters (pam)	3 clusters
+25 % outliers	MR = 0.90921	MR = 0.93298
Cassini	2 clusters (SClust)	3 clusters
+3 noisy variables	MR = 0.32009	MR = 0.89987
Spirals	2 clusters (pam)	2 clusters
+20 % outliers	MR = 0.54092	MR = 0.97630
Smiley	3 clusters (pam)	4 clusters
+1 noisy variable	MR = 0.43331	MR = 0.902134

Besides artificial data sets also one real data sets was applied. The car symbolic interval data set consists of 33 objects (car models) described by 8 interval-valued variables, 2 categorical multi-nominal variables and 1 nominal variable (see: de Carvalho et al. 2006). In this application, only 8 interval-valued variables—*Price*, *Engine Capacity*, *Top Speed*, *Acceleration*, *Step*, *Length*, *Width* and *Height*—were considered for brand segmentation purposes.

Car data set was clustered 20 times with single clustering methods (with application of Ichino and Yaguchi unnormalized distance). Best results were reached for 3 clusters (pam clustering)—mean Rand index was equal to 0.787362. In ensemble approach this data set was clustered 15 times (also with application of Ichino and Yaguchi unnormalized distance) with application of different clustering methods. Number of clusters was chosen at random from the interval [2; 20]. Finally also a 4 cluster solution was detected (mean Rand index was equal to 0.899823). First cluster contains mainly compact cars, there are also two sedans in this cluster. Second class contains only sedans. Third class contains only "flagships" (like Passat, BMW 7 series or Aston Martin). The fourth cluster contains mostly sport cars and one "flagship"—(Ferrari).

5 Final Remarks

Different ensemble clustering methods can be quite easily adapted to symbolic data situation. Ensemble learning methods can be applied for symbolic data analysis in all types of marketing problems. They usually reach better solutions (in terms of mean adjusted Rand index) than any of single clustering methods. They seem to be a good solution for marketing problems, as symbolic data tends to form not well separated clusters of many different shapes. When using symbolic data single clustering methods not always are able to detect correct number of clusters.

Simulation studies have shown that symbolic ensemble clustering based on co-association matrix reaches better results than single clustering methods—especially when dealing not well separated cluster structures, data sets with noisy variables and/or outliers.

While dealing real marketing problems ensemble clustering for symbolic data also reached better results than single clustering method. What is more it detected four quite homogeneous clusters—compact cars, sedans, "flagships" and sport cars.

Ensemble clustering can be applied to solve following marketing tasks (problems): market segmentation (for example brands, consumers, goods or services). Identification of homogeneous groups of consumers in order to understand better their behavior. Brand or company positioning. Determination of market structure. Dimension reduction. Identification of homogeneous groups of consumers to test new products (see for example Gatnar and Walesiak 2004).

References

Billard, L., & Diday, E. (2006). *Symbolic data analysis. Conceptual statistics and data Mining*. Chichester: Wiley.

Bock, H.-H., & Diday, E. (Eds.). (2000). *Analysis of symbolic data. Explanatory methods for extracting statistical information from complex data*. Berlin: Springer.

De Carvalho, F. A. T., Lechevallier, Y., & De Melo, F. M. (2012). Partitioning hard clustering algorithms based on multiple dissimilarity matrices. *Pattern Recognition 45*(1), 447–464.

De Carvalho, F. A. T., Souza, R. M. C. R., Chavent, M., & Lechevalier, Y. (2006). Adaptive Hausdorff distances and dynamic clustering of symbolic data. *Pattern Recognition Letters 27*(3): 167–179.

Fred, A. L. N. (2001). Finding consistent clusters in data partitions. In J. Kittler & F. Roli (Eds.), *Multiple classifier systems, Vol. 1857 of Lecture Notes in Computer Science* (pp. 78–86). Berlin: Springer.

Fred, A. L. N., & Jain, A. K. (2005). Combining multiple clustering using evidence accumulation. *IEEE Transaction on Pattern Analysis and Machine Intelligence, 27*, 835–850.

Gatnar, E., & Walesiak, M. (2004). *Metody statystycznej analizy wielowymiarowej w badaniach marketingowych. [Multivariate statistical methods in marketing researches]*. Wrocław University of Economics, Wrocław.

Ghaemi, R., Sulaiman, N., Ibrahim, H., & Mustapha, N. (2009). A survey: Clustering ensemble techniques. *Proceedings of World Academy of Science, Engineering and Technology, 38*, 636–645.

Gordon, A. D. (1999). *Classification*. Boca Raton: Chapman and Hall/CRC.

Hornik, K. (2005). A CLUE for CLUster ensembles. *Journal of Statistical Software, 14*, 65–72.

Hubert, L., & Arabie, P. (1985). Comparing partitions. *Journal of Classification, 1*, 193–218.

Jain, A. K., Murty, M. N., & Flynn, P. J. (1999). Data clustering: A review. *ACM Computational Surveys, 31*(3), 264–323.

Leisch, F., & Dimitriadou, E. (2010). The `mlbench` package. http://www.R-project.org/.

Milligan, G. W., & Cooper, M. C. (1985). An examination of procedures for determining the number of clusters in a data set. *Psychometrika, 50*(2), 159–179.

Pelka, M. (2010). Symbolic multidimensional scaling versus noisy variables and outliers. In H. Locarek-Junge & C. Weihs (Eds.), *Classification as a tool for research* (pp. 341–350). Berlin: Springer.

Qiu, W., & Joe, H. (2006). Generation of random clusters with specified degree of separation. *Journal of Classification, 23*, 315–334.
Rousseeuw, P. J. (1987). Silhouettes: A graphical aid to the interpretation and validation of cluster analysis. *Computational and Applied Mathematics, 20*, 53–65.
Stehl, A., & Gosh, J. (2002). Cluster ensembles – A knowledge reuse framework for combining multiple partitions. *Journal of Machine Learning Research, 3*, 583–618.
Walesiak, M., & Dudek, A. (2011). The `clusterSim` package. http://www.R-project.org/.

Image Feature Selection for Market Segmentation: A Comparison of Alternative Approaches

Susanne Rumstadt and Daniel Baier

Abstract The selection of variables (e.g. socio-demographic or psychographic descriptors of consumers, their buying intentions, buying frequencies, preferences) plays a decisive role in market segmentation. The inclusion as well as the exclusion of variables can influence the resulting classification decisively. Whereas this problem is always of importance it becomes overwhelming when customers should be grouped on the basis of describing images (e.g. photographs showing holidays experiences, usually bought products), as the number of potentially relevant image features is huge. In this paper we apply several general-purpose approaches to this problem: the heuristic variable selection by Carmone et al. (1999) and Brusco and Cradit (2001) as well as the model-based approach by Raftery and Dean (2004). We combine them with k-means, fuzzy c-means, and latent class analysis for comparisons in a Monte Carlo setting with an image database where the optimal market segmentation is already known.

1 Introduction

1.1 Traditional Market Segmentation

Market segmentation has always been the "key to market success" (Weinstein 1994). Its purpose is the definition and treatment of target markets. Important elements are data collection and the segmentation of the data, e.g. with cluster

S. Rumstadt (✉) · D. Baier
Chair of Marketing and Innovation Management, Institute of Business Administration and Economics, Brandenburg University of Technology Cottbus, Postbox 101344, D-03013 Cottbus, Germany
e-mail: susanne.rumstadt@tu-cottbus.de; daniel.baier@tu-cottbus.de

M. Spiliopoulou et al. (eds.), *Data Analysis, Machine Learning and Knowledge Discovery*, Studies in Classification, Data Analysis, and Knowledge Organization, DOI 10.1007/978-3-319-01595-8_24,

analysis (Wedel and Kamakura 2000). Commonly data collection is realized by asking consumers to reveal their actions, interests and opinions (AIO; Wells and Tiggert 1971) in a survey. But there are always problems with questionnaires, e.g. social desirability effects (Groves 1989). Therefore market segmentation on the basis of individual photographs, that describe their AIOs in a much more indirect way, seems to be a promising alternative.

1.2 New Approach: Grouping Consumers by Images

Our new approach is segmentation of consumers by their uploaded images—in social networks, on holiday rating pages, image sharing platforms and so on. The assumption is that these shared images with others reflect the way they want to be seen. And that they reflect their AIOs much more unbiased than their responses to questionnaires. So maybe these images contain more and more realistic information, which were not used in market analysis and marketing up to now. In prior research by Baier and Daniel (2012) and Baier et al. (2012) the research questions, whether uploaded images really reflect the consumers activities, interests and opinions and if clustering algorithms are able to reproduce expected grouping only using the uploaded images and their digital information, were analysed. Both questions were answered positively. This is advantageous, because more and more realistic data can be collected for market analysis. The questionnaires can be designed in a more attractive way for the respondents, because the form is more open. Further references can be found in Baier et al. (2012). But considering images as a data base for market segmentation, new problems arouse. In fact, out of images a huge amount of information can be extracted. On one hand, large mounds of data can lead to failure of the state-of-the-art methods in classification. On the other hand, if these methods do not fail at the task, it is possible, they eventually do not deliver satisfying results in an appropriate computing time. This is the reason why it is necessary to distinguish between relevant, irrelevant and redundant information. Every classification method can work more efficient if only relevant data is processed, of course. Hence the aim is to reduce noise in the extracted image data. And the reduction of dimensions is a wide-spread technique in pattern recognition, and consequently expedient for classification as well.

2 Methodology

The methods we compared are made up of a clustering method combined with a variable selection approach. More precisely, the clustering methods are presented in Sect. 2.1 and the variable selection approaches are explained in particular in Sect. 2.2. Because clustering describes unsupervised grouping procedures, it is not possible to combine these methods with feature selection techniques which are made for classification task, which means supervised grouping procedures. Furthermore, hierarchical clustering methods are not considered in this framework.

2.1 Clustering Methods

The clustering methods we took into consideration are first the *l*atent *c*lass *a*nalysis (LCA) as a model-based approach. Considering a given data matrix $\mathcal{Y} = (\mathbf{y_i})$, where $\mathbf{y_i}$ represents the real-valued data vector of object i, the LCA has the following objective function

$$\log J = \sum_{i,j} z_{ij} \cdot \log[\alpha_j \cdot p(\mathbf{y_i}|\theta_j)]. \tag{1}$$

(1) represents the loglikelihood function, where the z_{ij} are binary and the values 0 and 1 represent the assignment of object i to cluster j. The α_j's are the weights of the clustering components respectively the classes. $p(\cdot|\theta_j)$ stands for the density of cluster j with the parameters θ_j. Furthermore the *fuzzy c-means* clustering with the objective function

$$J = \sum_{i,j} z_{ij}^m \cdot \|\mathbf{y_i} - \mu_j\|^2 \tag{2}$$

is taken into comparison. The notation are similar to (1) with the difference that the z_{ij} are fuzzy and not binary, hence they can take values from 0 to 1. The μ_j represents the cluster centroid of the cluster j. $\|\cdot\|$ stands for the Euclidean distance. The fuzzy parameter m defines the degree of fuzzification in the calculation. The *k-means* (3) clustering is a special case of the fuzzy c-means clustering with z_{ij} from the binary set $\{0, 1\}$ and with fuzzifier $m = 1$.

$$J = \sum_{i,j} z_{ij} \cdot \|\mathbf{y_i} - \mu_j\|^2 \tag{3}$$

2.2 Variable Selection Methods

In the model-based variable selection approach *(clustvarsel)* by Raftery and Dean (2004) a statistical model selection is combined with a greedy search. It is supported by the R-package *c*lusteva*rs*el and hereafter referred to as CVS. The data set of all variables $\mathcal{Y}$ is divided into three disjunct sets of variables, where $\mathcal{Y}^1$ is the set of the selected variables, in $\mathcal{Y}^2$ are the candidate variable(s) and $\mathcal{Y}^3$ contains the remaining variables. By comparing the two models M_1 (4) and M_2 (5) candidates for addition and removal are chosen and evaluated (Raftery and Dean 2004). $\mathbf{z}$ stands for the cluster membership. Model M_1 is a model where $\mathcal{Y}^2$ is conditionally independent given $\mathcal{Y}^1$ means $\mathcal{Y}^2$ does not give additional information in relation to the clustering. Unlike M_2 where $\mathcal{Y}^2$ gives additional information after $\mathcal{Y}^1$ has been observed.

$$M_1 : p(\mathcal{Y}|\mathbf{z}) = p(\mathcal{Y}^1, \mathcal{Y}^2, \mathcal{Y}^3|\mathbf{z}) = p(\mathcal{Y}^3|\mathcal{Y}^2, \mathcal{Y}^1)p(\mathcal{Y}^2|\mathcal{Y}^1)p(\mathcal{Y}^1|\mathbf{z}) \quad (4)$$

$$M_2 : p(\mathcal{Y}|\mathbf{z}) = p(\mathcal{Y}^1, \mathcal{Y}^2, \mathcal{Y}^3|\mathbf{z}) = p(\mathcal{Y}^3|\mathcal{Y}^2, \mathcal{Y}^1)p(\mathcal{Y}^2, \mathcal{Y}^1|\mathbf{z}) \quad (5)$$

The first two variables are chosen to be the most likely to form a univariate clustering respectively a bivariate clustering in combination with the first variable chosen. Afterwards, an additional variable is proposed to be the most likely to form a multivariate clustering with the already chosen ones and evaluated and another selected variable is proposed for removal and evaluated. The proposition and evaluation steps are repeated until two propositions in series are rejected and the algorithm stops.

Other methods for variable selection are heuristics. Heuristics are simplified solutions for complex optimization problems. Heuristics do not demand finding the best solution, because finding the best solution is either impossible nor it can be found in an appropriate computing time. They provide proportionally fast and relatively high quality solutions for optimization problems. The first heuristic presented in this framework is *HINoV*, which stands for *h*euristic *i*dentification of *no*isy *v*ariables. The advantage of HINoV is the fast computing performance, the disadvantage is that this heuristic does not consider combinations of the different variables. But this is on the other hand the reason for its fast performance and the computational complexity is linear. Carmone et al. (1999) presented this heuristic for the purpose of variable selection and hence the reduction of dimensions as well. The HINoV-procedure works as follows: At first, for each variable the selected clustering method is executed separately. Therefore a single classification is build for every variable. These classifications can be compared using the *a*djusted *R*and *i*ndex (ARI; Hubert and Arabie 1985).[1] Summing up all these adjusted Rand indices, so called *to*tal *p*airwise adjusted *R*and *i*ndices (TOPRI) are calculated. And sorting these TOPRI-values in a decreasing order and cutting them down by a ratio rule leads to the set of the selected variables.

The *v*ariable-*s*election heuristic for k-means clustering *(VS)* was developed by Brusco and Cradit (2001) especially for the k-means clustering algorithm and is more complex than the HINoV. The VS heuristic considers the impact of the combination of variables. This is on one hand an advantage but it is disadvantageous in computing time issues. The computing complexity is quadratic and is hence not as fast as the HINoV. The procedure of VS is as follows: The VS works based on sets, so two sets were initialized, S for the set of selected variables is an empty set in the beginning and U, the set which contains all unselected variables. The clustering method is processed with every variable and every pair of variables

[1]Further information concerning the ARI can be found in Sect. 3.1.

treated as a bivariate data input. The result therefore are one-variable-classifications and two-variable-classifications. In case of the one-variable-classifications the ARI-values were calculated from pairwise combination. For every combination of two-variable-classifications the accounted for variance (VAF) is calculated and the pair which has the maximal VAF under the condition that the analogous ARI-values exceed a certain threshold, are picked as the first two selected variables. If two variables are selected the algorithm adds further variables in case the addition of these variables improves the clustering result. In every step for all the unselected respectively remaining variables ARI values are calculated. The full procedure can be followed in Brusco and Cradit (2001).

3 Evaluation

3.1 Synthetic Data

The concept of our proceeding in the empirical part is as follows: At first, we test all the combinations (LCA-CVS, LCA-HINoV, LCA-VS, Fuzzy-HINoV, Fuzzy-VS, K-means-HINoV, K-means-VS) with synthetic data we generated for this purpose, in order to show how the methods and variable selection approaches perform in general. Therefore we generated data according to Gaussian mixtures with randomly generated "true" parameters α^{true}, the number of classes K^{true}, and distribution parameters θ^{true}. And the clustering methods were supposed to find the grouping according to the "true" data structure we forced. Based on the "true" data structure, we have a so-called ground truth and are able to evaluate the performances exactly by applying the ARI. The ARI is a measure of similarity of two partitions. The original Rand index (Rand 1971) has been adjusted to chance by Hubert and Arabie (1985).

3.2 Real Data

In the experimental part with real data, we chose three color images, which come from the following three categories: mountain view, sunsets and city lights (Figs. 1–3). These three images have been edited in 19 different ways (brightness (+/−), clarity (+), contrast (+/−), cross processing (blue/green/red), film grain (black/white), high dynamic range, highlight, saturation (+/−), shadow, sharpen, soften, temperature (+/−)). Hence we have 20 versions of each image, 1 original and 19 derivatives. This leads to a set of 60 images, from which the following digital low level features (consisting of several variables) that are described in image processing literature (Del Bimbo 1999; Shapiro and Stockman 2001; Chatzichristofis and Boutalis 2008a,b; Tamura et al. 1978) were extracted: color and edge directivity

Fig. 1 Mountain view

Fig. 2 Sunset

Fig. 3 City lights

descriptor CEDD, color histogram, color structure, edge histogram, fuzzy color and texture histogram FCTH, Gabor, region shape, scalable color and Tamura.

4 Results

4.1 Synthetic Data

We have evaluated the data with 0, 50 and 80 % of noisy variables in the data. As it can be seen in Table 1, the quality of the results is decreasing with the increasing percentage of noise. This applies to all methods. Partially the increasing percentage of noise effects the computing time. In the second part of the evaluation we took the 80 % noisy data in order to show the general performing characteristics of the variable selection and weighting algorithms. Every one of the chosen variable selection procedures worked reliably for the tested cases. Only the computing times differ.

Table 1 Mean computation time (in seconds) and consensus with true classification (as ARI value) for LCA, fuzzy and k-means clustering with 0 %, 50 % and 80 % noise and with CVS, HINoV and VS

	LCA		Fuzzy		K-means	
	Time	ARI	Time	ARI	Time	ARI
0 % noise	115.57	1.00	261.35	0.94	141.86	0.94
50 % noise	377.16	0.42	266.94	0.49	138.75	0.38
80 % noise	680.11	0.49	259.78	0.42	144.37	0.36
CVS	795.56	1.00				
HINoV	157.30	0.99	988.14	0.99	139.84	0.99
VS	5729.9	0.99	14896.	0.99	6024.6	0.99

Table 2 Consensus with true classification (as ARI value) for LCA, fuzzy and k-means clustering with the extracted image features and with CVS, HINoV and VS (italic values are line by line and column by column means)

Features (number of variables)	LCA	Fuzzy	K-means	
CEDD (144)	0.261	0.479	0.367	*0.369*
ColorHistogram (512)	0.478	0.121	0.183	*0.261*
ColorStructure (128)	0.288	0.571	0.518	*0.459*
EdgeHistogram (80)	0.313	0.571	0.561	*0.482*
FCTH (192)	0.602	0.642	0.460	*0.568*
Gabor (60)	0.390	0.456	0.424	*0.423*
RegionShape (35)	0.377	0.695	0.323	*0.465*
ScalableColor (64)	0.422	0.478	0.681	*0.527*
Tamura (18)	0.626	0.905	0.821	*0.784*
	0.417	*0.546*	*0.482*	
All features (1,233)	0.821	0.859	0.859	*0.846*
with the following methods				
CVS	0.951	0.873	0.860	*0.895*
HINoV	0.821	0.951	0.905	*0.892*
VS	0.784	0.951	0.911	*0.882*
	0.844	*0.920*	*0.892*	
HSMatrix (16)	0.904	0.904	0.465	*0.758*
HSVCube (64)	0.379	0.859	0.456	*0.565*
LABCube (64)	0.491	0.859	0.423	*0.591*
RGBCube (64)	0.491	0.862	0.862	*0.738*
RGBMatrix (192)	0.414	0.416	0.327	*0.386*
	0.536	*0.780*	*0.507*	

4.2 Real Data

The real image data clustering by only the single feature data (upper part of Table 2) did not lead to satisfying result, although some appropriate features for clustering could be identified, e.g. Tamura or FCTH. If all features were taken into the experiment the resulting ARIs increase (middle part of Table 2). And they increase even more with applying the variable selection procedures. Outstanding good result were accomplished by the combinations of LCA-CVS, Fuzzy-HINoV and Fuzzy-VS. In the bottom part of the table, we investigated color histograms from the different color spaces and found out the HSMatrix gave good data to clust and the fuzzy-c-means procedure was the most successful procedure. Variable selection techniques were not applied in this case. The italic values are line by line and column by column means of the certain parts of the table and are stated to give an idea of the overall performance of the procedures and the applicability of the features.

4.3 Conclusion and Outlook

The experiment shows that clustering images only on the basis of their extracted digital data, which were all low level features, is possible and promising for further research. Single features are not convincingly successful in this task, but with the support of variable respectively feature selection methods the results increased up to an ARI value of 0.951. Of course, this is not the ultimate proof, hence further research has to be done. The basis of images will be expanded and the image processing respectively modifying techniques will vary more. There will be an investigation which low level feature variations cause the problems in clustering processes and which feature does not effect clustering in general. But the main aim of this paper, to show that unsupervised grouping of images can be done with low-level features and feature selection techniques, was successful. Later further features, e.g. the Exif data or high level features like manually added tags, can be included in the clustering behaviour investigations of images.

References

Baier, D., & Daniel, I . (2012). Image clustering for marketing purposes. In *Challenges at the interface of data analysis, computer science, and optimization. Proceedings of the Annual Conference of the Gesellschaft für Klassifikation e. V., Karlsruhe, July 21-23, 2010* (pp. 487–494). Berlin: Springer.

Baier, D., Daniel, I., Frost, S., & Naundorf, R. (2012). Image data analysis and classification in marketing. *Advances in Data Analysis and Classification, 6*(4), 253–276.

Brusco, M. J., & Cradit, J. D. (2001). A variable-selection heuristic for K-means clustering. *Psychometrika, 66*, 249–270.

Carmone, F. J., Kara, A., & Maxwell, S. (1999). HINoV: A new model to improve market segment definition by identifying noisy variables. *Journal of Marketing Research, 36*, 501–509.

Chatzichristofis, S. A., & Boutalis, Y. S. (2008a). CEDD: Color and edge directivity descriptor. A compact descriptor for image indexing and retrieval. In *ICVS'08 6th International Conference on Computer Vision Systems* (pp. 312–322), Santorini.

Chatzichristofis, S. A., & Boutalis, Y. S. (2008b). FCTH: Fuzzy color and texture histogram – A low level feature for accurate image retrieval. In *WIAMIS'08 9th International Workshop on Image Analysis for Multimedia Interactive Services* (pp. 191–196), Klagenfurt.

Del Bimbo, A. (1999). *Visual information retrieval*. San Francisco: Morgan Kaufman.
Groves, R. M. (1989). *Survey errors and survey costs. Wiley Series in probability and statistics: Probability and mathematical statistics*. New York: Wiley.
Hubert, L., & Arabie, P. (1985). Comparing partitions. *Journal of Classification, 2*, 193–218. Japan Electronics and Information Technology Industry.
Raftery, A. E., & Dean, N. (2004). *Variable Selection for Model-Based Clustering*. Technical Report No. 452, Department of Statistics, University of Washington, Seattle.
Rand, W. (1971). Objective criteria for the evaluation of clustering methods. *Journal of the American Statistical Association, 66*, 846–850.
Shapiro, L. G., & Stockman, G. C. (2001). *Computer vision*. Upper Saddle River: Prentice Hall.
Tamura, H., Mori, S., & Yamakawi, T. (1978). Textural features corresponding to visual perception. *IEEE Transactions on Systems, Man, and Cybernetics, 8*(6), 460–472.
Wedel, M., & Kamakura, W. A. (2000). *Market segmentation: Conceptual and methodological foundations*. Boston: Kluwer.
Weinstein, A. (1994). *Market segmentation. Using demographics, psychographics and other niche marketing techniques to predict and model consumer behaviour*. Burr Ridge: Irwin.
Wells, W. D., & Tiggert, D. J. (1971). Activities, interests and opinions. *Journal of Advertising Research, 11*(4), 27–35.

The Validity of Conjoint Analysis: An Investigation of Commercial Studies Over Time

Sebastian Selka, Daniel Baier, and Peter Kurz

Abstract Due to more and more online questionnaires and possible distraction—e.g. by mails, social network messages, or news reading during the processing in an uncontrolled environment—one can assume that the (internal and external) validity of conjoint analyses lowers. We test this assumption by comparing the (internal and external) validity of commercial conjoint analyses over the last years. Research base are (disguised) recent commercial conjoint analyses of a leading international marketing research company in this field with about 1.000 conjoint analyses per year. The validity information is analyzed w.r.t. research objective, product type, period, incentives, and other categories, also w.r.t. other outcomes like interview length and response rates. The results show some interesting changes in the validity of these conjoint analyses. Additionally, new procedures to deal with this setting will be shown.

1 Introduction

Conjoint Analysis (CA) is a wide used and well established method for measuring consumer preferences (see Green and Srinivasan (1990); Sattler and Hartmann (2008)). Even today, after more than 40 years of research in the context of marketing science, CA is still an object of investigation in this field of research

---

S. Selka (✉) · D. Baier
Chair of Marketing and Innovation Management, Institute of Business Administration and Economics, Brandenburg University of Technology Cottbus, Postbox 101344, 03013 Cottbus, Germany
e-mail: selkaseb@tu-cottbus.de; baier@tu-cottbus.de

P. Kurz
TNS Infratest GmbH, Arnulfstrasse 205, Munich, Germany
e-mail: peter.kurz@tns-infratest.com

M. Spiliopoulou et al. (eds.), *Data Analysis, Machine Learning and Knowledge Discovery*, Studies in Classification, Data Analysis, and Knowledge Organization, DOI 10.1007/978-3-319-01595-8_25,

(e.g., Green and Rao (1971); Green et al. (2001); Meissner and Decker (2010); Pelz (2012); Sänn and Baier (2012); Selka et al. (2012); Toubia et al. (2012)). Therefore, an extensive view on conjoint analyses (CAs) seems to be productive. Nowadays, with a lot of distraction potentials due to the ubiquitous availability of information in the internet in general (e.g., online newspapers, Twitter) or social network services (e.g., Facebook, LinkedIn) a lower validity for computer-assisted self interviews (CASI) on the web could be expected. Further the dynamic nature of consumers' preferences, measurement errors due to fatigue and boredom as well as the often stated learning effects of respondents can have negative impacts on CA' validity (see Mccullough and Best (1979); Desarbo et al. (2005); Netzer et al. (2008)). Information overload is also stated as potential source of lower internal and external validity (e.g., Jacoby (1984); Chen et al. (2009)). In contrast to this background, some intuitive presumptions, based on methodological and technological development in the past, will be given and completed with an analysis of the mean validity and the validity variance of conjoint analysis over time.

2 Intuitive Presumptions and Research Questions

Due to the evolutional development of conjoint analytical data collection methods, many different methods and approaches arose. There are traditional methods to avoid information overload by shrinking potential attribute combinations through orthogonal designs (Addelman 1962) or modern methods to collect as many data as possible by each respondent by asking as few questions as possible (Johnson 1987; Green and Krieger 1996; Netzer et al. 2008). Besides them, there are other approaches, which produce much better validity values due to the usage of modern computer technology (e.g., Allenby et al. (1995); Johnson and Orme (2007)). Additionally there is an intuitive presumption of learning effects of research company employees regarding the question of "How to do a good CA?"

Against this background and even with the potential negative impacts due to the distraction potentials of the internet, a validity gain over time in the conjoint analytical research area is expected. On the one hand, this expectation is based on the developments in technology and on the other hand on the previously given methodological developments in the past. Both expectations are probably compensating given negative impacts mentioned in Sect. 1. Furthermore, support for this presumption could be found in other scientific areas (see Day and Montgomery 1983; Landeta 2006). Insofar, the research question could be given as "Is there a validity gain in CA over time?" and potential intuitive answers to this research question can be derived as hypotheses in such a manner:

H_1: The mean validity of CAs is increasing over time.
H_2: The mean validity variance CAs is decreasing over time.

By following these hypotheses, a brief introduction into validity measures should be given here. Just to clarify the understanding of them within the scope of this paper. Typically, validity in CA context is measured by internal and external validity values. Internal validity values are represented through the Root Likelihood (RLH) value, which is measuring the correlation between estimated respondent answers and the given ones. Other correlation coefficients are also possible (and typical; e.g., R^2), but not within the scope of this paper. Typically the RLH vlaues are measuring the validity of hierarchical bayes (HB) estimation models of CAs, which were applied to all data sets here. The Mean Absolute Error (MAE) measures the errors between the estimated data model and the prior calculated simulation results. Therefore, the MAE values are also a measure for the internal model fit. For measuring the external validity, here, First-Choice Hit-Rates (FCHR) are given. Upcoming analyses here are based on linear regression models and F-Statistics (to test hypothesis H_1) and on Breusch-Pagan- (BPT) and Goldfeld-Quandt-Tests (GQT) to check for heteroscedasticity (and therefore to test hypothesis H_2).

3 Database of Recent Commercial CAs

To investigate the research question and to support or reject the given hypotheses, a database containing 2,093 data sets of commercial CAs over a time (1996–2011) was analyzed. The data sets are provided from a german market research institute. Therefore, this database just contains data about german CAs. To come up with the meta-data examination, it should be summarized here, that the given data contains information about…

- …the end date, the topic, the user amount, the drop-off rate,
- the representativeness of the study, the questionnaire duration,
- the purpose, the usage of incentives, the multimedia usage,
- the questionnaire type (CASI, CAPI), the used approach (e.g. ACA, CBC),
- the features and levels and (of course)
- some validity values (RLH, MAE, FCHR) for each CA.

In a first analysis step, a brief overview of the meta-data information is given in an aggregated manner in Table 1. The given summary is following the given characteristics in Wittink et al. (1994). The given results showing w.r.t. the application context a similar result, as given by Green et al. (1981); Louviere and Woodworth (1983), and Green and Srinivasan (1990). With respect to the purpose, the results showing a similar result as given in Cattin and Wittink (1982), Wittink and Cattin (1989), Wittink et al. (1994), and Sattler and Hartmann (2008). Therefore, with these given distributions and results, it could be stated, that the upcoming results here probably can be transferred to other international CAs as well. Even if the database here is just taken from the German market. Furthermore, it should be outlined, that the data examination shows an meaningful result regarding the usage of incentives and the usage of CASI over time. Both usages increased highly significant over time ($p < 0.001$).

Table 1 Characteristics of given database

Category	% of usage	Purpose	% of usage
Consumer goods (to expend)	29.35%	Product optimization	52.41%
Consumer goods (to utilize)	20.79%	Pricing	31.49%
Telco services	17.40%	Drug admission	6.31%
Financial services	10.95%	New product development	3.15%
Medical services	9.70%	Distribution	1.82%
Other services	7.93%	Other	4.83%
Other	3.87%		
CA method	% of usage	Avg. duration	% of usage
CBC	89.10%	Up to 20 min	7.03%
ACA	5.73%	Up to 30 min	33.56%
HCA	3.20%	Up to 45 min	31.60%
CVA	1.96%	Up to 60 min	27.82%
		Up to 90 min	1.82%
Computer-aided collection type	% of usage	Usage of incentives	% of usage
CAWI	66.65%	Incentives used	69.47%
CAPI	25.99%	No incentives used	30.53%
other	7.36%		

ACA Adaptive CA, *CBC* Choice Based CA, *HCA* Hierarchical CA, *CVA* Conjoint Value Analysis, *CAWI/CAPI* Computer-assisted (Web/Personal) Interview, *Avg.* Average

4 Validity and Variance Analysis of Recent Commercial CAs

For the upcoming data analysis it has to be repeated here, that all data sets are collected over a time period of 16 years. Due to technological and methodological development over time, the given results are not directly comparable (e.g., different estimation software versions, iteration amounts, etc.). Therefore, all recent commercial CAs were recalculated on the same computer, with the same software base and with the same estimation parameters (10k burn-in and estimation iterations, prior variance of 2, 5 degrees of freedom and an acceptance rate of 35 %) to create a homogeneous database.

As mentioned in Sect. 2, F-Statistics and linear regression models where used to analyze the internal validity values. Given p-values indicating the significance levels and the regression coefficient b will be used to indicate the validity development of the specific dataset over time (positive b-values indicating a gain, negative values the opposite). The regression model here is given through the usual regression formula:

$$\hat{y}_i = b \cdot x_i + \alpha$$

The first investigation is about the CBC and ACA approaches in general. Table 2 shows the results in an appropriate manner.

Table 2 Summary of validity development over time for different time frames

Time frame	ACA	CBC
1996–2011	$N = 119; b = -0.035; p < 0,01$	$N = 1772; b = -0,014; p < 0,001$
2002–2011	$N = 28$; data set to small	$N = 1683; b = -0,007; p < 0,01$
2006–2011	$N = 8$; data set to small	$N = 1382; b = 0,003644; p > 0,1$

E Exponent

Table 3 Internal and external validity development in CA

CA approach	Validity criterion	Result
Time frame: 1996–2002		
ACA	R^2 ($N = 89$)	$b = -2.2E^{-6}; p > 0.1$
	MAE ($N = 92$)	$b = 1.8E^{-4}; p > 0.1$
	$FCHR$ ($N = 92$)	$b = -6.7E^{-4}; p > 0.1$
CBC	R^2 ($N = 88$)	$b = -3.7E^{-6}; p < 0.1$
	MAE ($N = 140$)	$b = 4.7E^{-4}; p > 0.1$
	$FCHR$ ($N = 140$)	$b = 1.8E^{-5}; p > 0.1$
Time frame: 2003–2011		
ACA	Dataset too small, for research implications ($N = 28$)	
CBC	RLH ($N = 1770$)	$b = 4.5E^{-4}; p < 0.1$
	MAE ($N = 1719$)	$b = 1.1E^{-3}; p < 0.001$
	$FCHR$ ($N = 1719$)	$b = -4.6E^{-3}; p < 0.001$

E Exponent

Given results showing no significant validity gain over time rather the opposite is shown over the whole time period. A more deep view on further given validity values is inline with the given results above (see Table 3)—No validity gain over time.

Therefore, hypothesis H_1 has to be rejected. No significant increase of CA' validity values over time could be detected. In fact, the opposite development was found for some validity values. To complete these findings, a variance analysis was applied to check for result dispersons within the data set. On this, all RLH values got regressed and tested for homo- and heteroscadisticity by using GQT and (studentized) BPT. The statistical test and p-values are summarized in Table 4. Compared to the results above, they are showing a consistent result. The validity variance of CAs over time was not decreased. Significant p-values there indicating a heteroscedasticity, what has to be interpreted as an variance increase over time. Non significant values indicating homoscedasticity, what has to be interpreted as constant variance over time. The CBC validity values (1996–2011) from Table 4 were also plotted in Fig. 1 and using the example of the FCHR. A visual analysis of figure's shape shows the increased variance. Hypothesis H_2 could not be accepted either. A validity gain based on validity variance could not be detected and therefore, support for both hypotheses could not be found.

Table 4 Validity variance values and variance development in CA

CA approach	VC	Studentized BPT	BPT	GQT
Time frame: 1996–2011				
ACA ($N = 119$)	*RLH*	$BP = 0.085$	$BP = 0.252$	$GQ = 1.334$
	MAE	$BP = 3.59$*	$BP = 1.81$	$GQ = 0.97$
	FCHR	$BP = 0.202$	$BP = 0.087$	$GQ = 1.102$
CBC ($N = 1772$)	*RLH*	$BP = 8.003$***	$BP = 3.854$**	$GQ = 0.93$
($N = 1859$)	*MAE*	$BP = 4.353$**	$BP = 2.178$	$GQ = 0.889$
($N = 1859$)	*FCHR*	$BP = 7.33$***	$BP = 3.757$*	$GQ = 1.16$**
Time frame: 2003–2011				
ACA	Dataset too low, for research implications ($N = 41$)			
CBC ($N = 1683$)	*RLH*	$BP = 0.238$***	$BP = 0.108$**	$GQ = 0.967$
($N = 1719$)	*MAE*	$BP = 3.253$*	$BP = 1.6$	$GQ = 0.932$
($N = 1719$)	*FCHR*	$BP = 12.921$****	$BP = 6.521$**	$GQ = 1.134$**
Time frame: 2006–2011				
CBC ($N = 1382$)	*RLH*	$BP = 0.193$***	$BP = 0.085$**	$GQ = 0.957$
	MAE	$BP = 6.694$***	$BP = 3.287$*	$GQ = 0.905$
	FCHR	$BP = 5.415$**	$BP = 2.802$*	$GQ = 1.057$

E Exponent, *BPT* Breusch-Pagan-Test, *GQT* Goldfeldt-Quandt-Test, *BP* Breusch-Pagan's value of test statistic, *GQ* Goldfeld-Quandt's value of test statistic, *VC* Validity Criterion
$*p < 0,1$; $**p < 0,05$; $***p < 0,01$; $****p < 0,001$

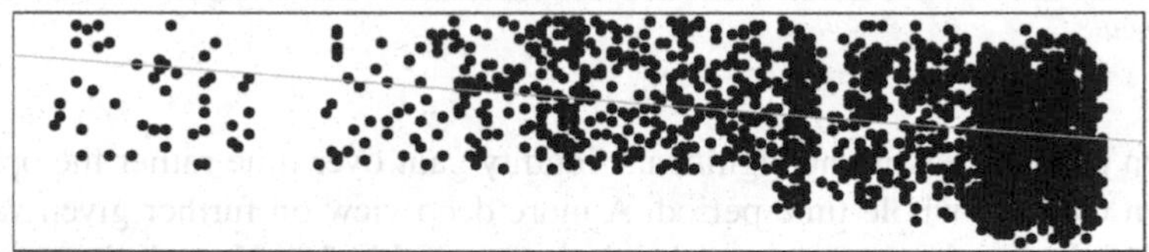

Fig. 1 Graphical representation of CBC's first-choice-hit-rates over time

5 Conclusion and Outlook

The given analysis here has not proven the intuitive presumption of a validity gain over time. Neither the internal and external validity values nor the validity variance of more than 2,000 analyzed commercial CAs from the last 16 years have shown support for the presumptions and the derived hypotheses. Even modern technological and methodological developments in the past seeming to have no positive effects on CA' validity values. Maybe the mentioned negative impacts in Sect. 1 are cutting through, which means, that the negative impacts of modern internet on respondent's distraction are compensating the potential positive impact of such new modern approaches and technological enhancements. This should be tested and investigated in future.

References

Addelman, S. (1962). Orthogonal main-effect plans for asymmetrical factorial experiments. *Technometrics*, *4*, 21–46.

Allenby, G. M., Arora, N., & Ginter, J. L. (1995). Incorporating prior knowledge into the analysis of conjoint analysis. *Journal of Marketing Research*, *32*, 152–162.

Cattin, P., & Wittink, D. R. (1982). Commercial use of conjoint analysis: a survey. *Journal of Marketing*, *46*, 44–53.

Chen, Y.-C., Shang, R.-A., & Kao, C.-Y. (2009). The effects of information overload on consumers' subjective state towards buying decision in the internet shopping environment. *Electronic Commerce Research and Applications. Elsevier Science Publishers B. V.*, *8*, 48–58.

Day, G. S., & Montgomery, D. B. (1983). Diagnosing the experience curve. *Journal of Marketing*, *47*, 44–58.

Desarbo, W., Font, D., Liechty, J., & Coupland, J. (2005). Evolutionary preference/utility functions: a dynamic perspective. *Psychometrika*, *70*, 179–202.

Green, P. E., Carroll, J. D., & Goldberg, S. M. (1981). A general approach to product design optimization via conjoint analysis. *Journal of Marketing*, *45*, 17–37.

Green, P. E., & Krieger, A. M. (1996). Individualized hybrid models for conjoint analysis. *Management Science*, *42*, 850–867.

Green, P. E., Krieger, A. M., & Wind, Y. (2001). Thirty years of conjoint analysis: reflections and prospects. *Interfaces, INFORMS*, *31*, 56–73.

Green, P. E., & Rao, V. R. (1971). Conjoint measurement for quantifying judgmental data. *Journal of Marketing Research*, *8*, 355–363.

Green, P. E., & Srinivasan, V. (1990). Conjoint analysis in marketing: new developments with implications for research and practice. *The Journal of Marketing*, *54*, 3–19.

Jacoby, J. (1984). Perspectives on information overload. *Journal of Consumer Research*, *4*, 432–435.

Johnson, R. M. (1987). Adaptive conjoint analysis. *Conference Proceedings of Sawtooth Software Conference on Perceptual Mapping, Conjoint Analysis, and Computer Interviewing*, pp. 253–265.

Johnson, R. M., & Orme, B. K. (2007). A new approach to adaptive CBC. *Sawtooth Software Inc*, 1–28.

Landeta, J. (2006). Current validity of the delphi method in social sciences. *Technological Forecasting and Social Change*, *73*, 467–482.

Louviere, J. J., & Woodworth, G. (1983). Design and analysis of simulated consumer choice or allocation experiments: an approach based on aggregate data. *Journal of Marketing Research*, *20*, 350–367.

Mccullough, J., & Best, R. (1979). Conjoint measurement: temporal stability and structural reliability. *Journal of Marketing Research*, *16*, 26–31.

Meissner, M., & Decker, R. (2010). Eye-tracking information processing in choice-based conjoint analysis. *International Journal of Market Research*, *52*, 591–610.

Netzer, O., Toubia, O., Bradlow, E., Dahan, E., Evgeniou, T., Feinberg, F., et al. (2008). Beyond conjoint analysis: advances in preference measurement. *Marketing Letters*, *19*, 337–354.

Pelz, J. R. (2012). *Aussagefähigkeit und -willigkeit aus Sicht der Informationsverarbeitungstheorie Aussagefähigkeit und Aussagewilligkeit von Probanden bei der Conjoint-Analyse*. Wiesbaden: Gabler.

Sänn, A., & Baier, D. (2012). Lead user identification in conjoint analysis based product design. In W. A. Gaul, A. Geyer-Schulz, L. Schmidt-Thieme, & J. Kunze (Eds.), *Challenges at the interface of data analysis, computer science, and optimization* (pp. 521–528). Studies in classification, data analysis, and knowledge organization, vol. 43. Berlin, Heidelberg: Springer.

Sattler, H., & Hartmann, A. (2008). Commercial use of conjoint analysis. In M. Höck, & K. I. Voigt (Eds.), *Operations management in theorie und praxis* (pp. 103–119). Wiesbaden: Gabler.

Selka, S., Baier, D., & Brusch, M. (2012). Improving the validity of conjoint analysis by additional data collection and analysis steps. In W. A. Gaul, A. Geyer-Schulz, L. Schmidt-Thieme & J. Kunze (Eds.), *Challenges at the interface of data analysis, computer science, and optimization* (pp. 529–536). Studies in classification, data analysis, and knowledge organization, vol. 43. Berlin, Heidelberg: Springer.

Toubia, O., De Jong, M. G., Stieger, D., & Füller, J. (2012). Measuring consumer preferences using conjoint poker. *Marketing Science*, *31*, 138–156.

Wittink, D. R., & Cattin, P. (1989). Commercial use of conjoint analysis: an update. *Journal of Marketing*, *53*, 91–96.

Wittink, D. R., Vriens, M., & Burhenne, W. (1994). Commercial use of conjoint analysis in Europe: results and critical reflections. *International Journal of Research in Marketing*, *11*, 41–52.

Solving Product Line Design Optimization Problems Using Stochastic Programming

Sascha Voekler and Daniel Baier

Abstract In this paper, we try to apply stochastic programming methods to product line design optimization problems. Because of the estimated part-worths of the product attributes in conjoint analysis, there is a need to deal with the uncertainty caused by the underlying statistical data (Kall and Mayer, 2011, Stochastic linear programming: models, theory, and computation. *International series in operations research & management science*, vol. 156. New York, London: Springer). Inspired by the work of Georg B. Dantzig (1955, Linear programming under uncertainty. *Management Science*, *1*, 197–206), we developed an approach to use the methods of stochastic programming for product line design issues. Therefore, three different approaches will be compared by using notional data of a yogurt market from Gaul and Baier (2009, Simulations- und optimierungsrechnungen auf basis der conjointanalyse. In D. Baier, & M. Brusch (Eds.), *Conjointanalyse: methoden-anwendungen-praxisbeispiele* (pp. 163–182). Berlin, Heidelberg: Springer). Stochastic programming methods like chance constrained programming are applied on Kohli and Sukumar (1990, Heuristics for product-line design using conjoint analyses. *Management Science*, 36, 1464–1478) and will be compared to its original approach and to the one of Gaul, Aust and Baier (1995, Gewinnorientierte Produktliniengestaltung unter Beruecksichtigung des Kundennutzens. *Zeitschrift fuer Betriebswirtschaftslehre*, *65*, 835–855). Besides the theoretical work, these methods will be realized by a self-written code with the help of the statistical software package R.

S. Voekler (✉) · D. Baier
Institute of Business Administration and Economics, Brandenburg University of Technology Cottbus, Postbox 101344, D-03013 Cottbus, Germany
e-mail: sascha.voekler@tu-cottbus.de; daniel.baier@tu-cottbus.de

M. Spiliopoulou et al. (eds.), *Data Analysis, Machine Learning and Knowledge Discovery*, Studies in Classification, Data Analysis, and Knowledge Organization, DOI 10.1007/978-3-319-01595-8_26,

1 Introduction

In Marketing, product line design optimization holds an important role. Companies want to place new products in special markets, so they have to know which products are the most promising ones to minimize costs or to maximize market share, profits and sales (Green et al. 1981; Gaul et al. 1995). For measuring the preferences of the customers for that purpose, conjoint analysis is very well suited (Green and Srinivasan 1978). Based on the part-worths measured and estimated by conjoint analysis, it is possible to compute the optimal product line via different approaches. All approaches to these optimization problems have the same weakness: the optimization is based on uncertain statistical data. As one can see, there is a need to deal with these uncertainties. One possibility is to solve product line design problems with the techniques of stochastic programming (Dantzig 1955). In this paper we applied a special kind of stochastic programming: chance constrained programming (Kall and Mayer 2011) which was incorporated to the optimization approach of Kohli and Sukumar (1990) and afterwards compared to the approach of Gaul et al. (1995). Therefore in Sects. 2.1 and 2.2, we introduce the two approaches. Afterwards in Sect. 3, we give the reader some theory of chance constrained programming techniques with the aim to understand the application of these techniques to the approach of Kohli and Sukumar (1990) in Sect. 5.

2 Product Line Design Optimization

In product line design optimization, there are several different approaches to get the optimal product line from the estimated part-worths and other additional conditions. To test chance constrained programming, we consider two models which are introduced below.

On a market with J products, a company plans to add R new products to the product line. With K the number of attributes and L_k ($k = 1, 2, \ldots, K$) the number of attribute levels, I the number of segments, the utility of product j ($j = 1, 2, \ldots, J$) in segment i ($i = 1, 2, \ldots, I$) is denoted by u_{ij}. The part-worth of an attribute level l ($l = 1, 2, \ldots, L_k$) from attribute k in segment i is described by β_{ikl}. If a product j with attribute k possesses attribute level l, the decision variable x_{jkl} takes the value 1 and 0 else. The amount of coverage for product j with attribute level l of attribute k is d_{jkl}.

2.1 Approach of Kohli and Sukumar (1990)

The number of possible product configurations is $J = \prod_{k=1}^{K} L_k$. Because that number can be very high, one choose just R out of J products to add to the product line. The approach of Kohli and Sukumar (1990) uses the first choice decision rule. Here, p_{ij} denotes the probability of choosing product j over product j' if the utility u_{ij} of product j in segment i exceeds the utility $u_{ij'}$ of product j in segment i ($p_{ij} = 1$ and $p_{ij} = 0$, else). Then, the optimization problem can be formulated as follows (Gaul and Baier 2009):

$$\sum_{i=1}^{I}\sum_{j=1}^{R}\sum_{k=1}^{K}\sum_{l=1}^{L_k} d^r_{ikl} x_{ijkl} y_i \longrightarrow \max!$$

subject to

$$\sum_{j=1}^{R}\sum_{l=1}^{L_k} x_{ijkl} = 1 \qquad \forall i,k, \tag{1}$$

$$\sum_{l=1}^{L_k} x_{ijkl} - \sum_{l=1}^{L_{k'}} x_{ijk'l} = 0 \qquad k' < k, \forall i,j,k, \tag{2}$$

$$x_{ijkl} + x_{i'jkl'} \leq 1 \qquad \forall i < i', l < l', j, k, \tag{3}$$

$$\sum_{j=1}^{R}\sum_{k=1}^{K}\sum_{l=1}^{L_k} \beta_{ikl}(x_{ijkl} - x_{i'jkl}) \geq 0 \qquad \forall i \neq i', \tag{4}$$

$$y_i \sum_{j=1}^{R}\sum_{k=1}^{K}\sum_{l=1}^{L_k} \beta_{ikl} x_{ijkl} \geq y_i(u_{i0} + \epsilon) \qquad \forall i, \tag{5}$$

$$y_i, x_{ijkl} \in \{0,1\} \qquad \forall i,j,k,l. \tag{6}$$

The objective function represents the profit for the supplier and should be maximized subject to the constraints. The constraints (1)–(3) ensure that for each individual just one new product is chosen. By (4) and (5) the first choice decision rule is modeled. Equation (6) represents the binary restriction for the decision variables y_i (= 1, if individual i buys a product out of the product line;= 0 else) and x_{ijkl}.

2.2 *Approach of Gaul/Aust/Baier (1995)*

The optimization model of Gaul et al. (1995) uses the BTL (Bradley–Terry–Luce) choice decision rule (Bradley and Terry 1952). An individual i chooses product j with probability p_{ij} proportional to its utility for product j:

$$p_{ij} = \frac{u_{ij}^{\alpha}}{\sum_{j'=1}^{J} u_{ij'}^{\alpha}},$$

The parameter $\alpha \geq 0$ is for adjusting the simulation on the basis of market information. For $\alpha = 0$ the BTL rule leads to uniformly distributed choices and if $\alpha \rightarrow \infty$ the BTL rule becomes the first choice decision rule.

Consider a market with $j = 1, \ldots, F$ external products and $j = F+1, \ldots, F+E$ own products. The new products to add to the product line are represented by $j = F+E+1, \ldots, F+E+R$. This model is able to include part fixed costs f_{kl} for the attribute levels. However, the notation is the same like in the model above.

$$\sum_{k=1}^{K}\sum_{l=1}^{L_k}\left(\sum_{j=F+E+1}^{F+E+R} x_{ikl}\left(\sum_{i=1}^{I} p_{ij}\omega_i d_{ikl} - f_{kl}\right)\right.$$
$$\left. + \sum_{j=F+1}^{F+E} x_{ikl}\sum_{i=1}^{I} p_{ij}\omega_i d_{ikl}\right) \longrightarrow \max!$$

subject to

$$\sum_{l=1}^{L_k} x_{jkl} \leq 1 \qquad \forall k, j = F+E+1, \ldots, F+E+R, \tag{7}$$

$$p_{ij} = \frac{\left(\sum_{k=1}^{K}\sum_{l=1}^{L_k} \beta_{ikl} x_{jkl}\right)^{\alpha}}{\sum_{j'=1}^{F+E+R}\left(\sum_{k=1}^{K}\sum_{l=1}^{L_k} \beta_{ikl} x_{jkl}\right)^{\alpha}} \qquad \forall i, j, \tag{8}$$

$$\sum_{l=1}^{L_k} x_{jkl} = \sum_{l=1}^{L_{k+1}} x_{j(k+1)l} \qquad k = 1, \ldots, K-1, \tag{9}$$
$$j = F+E+1, \ldots, F+E+R,$$

$$x_{jkl} \in \{0, 1\} \qquad \forall k, l, j = F+E+1, \ldots, F+E+R. \tag{10}$$

Here, the objective function is the maximization of the amounts of coverage for the new and already established products less the fixed costs for the new products. Summarizing over all individuals, the probability of buying the products is considered as well. Equation (7) ensures that every attribute of a product has just one attribute level. Constraint (8) represents the BTL choice decision rule. while (9) assures that each product is offered with a full set of attributes. The binary restriction of the decision variables x_{jkl} is shown in constraint (10).

3 Stochastic Programming

The first stochastic programming techniques were proposed by Dantzig (1955). The purpose was to deal with uncertain parameters in (non-)linear programs. Since then, it proved its flexibility and usefulness in many fields where the incorporation of uncertainty is necessary, e.g. finance, operations research or meteorology (Shapiro et al. 2009). There are three different types of stochastic programs: single-stage programs (e.g. chance constrained programming), second-stage programs and multistage programs (Birge and Louveaux 2011).
Now, we want to give a very short introduction into the field of chance constrained programming according to Kall and Mayer (2011). Imagine an ordinary linear (or integer) program with fixed matrix A, fixed solution vector b, the parameter vector of the objective function c and the vector of the decision variables x:

$$\begin{aligned} \min \quad & c^T x \\ s.t. \quad & Ax \geq b, \\ & x \geq 0. \end{aligned}$$

Now, we assume that not every constraint in $Ax \geq b$ is fixed, but also some of them depending on uncertain data, like statistical parameters. Then with $T \subseteq A$, $h \subseteq b$ and a random variable ξ follows:

$$\begin{aligned} \min \quad & c^T x \\ s.t. \quad & Ax \geq b, \\ & T(\xi)x \geq h(\xi), x \geq 0. \end{aligned}$$

Because of the uncertain parameters in $T(\xi)x \geq h(\xi)$, one cannot ensure that every constraint in this system holds, so that no optimal solution can be found. To avoid that, there is a need to incorporate a probability function P:

$$\begin{aligned} \min \quad & c^T x \\ s.t. \quad & Ax \geq b, \\ & P(T(\xi)x \geq h(\xi)) \geq p, x \geq 0. \end{aligned}$$

That means, that the constraints $T(\xi)x \geq h(\xi)$ are holding with high probability for a certain probability p. After transformation of these constraints, the chance constrained problem becomes a linear (or integer) program again.

4 Chance Constrained Programming Applied to Kohli and Sukumar (1990)

Remember constraint (5) from the optimization model of Kohli and Sukumar (1990). This constraint is chosen for the stochastic program because in that a very important decision for choosing new products to the product line is made. A new product will be taken over the status quo products if the utility of the new product exceeds the utility of the status quo products ($u_{i0}, \forall i$) of a free chosen parameter ϵ. We will adjust ϵ that way, that the uncertainty of the estimated part-worths β_{ikl} will be incorporated in the decision process. The advantage is that we get a parameter which considers the quality and uncertainty of the conjoint part-worths estimation. Therefore, we used the standard error of the estimated part-worths for each segment i evolving from their variances:

$$Var(\beta_{ikl}) = \sigma_i^2.$$

Then, the standard error is defined as:

$$\sigma^2 = \frac{1}{I^2} \sum_{i=1}^{I} \sigma_i^2$$

The standard error (the uncertainty of the model) will now be incorporated into the optimization model through the manipulation of the parameter ϵ which we assume is normal distributed with the expectation value 0 and the variance σ^2: $\epsilon \sim N(0, \sigma^2)$. Then, we get:

$$P\left(y_i \sum_{j=1}^{R} \sum_{k=1}^{K} \sum_{l=1}^{L_k} \beta_{ikl} x_{ijkl} \geq y_i (u_{i0} + \epsilon) \right) \geq p \qquad \forall i,$$

$$P\left(y_i \left(\sum_{j=1}^{R} \sum_{k=1}^{K} \sum_{l=1}^{L_k} \beta_{ikl} x_{ijkl} - \sum_{k=1}^{K} \sum_{l=1}^{L_k} \beta_{0kl} x_{0jkl} \right) \geq y_i \epsilon \right) \geq p \qquad \forall i.$$

Transform ϵ into a random variable with a standard normal distribution:

$$\eta := \frac{\epsilon}{\sigma} \sim N(0, 1)$$

We get:

$$P\left(\frac{1}{\sigma} \cdot y_i \left(\sum_{j=1}^{R} \sum_{k=1}^{K} \sum_{l=1}^{L_k} \beta_{ikl} x_{ijkl} - \sum_{k=1}^{K} \sum_{l=1}^{L_k} \beta_{0kl} x_{0jkl} \right) \geq y_i \cdot \eta \right) \geq p \qquad \forall i.$$

Inversing the probability function P and let q_p be the p-quantile of the standard normal distribution leads to:

$$y_i \left(\sum_{j=1}^{R} \sum_{k=1}^{K} \sum_{l=1}^{L_k} \beta_{ikl} x_{ijkl} - \sum_{k=1}^{K} \sum_{l=1}^{L_k} \beta_{0kl} x_{0jkl} \right) \geq y_i \cdot \sigma \cdot q_p \qquad \forall i. \tag{11}$$

We have a simple integer program again.

5 Results and Application of the New Approach

For our first test of the model, we took a small example from Gaul and Baier (2009) with given part-worths of the attribute levels of natural yogurts, the corresponding amounts of coverage and four segments with their segment weights like shown in Table 1. On the status quo market, there are three suppliers with five products. For the simulation, the original and the chance constrained approach of Kohli and Sukumar (1990) as well the approach of Gaul et al. (1995) were coded with the statistical software package *R* and solved by complete enumeration.

In this example we choose $p = 0.95$, because we want that the constraints hold with a probability of 95 %. This is a good trade off between the relaxation of constraint (5) and the need that this constraint is fulfilled. Then, the p-quantile of the standard normal distribution is $q_p = 1.645$ and the root of the standard error is $\sigma = 0.0091$. With these values we can run the simulation. In Tables 2 and 3 are the results of the simulation with the original approach of Kohli and Sukumar (1990) with the proposed $\epsilon = 0$ and the chance constrained approach with $\epsilon = \sigma q_p$. It shows that the solution with $\epsilon = 0$ is very bad, because these are two products (p,m,0.4) and (p,s,0.5) who are already present on the status quo market. With an ϵ depending on the standard error one get products (g,m,0.5) and (p,s,0.4) which are not on the market yet and thereby much better results. The solution after Gaul et al. (1995) is shown in Table 4. In this approach the two products (p,s,0.4) and (p,m,0.7) give the optimal solution for the problem. To compare the solutions found with the chance constrained approach applied to Kohli and Sukumar (1990) with the solutions of the approach of Gaul et al. (1995), we inserted them into the objective function of Gaul et al. (1995). For the product (p,s,0.4) there is a yield of 19.43 PMU/TU (part money unit per time unit) and for (g,m,0.5) we get 7.77 PMU/TU, so combined a total amount of 27.20 MU/TU for the chance constrained approach. With the solutions of Gaul et al. (1995) we have a yield of 21.32 PMU/TU for (p,s,0.4) and a yield of 18.31 PMU/TU which makes a summarized yield of 39.63 MU/TU.

Table 1 Illustration of the segment specific part-worths (Source: Own illustration according to Gaul and Baier (2009))

		Part-worths β_{ikl}						
		Wrapping $k = 1$		Taste $k = 2$		Price $k = 3$		
Segment i	Weight of segment w_i	Plastic $l = 1$	Glass $l = 2$	Sour $l = 1$	Mild $l = 2$	0.4 MU $l = 1$	0.5 MU $l = 2$	0.7 MU $l = 3$
1	20000 MU/TU	0.00	0.06	0.00	0.44	0.33	0.17	0.00
2	10000 MU/TU	0.00	0.06	0.44	0.00	0.33	0.17	0.00
3	5000 MU/TU	0.05	0.00	0.14	0.00	0.55	0.27	0.00
4	10000 MU/TU	0.00	0.53	0.00	0.27	0.13	0.07	0.00

Table 2 Solution after Kohli/Sukumar with proposed $\epsilon = 0$ with the two products (p,m,0.4) and (p,s,0.5).

i	y_i	x_{i11}	x_{i12}	x_{i21}	x_{i22}	x_{i31}	x_{i32}	x_{i33}
1	1	1	0	0	1	1	0	0
2	1	1	0	1	0	0	1	0
3	1	1	0	0	1	1	0	0
4	0	1	0	0	1	1	0	0

Table 3 Solution after Kohli/Sukumar with chance constrained $\epsilon = \sigma q_p$ with the two products (g,m,0.5) and (p,s,0.4).

i	y_i	x_{i11}	x_{i12}	x_{i21}	x_{i22}	x_{i31}	x_{i32}	x_{i33}
1	0	0	1	0	1	0	1	0
2	1	1	0	1	0	1	0	0
3	1	1	0	1	0	1	0	0
4	1	0	1	0	1	0	1	0

Table 4 Solution after Gaul et al. with the products (p,s,0.4) and (p,m,0.7).

j	x_{j11}	x_{j12}	x_{j21}	x_{j22}	x_{j31}	x_{j32}	x_{j33}
1	1	0	1	0	1	0	0
2	1	0	0	1	0	0	1

6 Conclusion and Outlook

Nevertheless the solution of the chance constrained approach is not that good as the approach of Gaul et al. (1995), we managed to make the original approach of Kohli and Sukumar (1990) better. Now, one can consider the failure that inevitably occur in the conjoint estimation process. That is an advantage, especially when you have bigger problems than above and cannot adjust the parameter ϵ manually. However, stochastic programming (chance constrained programming) showed its power to product line design optimization problems and will be studied further. The next step is to design new product optimization models with two- or multi-stage stochastic programming. Stochastic programming seems to be very promising to these kind of problems.

References

Birge, J. R., & Louveaux, F. (2011). Introduction to stochastic programming. *Springer series in operations research and financial engineering*. New York, London: Springer.

Bradley, R. A., & Terry, M. E. (1952). Rank analysis of incomplete block designs: the method of paired comparisons. *Biometrika, 39*, 324–345.

Dantzig, G. B. (1955). Linear programming under uncertainty. *Management Science, 1*, 197–206.

Gaul, W., & Baier, D. (2009). Simulations- und optimierungsrechnungen auf basis der conjointanalyse. In D. Baier, & M. Brusch (Eds.), *Conjointanalyse: methoden-anwendungen-praxisbeispiele* (pp. 163–182). Berlin, Heidelberg: Springer.

Gaul, W., Aust, E., & Baier, D. (1995). Gewinnorientierte Produktliniengestaltung unter Beruecksichtigung des Kundennutzens. *Zeitschrift fuer Betriebswirtschaftslehre, 65*, 835–855.

Green, P. E., Carroll, J. D., & Goldberg, S. M. (1981). A general approach to product design optimization via conjoint analysis. *Journal of Marketing, 45*, 17–37.

Green, P. E., & Srinivasan, V. (1978). Conjoint analyses in consumer research: Issues and outlook. *Journal of Consumer Research, 5*, 103–123.

Kall, P., & Mayer, J. (2011). Stochastic linear programming: models, theory, and computation. *International series in operations research & management science*, vol. 156. New York, London: Springer.

Kohli, R., & Sukumar, R. (1990). Heuristics for product-line design using conjoint analyses. *Management Science, 36*, 1464–1478.

Shapiro, A., Dentcheva, D., & Ruszczynski, A. (2009). Lectures on stochastic programming: Modeling and theory. *MPS/SIAM Series on Optimization, 9*, xvi–436.

Part IV
AREA Data Analysis in Finance

On the Discriminative Power of Credit Scoring Systems Trained on Independent Samples

Miguel Biron and Cristián Bravo

Abstract The aim of this work is to assess the importance of independence assumption in behavioral scorings created using logistic regression. We develop four sampling methods that control which observations associated to each client are to be included in the training set, avoiding a functional dependence between observations of the same client. We then calibrate logistic regressions with variable selection on the samples created by each method, plus one using all the data in the training set (biased base method), and validate the models on an independent data set. We find that the regression built using all the observations shows the highest area under the ROC curve and Kolmogorv–Smirnov statistics, while the regression that uses the least amount of observations shows the lowest performance and highest variance of these indicators. Nevertheless, the fourth selection algorithm presented shows almost the same performance as the base method using just 14 % of the dataset, and 14 less variables. We conclude that violating the independence assumption does not impact strongly on results and, furthermore, trying to control it by using less data can harm the performance of calibrated models, although a better sampling method does lead to equivalent results with a far smaller dataset needed.

M. Biron (✉)
Department of Industrial Engineering, Universidad de Chile. República 701, 8370439 Santiago, Chile
e-mail: mbiron@ing.uchile.cl

C. Bravo
Finance Center, Department of Industrial Engineering, Universidad de Chile. Domeyko 2369, 8370397 Santiago, Chile
e-mail: cbravo@dii.uchile.cl

M. Spiliopoulou et al. (eds.), *Data Analysis, Machine Learning and Knowledge Discovery*, Studies in Classification, Data Analysis, and Knowledge Organization, DOI 10.1007/978-3-319-01595-8_27,

1 Introduction

Behavioral scorings are a well-known statistical technique used by financial institutions to assess the probability that existing customers default a loan in the future. Various statistical models are used to measure this probability, with logistic regression the most popular among them, used by 95 % of financial institutions worldwide according to Thomas et al. (2002).

The aim of this study is to empirically measure the importance of violating the independence assumption needed to calibrate the parameters of a logistic regression using standard maximum likelihood estimation (Hosmer and Lemeshow 2000), in the context of behavioral scoring development. The situation has been documented in the literature (Medema et al. 2007), and there is strong evidence on the perverse effects of ignoring its effects, particularly in the case of reject inference, where it was concluded that sample bias leads to different classification variables, and to an overall reduction in predictive capability as Verstraeten and Van Der Poel (2005) show. Another, very disruptive, consequence of dependent events is that common goodness-of-fit tests applied give biased and incorrect results, resulting in the commonly overlooked need of applying ad-hoc tests such as the one developed by Archer et al. (2007), or measuring the model for misspecification using a statistical test alike the one introduced by White and Domowitz (1984).

In this paper, we propose several sampling methods in order to avoid using dependent data as input when training logistic regressions, focusing on eliminating or reducing sampling bias by different sub-sampling methodologies in temporal, panel-like, databases. This problem, very common consumer credit risk, is also of relevance whenever there is a panel structure of data with a high rate of repeated cases, such as predictive modeling in marketing campaign response.

The paper is structured as follows: First, we show how is that a dependence structure arises when monthly account data and a standard definition of default. Second, we describe the four sampling techniques implemented to overcome the latter issue. Third, we calibrate logistic regressions using these techniques, and compare the discrimination power of these regressions against one calibrated using all the data points. The paper ends with conclusions and practical recommendations about the importance of the issue.

2 Dependence Structures

The data commonly used in developing behavioral scoring models corresponds to observations at the end of every month for every client currently on an active relationship with the financial institution. Usually, these databases contain information about the products that every client has, so there will normally be one record for every product—client, each month. Because the risk that is being studied is the one associated with the client and not with a single product, these tables are usually consolidated to show the credit behavior of the clients, in each particular month. Finally, these tables are cross-referenced with other sources to obtain additional information, normally consisting of demographic (in the case of retail lending) and external credit information.

The dependent variable associated with the client behavior that is going to be modeled through a logistic regression is constructed using the days in arrears information. So, for client i on time t, the dependent variable y_{it} is obtained as follows:

$$y_{it} = \begin{cases} 1 \; \max_{s=1,\dots,h} d_{i,t+s} \geq 90 \\ 0 \qquad\qquad \text{otherwise} \end{cases} \tag{1}$$

where d_{it} corresponds to the days in arrears that client i shows on time t, and h is a parameter of the model and corresponds to the length of the time frame to include on an observation of future behavior. This means that if the client becomes 90 days in arrears or more in any of the h subsequent months, that client is marked on time t as showing a default from that record onward. We use here the standard definition of default accepted by the Bank for International Settlements and the Basel Committee on Banking Supervision (2006).

With this definition of default, a 0 means that none of the following h months show more than 89 days in arrears. This is where the independence issue arises. Consider a client that never shows a delay in his/her monthly payments. Now, suppose that y_{it} and $y_{i,t+h}$ are known and that both equal 0. Then, it is possible to deduce the value of the dependent variable for every month in this interval of time. Indeed, $y_{it} = 0$ means that

$$d_{i,t+s} < 90 \quad \forall s = 1, \dots, h \tag{2}$$

and $y_{i,t+12} = 0$ means that

$$d_{i,t+12+s} < 90 \quad \forall s = 1, \dots, h \tag{3}$$

Of course, together these two equations imply that:

$$d_{i,t+s} < 90 \quad \forall s = 1, \dots, 2h \tag{4}$$

We conclude that every indicator of default between t and $t + h$ has to be equal to 0. Therefore, there is a dependence structure between y_{it}, $y_{i,t+h}$, and every point in between.

3 Sampling Algorithms

We develop four sampling algorithms that choose data in a way so that dependence structures are eliminated. The motivation behind the use of algorithms is to "simulate" a data generation process that does not produce dependent observations

Algorithm 1 Algorithm 1: One-month-per-client sampling

```
S = ∅ {Reduced sample set}
for t = to t do
    for i = 1 to N do
        Identify the subset D_i ⊂ {1,...,T} in which the client is observed.
        Choose a random date t' ∈ D_i.
        Include the observation (i,t') in the sample S, S = S ∪ (i,t').
    end for
end for
return S
```

Algorithm 2 Algorithm 2: One-month-per-client sampling 2

```
S = ∅ {Reduced sample set}
for i = 1 to N do
    Select random time t' ∈ {1,...,T}.
    if Pair (i,t') exists in full sample. then
        S = S ∪ (i,t')
    end if
end for
return S
```

in the sense explained in the above section, so that the estimation can be applied in a straightforward manner. Of course, this imposition implies that less data is going to be included on model calibration. We assume that there is a full sample set index F such that a pair (i,t) is in F if both a given customer i is in the books (and therefore has data available) at time t.

Algorithms 1–3 construct small samples of data, because they select at most 1 observation per client. Therefore, to assure consistency of the estimates, parameters obtained on each sample are then averaged to obtain the final estimate, in a bootstrapping approach. Algorithm 1 corresponds to stratified sampling, regarding each client as a stratum. It selects, for every i, one random date t in $F_i = (c,t) \in F : c = i$. This approach distorts the distribution of the data, because it tends to over represent "occasional clients": individuals who have little history with the institution.

Algorithm 2 tries to control for the distortion previously described. It can be regarded as a cluster sampling, in which each client (with all of its corresponding observations) is regarded as a cluster. First, each cluster i is selected with probability $\frac{|F_i|}{|F|}$. Then, one observation from each individual is selected at random from the dates available for it in the dataset.

Algorithm 3 builds on the fact that, by definition, clients appear only once each month. Thus, it constructs T different samples, labeled $F_t, t = 1 \ldots T$, so that $F_t = (i,s) \in F : s = t$. This method assumes that the credit behavior of the population does not change heavily from one month to the other.

Finally, the aim of Algorithm 4 is to try to include the most data associated with each client available set of data F_i, controlling that every pair of data is separated for more than h months, and therefore independent with respect to the framework

Algorithm 3 Algorithm 3: Each month as different sample

```
S = ∅
Select random time t'.
S = {i : (i, t') ∈ F}
return S
```

Algorithm 4 Algorithm 4: mod h Sampling

```
Construct a reduced sample Z' using (for example) algorithm 1.
for Clients i in Z' do
  F_i = {t : (i, t) ∈ Z'}
  FLAG = true.
  for f = (i, t) ∈ F_i do
    S_i = {s : (i, t) ∈ S
    for s ∈ S_i do
      if |s − t| mod_h ≠ 0 then
        FLAG = false.
      end if
    end for
    if FLAG= true then
      S = S ∪ f
    end if
  end for
end for
return S
```

presented earlier. In other words, the aim is to solve for each client i:

$$\begin{aligned} \max_{S_i \in Z} \quad & Card(S_i) \\ \text{s.t.} \quad & t_2 - t_1 \geq h \quad \forall t_2 > t_1 . t_j \in S_i, j = \{1, 2\}. \end{aligned} \tag{5}$$

The feasible set is non-empty (one can always use just one datum), so the solution exists; although, it may not be unique, depending on the data available. Algorithm 4 is a heuristic approach to solve this problem.

4 Experimental Results

To illustrate the difference in performance of the sampling methods, we implemented them on a real financial institution database. The data comes from monthly observations of the behavior of the consumer retail business line. The list of clients was split 70–30 % to create a training set and a validation set of data.

The first step to build the logistic regression models was to perform univariate analysis of the variables included, resulting on a predictive power ranking of variables. The level association of every variable with the target was calculated using CHAID trees, based on the chi square statistic. The optimal splitting constructed by

Table 1 Description of datasets used per algorithm

Algorithm	No. of sets	Avg. size	Avg. bad rate	No. variables
1	50	19,063	2.55 %	12
2	50	10,738	2.01 %	5
3	50	10,733	2.05 %	4
4	1	57,038	2.08 %	25
Base	1	407,844	2.07 %	39

CHAID trees was used to define sets of dummy indicators for each variable. Then, a correlation filter was applied to the dummies, with a tolerance of 50 %, assuring that whenever a conflict appeared, the dummy associated with the variable that was higher on the univariate ranking was preserved.

The next step was to construct sample using the above mentioned algorithms. Table 1 shows statistics of the sets built. The base algorithm refers to using all the data points in the training set associated with one client. Due to the fact that the sample size of sets created by Algorithms 1, 2 and 3 was less than 5 % of the available data (represented by the set size of the Base Algorithm), 50 samples were extracted so that the logistic regression model would be obtained by averaging coefficients obtained across the 50 sets, in order to assure the robustness of the results.

Using the above mentioned sets, forward selection with logistic regression was applied to reduce the number of selected variables. For algorithms 1, 2 and 3, this process was run for every one of the 50 sets constructed, so 50 lists of selected dummies were obtained. The final set of dummies for each method was the one that appeared on at least 90 % of the lists. The same table also shows the number of variables selected with each method, and it is clear that the algorithms that produced the smaller sets of data tend to include the least number of dummies.

The last step was to calibrate logistic regressions using just the list of dummies selected under each sampling approach. Again, for Algorithms 1, 2 and 3, logistic models were fitted on each of the samples, and then the final coefficients of the model were calculated as the average across all sets.

In order to validate the results of each model, Kolmogorv–Smirnov and Area Under ROC statistics were calculated in the validation set. These statistics were calculated for each month of the set, so as to assure that just one observation per client would be used. Table 2 shows the average and standard deviation for both statistics, across all months on the validation set.

The results show that the Base Algorithm is the one that shows the best performance under both measures. Nevertheless, Algorithm 4 shows less volatility on the KS statistic (although this does not occur with AUC), and achieves almost the same performance. The worst results are shown by Algorithms 2 and 3.

To investigate if the differences in performance of Algorithm 4 and Base were due to model misspecification, a neural network model was calibrated and validated with the same approach as before. Table 3 shows these results.

Table 2 Results per algorithm

	KS		AUC	
Algorithm	Average	Std. Dev.	Average	Std. Dev.
1	71.39 %	3.49 %	91.58 %	1.78 %
2	68.95 %	4.06 %	87.92 %	1.71 %
3	68.68 %	3.98 %	87.00 %	2.07 %
4	74.39 %	2.85 %	93.11 %	1.27 %
Base	74.89 %	3.02 %	93.55 %	1.20 %

Table 3 K–S statistics, and AUC for each algorithm

	KS		AUC	
Algorithm	Average	Std. Dev.	Average	Std. Dev.
4	69.46 %	2.92 %	91.16 %	1.47 %
Base	71.08 %	3.04 %	91.66 %	1.39 %

It is clear that the same results are obtained using neural networks instead, and it can be concluded that the most likely reason behind the difference in performance across the methods is the number of data points in the sets created by them. Indeed, Algorithm Base and 4 show the largest size and the best performance. Conversely, Algorithms 2 and 3 show the worst performance and the smallest sample size. Therefore, the differences in performance would be explained by the fact that the quality of the models calibrated using more data is better, because more and more complex patterns can be discovered with more data.

In sum, although we couldn't find reductions in performance associated with the use of dependent data, we do find that one can achieve almost the same performance with less data points, using considerably less classification variables (as with Algorithm 4), which in turn reduces the volatility of the K–S statistic in time. Consequently, we believe that sampling algorithms that demand a minimum separation of h months between observations associated with the same client on each sample can deliver better performance results in behavioral scoring applications.

5 Conclusions

In this paper we aimed to empirically measure the importance of the independence assumption needed to use standard maximum likelihood calibration methods for logistic regressions, in the context of behavioral scoring model development. We showed how dependence structures arise when using monthly client behavior data, and a standard definition of target variable. Finally, sampling methods that eliminated these dependencies were developed, and testing them on real life data from a financial institution show how reducing the impact of bias can lead to worst predicting capabilities.

The experiments showed that when using all the observations of a client the best performance was achieved. In other words, trying to control the dependence in data caused model impairment. This situation was related with the smaller training set size that sampling implies. Nevertheless, Algorithm 4 achieves almost the same performance than Base Algorithm, but using 14 variables less and showing lower volatility of the KS statistic.

In sum, trying to control for dependence in data using sampling techniques could harm the performance of the calibrated models. Although it may be useful in order to lower the volatility of the model without sacrificing a lot of discriminating power, it should be used carefully, and trying to include the most data possible, using algorithms such as Algorithm 4.

Acknowledgements The work reported in this paper has been partially funded by the Finance Center, DII, Universidad de Chile, with the support of bank Bci.

References

Archer, K. J., Lemeshow, S., & Hosmer, D. W. (2007). Goodness-of-fit tests for logistic regression models when data are collected using a complex sampling design. *Computational Statistics & Data Analysis, 51*, 4450–4464.

Basel committee on banking supervision (2006). Basel II: International convergence of capital measurement and capital standards: A revised framework—comprehensive version. http://www.bis.org/publ/bcbsca.htm. Accessed 15 October 2011.

Hosmer, D., & Lemeshow, H. (2000). *Applied logistic regression*. New York: Wiley.

Medema, L., Koning, R. H., & Lensink, R. (2007). A practical approach to validating a PD model. *Journal of Banking and Finance, 33*, 701–708.

Thomas, L. C., Crook, J. N., & Edelman, D. B. (2002). *Credit scoring and its applications*. Philadelphia: SIAM.

Verstraeten, G., & van der Poel, D. (2005). The impact of sample bias on consumer credit scoring performance and profitability. *Journal of the Operational Research Society, 56*, 981–992.

White, H., & Domowitz, I. (1984). Nonlinear regression with dependent observations. *Econometrica, 52*, 143–162.

A Practical Method of Determining Longevity and Premature-Death Risk Aversion in Households and Some Proposals of Its Application

Lukasz Feldman, Radoslaw Pietrzyk, and Pawel Rokita

Abstract This article presents a concept on how to infer some information on household preference structure from expected trajectory of cumulated net cash flow process that is indicated by the household members as the most acceptable variant. Under some assumptions, financial planning style implies cumulated surplus dynamics. The reasoning may be inverted to identify financial planning style. To illustrate the concept, there is proposed a sketch of household financial planning model taking into account longevity and premature-death risk aversion, as well as bequest motive. Then, a scheme of a procedure to identify and quantify preferences is presented. The results may be used in life-long financial planning suited to the preference structure.

1 Introduction

Modern personal financial planning builds on achievements of many interrelating disciplines and specializations. Its theoretical backbone is life cycle theory, developed since the middle of the twentieth century, starting from seminal works by Ando and Modigliani (1957) and Yaari (1965), based on earlier achievements by Ramsey (1928), Fisher (1930) and Samuelson (1937) in the field of intertemporal choice. Under the influence of concepts developed by these authors it is often postulated that a life-long financial plan of a household should optimize time structure of consumptby maximization of expected discounted utility. To do it, one needs to formulate the utility function, estimate its parameters and define probabilities to be used in calculation of expected value. Most publications on this subject use utility of consumption, neglecting bequest motive. Longevity risk aversion is

L. Feldman (✉) · R. Pietrzyk · P. Rokita
Wroclaw University of Economics, Komandorska 118/120, Wroclaw, Poland
e-mail: lukasz.feldman@ue.wroc.pl; radoslaw.pietrzyk@ue.wroc.pl; pawel.rokita@ue.wroc.pl

M. Spiliopoulou et al. (eds.), *Data Analysis, Machine Learning and Knowledge Discovery*, Studies in Classification, Data Analysis, and Knowledge Organization,
DOI 10.1007/978-3-319-01595-8_28,

usually the only kind of risk aversion considered there. As regards the probabilities, life-cycle models are applied, assuming either deterministic (Yaari 1965), or, sometimes, stochastic mortality force (Huang et al. 2011). Existing life-cycle literature concentrates on models that support optimization of retirement capital distribution, rather than life-long financial planning. Moreover, these models are better suited for individuals than households. Consistently, they neglect premature-death and focus on longevity risk (for a one-person household premature death has no financial consequences). Another implication of that approach is treating the time when the life of the analyzed entity ends as the only stochastic element (unless stochastic mortality force is assumed). In real life, from financial point of view, it is important not only whether the household still exists, but also who of its members is alive (particularly when their incomes or variable costs differ substantially).

Also in the field of practical applications there is a lot of scope for improvements. The main financial goal that must be worked into personal financial plan is, certainly, retirement. This is because of its relatively big magnitude, as well as impossibility of post-financing it. Retirement planning professionals tend, however, to attach excessive weight to the following two rules, applied equally to all households, regardless their preferences: (1) retirement income of a household should be at least as high as total consumption of the household, and (2) individual retirement income of a household member should not be lower than his or her individual financial needs in retirement. Building retirement capital that guarantees this is a safe but expensive way of providing for old age. It may be inefficient, due to overlapping coverage of household fixed costs, and generates considerable unutilized surplus. Neglecting this surplus would, in turn, lead to overestimating retirement capital needs and, consistently, paying unnecessarily high contributions to private pension plans in pre-retirement period. It is possible to propose more risky, but less expensive solutions. They should be, however, suited to preference structure of household members.

In his article a modification to the existing approaches is presented. It is, so far, a basic version of the model, narrowed to retirement planning. The model is used here just as a tool facilitating assessment of financial-safety level demanded by household members for their retirement period, taking also into account their bequest motive. This will determine boundary conditions for any other financial plans. The results may be then used as an input to augmented versions of the model, that, in turn, may be applied as tools supporting life-long financial planning. The extensions might, for example, allow for multiple financial goals and a variety of financing sources.

The proposition encompasses a simplified household model, a concept of augmented preference-structure formal description and a general idea of a tool for assessing aversion to premature-death and longevity risk (with bequest motive taken into account). The underlying model of household cash flow is expressed in the form of cumulated net cash flow process (also referred to as cumulated surplus dynamics).

Under assumptions of Sect. 3, there is a strict dependence between financial plan and expected trajectory of cumulated net cash flow process. The focus is on the segment of the trajectory that corresponds to the retirement phase of the life cycle. One can identify just a couple of typical general shapes this expected trajectory may take on, each corresponding to some other level of premature death and longevity risk. Each has also its particular potential to generate bequest capital.

The core of this proposition is a simple technique of assessing risk aversion towards two types of length-of-life-related risk at once. It is shown how a scope of decision variants, together with information about their financial consequences for old age, might be visualized to household members, and how the feedback information received from the household may be interpreted and used in the model.

2 Definitions

The terms that are used here in specific meaning, suited for our approach, are: household, full and partial retirement and schemes of retirement gap filling.

Definition 1. Household
– an autonomous economic entity distinguished according to the criterion of individual property, whose members are two natural persons (*main members of the household*), fulfilling the following conditions: (1) jointly set and realize, in solidarity with each other, goals as regards majority of the most important needs of all members of this entity and (2) are supposed to intend to remain members of this entity, to the extent possible, throughout the whole period of its existence—together with other natural persons, who, in some phases of the entity life cycle, fulfill the condition 1 or are (at least economically) dependent on any of the main members.

Definition 1 uses some key elements of a broader definition proposed by Zalega (2007).

Definition 2. Full pension
– such retirement income of an individual that completely covers his/her variable costs and the whole amount of fixed costs.

Definition 3. Partial pension
– such individual retirement income that is lower than full pension.

Definition 4. General schemes of retirement gap filling
A scheme of retirement gap filling is one of the following three general classes of relations between retirement income and consumption needs:

1. "2 × Full"

$$In_t^{(1)} \geq FC + VC_t^{(1)}; \quad In_t^{(2)} \geq FC + VC_t^{(2)} \tag{1}$$

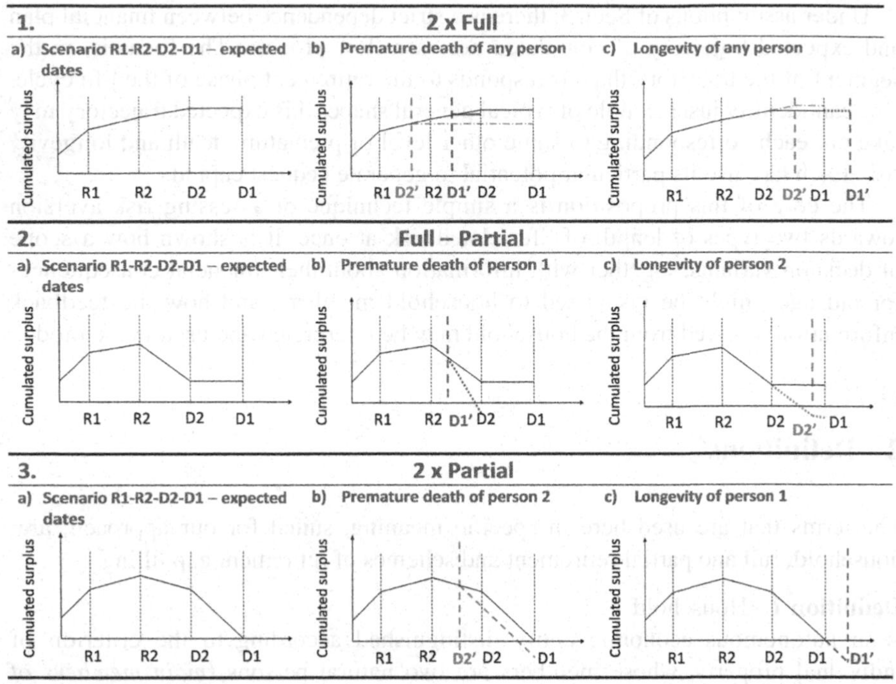

Fig. 1 Household's cumulated net cash flow in: (**1**) "2 × Full", (**2**) "Full–Partial", (**3**) "2 × Partial" scheme, for: (**a**) expected R1-R2-D2-D1 scenario, (**b**) premature death, (**c**) longevity

2. "Full–Partial"

$$In_t^{(1)} < VC_t^{(1)};\ In_t^{(2)} \geq FC + VC_t^{(2)};\ In_t^{(1)} + In_t^{(2)} < FC + VC_t^{(1)} + VC_t^{(2)} \quad (2)$$

3. "2 × Partial"

$$VC_t^{(1)} < In_t^{(1)} < FC + VC_t^{(1)};\quad VC_t^{(2)} < In_t^{(2)} < FC + VC_t^{(2)} \quad (3)$$

$$In_t^{(1)} + In_t^{(2)} < FC + VC_t^{(1)} + VC_t^{(2)}$$

Notation: $In_t^{(i)}$—i-th person's total income in period t, FC—household fixed costs, $VC_t^{(i)}$—variable costs assigned to the i-th person.

Graphical illustration of these general schemes present Fig. 1.

Moreover, the schemes "Full–Partial" and "2 × Full" have four specific variants (each). Together there are 9 possible schemes. More details are presented in Table 1.

Table 1 Income-cost structure and corresponding risks related to length of life

Ex.	In(1)	In(2)	PD(1)	Lgv.(1)	PD(2)	Lgv.(2)	Scheme
1	$In^{(1)} < VC^{(1)}$	$In^{(2)} < VC^{(2)}$	−	+	−	+	$2 \times Partial$
2		$VC^{(2)} < In^{(2)} < FC + VC^{(2)}$	−	+	+	+	$2 \times Partial$
3		$In^{(2)} > FC + VC^{(2)}$	−	+	+	−	$Full - Partial$
4	$VC^{(1)} < In^{(1)} < FC + VC^{(1)}$	$In^{(2)} < VC^{(2)}$	+	+	−	+	$2 \times Partial$
5		$VC^{(2)} < In^{(2)} < FC + VC^{(2)}$	+	+	+	+	$2 \times Partial$
6		$In^{(2)} > FC + VC^{(2)}$	−	+	+	−	$Full - Partial$
7	$In^{(1)} > FC + VC^{(1)}$	$In^{(2)} < VC^{(2)}$	+	−	−	+	$Full - Partial$
8		$VC^{(2)} < In^{(2)} < FC + VC^{(2)}$	+	−	−	+	$Full - Partial$
9		$In^{(2)} > FC + VC^{(2)}$	−	−	−	−	$2 \times Full$

Where:
Ex. – exposure type, In(i) – i-th person's income, PD(i), Lgv.(i) – existence (+) or nonexistence (−) of a threat to liquidity in case of premature death (PD(i)) or longevity (Lgv.(i)) of person i (premature-death and longevity risk)

3 Assumptions

The model takes on the following assumptions:

- Both main members intend to remain in the household until their death.
- Joint utility function of the whole household is considered.
- Utility function of the household is composed of two elements: utility of consumption and utility of bequest.
- Pre-retirement household income is constant in real terms (inflation indexed).
- Fixed real rate of return on private pension plan.
- Pension income constant in real terms (inflation indexed).
- Fixed replacement rate (but may be different for women and men).
- No other financial goals considered.
- Household members buy life annuity.
- Household consumption is fixed at a planned level, unless loosing liquidity.
- There is no "full pension" restriction

There are four key events in life cycle of a household. The times are denoted as:

- retirement of the first person (deterministic): $R1$,
- retirement of the second person (deterministic): $R2$,
- death of the second person (stochastic): $D2$,
- death of the first person (stochastic): $D1$.

If a typical case is taken into consideration, where the household is just a marriage and spouses are of similar age, then expected order of these events is just $R1, R2, D2, D1$ (or, put differently: $E(R1) < E(R2) < E(D2) < E(D1)$).

Are the assumptions listed above realistic? Together they make up an indeed simplified household finance model. This refers to the number of financial goals (retirement only), possible ways of financing, dynamics of labour income, consumption needs, etc. But this model is used just to produce general schemes of filling retirement gap, each connected to a corresponding level of risk and each showing somewhat different potential to generate bequest. The model also allows to represent the schemes in a simple graphical and tabular form, to be shown to household members. The choice of scheme by the household members should help assessing their attitude towards risk and bequest motive. The results may be then used in a final, more advanced, model, to support life-long financial planning. That other model should, in turn, allow for any number of financial goals and ways of financing them, as well as for more types of risk. This will require revising and relaxing some assumptions.

4 Optimization

Formally the goal function may be expressed with the following (4):

$$V(\mathbf{B};\gamma,\delta,\zeta) = \sum_{t=0}^{\infty}\left[\frac{1}{(1+\rho)^t}\cdot q_{t|\Psi_t}\cdot u(C_t(\mathbf{B});\gamma,\delta) + \frac{1}{(1+\rho)^t}\cdot \phi_{t|\Psi_t}\cdot v(CNCF_t(\mathbf{B});\zeta)\right] \tag{4}$$

for the optimization task defined as: $argmax_{\mathbf{B}} V(\mathbf{B};\gamma,\delta,\zeta)$.

where: C_t—consumption (in monetary terms) at t, $CNCF_t$—cumulated net cash flow at the moment t, γ—parameter describing longevity risk aversion, δ—parameter of premature death risk aversion, ζ—bequest motive parameter, $u(.)$—utility of consumption, $v(.)$—utility of bequest, ρ—interest rate used for discounting.

The decision variable **B** may be defined as (5):

$$\mathbf{B} = \begin{bmatrix} z_1 & x_1 \\ z_2 & x_2 \end{bmatrix} \tag{5}$$

where:

z_i is a fraction of variable costs of i-th person covered by her retirement income,
x_i is a fraction of fixed costs covered by retirement income of i-th person.

The feasible set is given by condition 6:

$$0 \leq z_i \leq 1; \quad \text{if} \quad z_i < 1, \quad \text{then} \quad x_i = 0; \quad \text{if} \quad z_i = 1, \quad \text{then} \quad x_i \geq 0 \tag{6}$$

The probabilities $q_{t|\Psi_t}$ and $q_{t|\Psi_t}$ used in (4) are not just conditional survival probabilities (like in Yaari (1965) or Huang et al. (2011)), but probabilities that consumption (or cumulated surplus) is of a given size at a given time. These probabilities depend on survival probabilities, but are not identical with them.

Under the assumption of Sect. 3 each of retirement gap filling schemes, represented by **B**, may be, ceteris paribus, realized by a unique combination of consumption, investments and unconsumed (but also uninvested) surplus. This, in turn, determines cumulated surplus dynamics. On this basis a cumulative stochastic process of surplus may be derived. Its will, certainly, depend on mortality model used.

Another task is to define utility functions $u(.)$ and $v(.)$. Even before this, it is possible, however, to present a general concept on how to estimate their parameters (γ, δ and ζ), which is the subject of this article.

At last, there is an interest rate (ρ), used for discounting. For now it is sufficient to say that it is a parameter, just set by the analyst. But in further research one should consider, what kind of risk premia the rate should include.

5 The Role of Bequest Motive

Choosing from amongst the schemes, a household chooses in fact a future dynamics of cumulated net cash flow. It implies corresponding combinations of premature-death and longevity risk. Each scheme may be also characterized with some level of expected bequest left. Two households with identical risk aversion may differently value available schemes if they differ in their attitude towards bequest.

This is why the goal function contains element reflecting bequest motive (4). Each row of Table 1 should be then considered twice: from the point of view of a household that is and is not interested in passing any bequest to their descendants.

6 Choice of Scheme and Model Calibration

Having defined all decision variants, one may start with goal function calibration. A sketch of rule (ordinal condition) for estimation results is presented in Table 2.

It is suggested that only nine levels of risk are considered (for both kinds of life-length risk) and risk aversion parameters γ and δ may take on just nine values (each). The values reflect ranks with respect to the level of risk, but in the inverted order (the higher risk aversion, the lower risk accepted). Risk aversion is measured here on an

Table 2 Income-cost structure and corresponding risks related to length of life

Ex.	In(1)	In(2)	PD(1)	Lgv.(1)	PD(2)	Lgv.(2)	B	Scheme	γ	δ	ζ
1.1		$In^{(2)} < VC^{(2)}$	−	+++++	−	++++	+	2 × *Partial*	0	9	1
1.2			−	+++++	−	++++	−		0	9	0
2.1	$In^{(1)} < VC^{(1)}$	$VC^{(2)} < In^{(2)} < FC + VC^{(2)}$	−	++++	+	+++	+	2 × *Partial*	2	8	1
2.2			−	++++	+	+++	−		2	8	0
3.1		$In^{(2)} > FC + VC^{(2)}$	−	++++	+++	−	+	*Full − Partial*	5	6	1
3.2			−	++++	+++	−	−		5	6	0
4.1		$In^{(2)} < VC^{(2)}$	+	+++	−	++	+	2 × *Partial*	4	8	1
4.2			+	+++	−	++	−		4	8	0
5.1	$VC^{(1)} < In^{(1)} < FC + VC^{(1)}$	$VC^{(2)} < In^{(2)} < FC + VC^{(2)}$	+	++	+	+	+	2 × *Partial*	6	5	1
5.2			+	++	+	+	−		6	5	0
6.1		$In^{(2)} > FC + VC^{(2)}$	−	++	++	−	+	*Full − Partial*	7	5	1
6.2			−	++	++	−	−		7	5	0
7.1		$In^{(2)} < VC^{(2)}$	+	−	−	++	+	*Full − Partial*	7	8	1
7.2			+	−	−	++	−		7	8	0
8.1	$In^{(1)} > FC + VC^{(1)}$	$VC^{(2)} < In^{(2)} < FC + VC^{(2)}$	+	−	−	+	+	*Full − Partial*	8	8	1
8.2			+	−	−	+	−		8	8	0
9.1		$In^{(2)} > FC + VC^{(2)}$	−	−	−	−	+	2 × *Full*	9	9	1
9.2			−	−	−	−	−		9	9	0

List of schemes, each considered twice, with and without bequest motive, and augmented by an attempt to assign values to model parameters or each scheme or now, yet, on an ordinal scale only; he values of the parameters are shown here to illustrate which decision variant is better suited for a household with higher and which for one lower risk aversion; bequest motive is measured on dichotonominal scale.

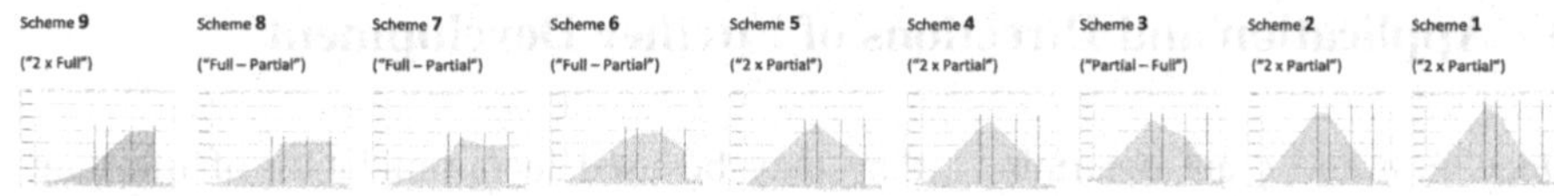

Fig. 2 Cumulated net cash flow of a household under various schemes of retirement gap filling; sorted from the safest (no. 9) to the most risky (no. 1)

ordinal scale only. Utility function should be then constructed in such a way that it makes sense of these values. A rule will be that the goal function with parameters γ_j, δ_j and ζ_j, corresponding to j-th row of Table 2, reaches its maximum on the range of values of **B** (Sect. 4, (5)) that fulfills income-cost conditions of the scheme j.

The procedure of model calibration boils down to presentation of graphical schemes and Table 1 to main members of the household, supplemented with additional explanation about threats in case dates of death deviate from their expected values, and requesting a feedback in the form of the choice of only one of these schemes. In addition, the information about household's bequest motive must be collected. Then values of γ, δ and ζ are read from the selected row of the table.

This is, certainly, not an estimation in statistic-econometric sense. No significance tests are possible. Here p-value would mean something like this: if a household has indeed no aversion to longevity risk ($\gamma = 0$), then what is the probability that it chooses, say, Scheme 3 ($\gamma = 5$)? Next, the probability might be compared to the test significance level. This is, however, beyond the scope of this article.

7 Various Gap-Filling Schemes: Numerical Example

Household parameters: a 38 year old woman (1) with income 4,000, and 39 year old man (2) with income 4,500. Scenario: R1-R2-D2-D1. Total expenses: (a) 6,450 (pre-retirement), (b) 5,950 ($R1$-$R2$), (c) 4,950 ($R2$-$D2$), (d) 2,975 ($D2$-$D1$). Let retirement age be 60 for women and 65 for men.

Cumulated surplus term structures for several schemes of retirement gap filling are presented in Fig. 2. The plots presented there show the relationship that is essential for the discussed problem. Namely, they illustrate a potential for investigating a trade-off between cumulated net cash flow on the day the first person retires and the same quantity close to the horizon of the plan.

8 Application and Directions of Further Development

Once γ, δ and ζ are determined, they may be used in optimization of life-long financial plan. In planning stage the model of cumulated net cash flow may be much more complex, due to new kinds of financial goals and ways of financing them.

The main challenge in further research will be defining such utility functions ($u(.)$ and $v(.)$ in (4)) that reflect properly preferences of a household. For a given set of values of risk-aversion and bequest-motive parameters, the goal function should reach its maximum just for the scheme corresponding to these values.

Another direction is classification of households with respect to life-length risk aversion.

References

Ando, A., & Modigliani, F. (1957). Tests of the life cycle hypothesis of saving: comments and suggestions. *Oxford Institute of Statistics Bulletin*, *19* (May), 99–124.

Fisher, I. (1930). *The theory of interest*. New York: Macmillan.

Huang, H., Milevsky, M. A., & Salisbury, T. S. (2011). Yaari's Lifecycle Model in the 21st Century: Consumption Under a Stochastic Force of Mortality. *URL:* http://papers.ssrn.com/sol3/papers.cfm?abstract_id=1816632[or:]http://dx.doi.org/10.2139/ssrn.1816632.Cited22March2012

Ramsey, F. P. (1928). A mathematical theory of saving. *The Economic Journal*, *38*(152), 543–559.

Samuelson, P. A. (1937). A note on measurement of utility. *Review of Economic Studies*, *4*, 155–161.

Yaari, M. E. (1965). Uncertain lifetime, life insurance and theory of the consumer. *The Review of Economic Studies*, *32*(2), 137–150.

Zalega, T. (2007). Gospodarstwa domowe jako podmiot konsumpcji. *Studia i Materiały*, Wydział Zarządzania UW

Correlation of Outliers in Multivariate Data

Bartosz Kaszuba

Abstract Conditional correlations of stock returns (also known as exceedance correlations) are commonly compared regarding downside moves and upside moves separately. The results have shown so far the increase of correlation when the market goes down and hence investors' portfolios are less diversified. Unfortunately, while analysing empirical exceedance correlations in multi-asset portfolio each correlation may be based on different moments of time thus high exceedance correlations for downside moves do not mean lack of diversification in bear market.

This paper proposes calculating correlations assuming that Mahalanobis distance is greater than the given quantile of chi-square distribution. The main advantage of proposed approach is that each correlation is calculated from the same moments of time. Furthermore, when the data come from elliptical distribution, proposed conditional correlation does not change, which is in opposition to exceedance correlation. Empirical results for selected stocks from DAX30 show increase of correlation in bear market and decrease of correlation in bull market.

1 Introduction

Correlation estimates play an important role in finance. It is especially visible in portfolio allocation, where the weight of an asset in a portfolio does not rely only on an estimated risk but also depends on an estimated correlation. Many previous studies (Longin and Solnik 2001; Ang and Bekaert 2002; Ang and Chen 2002; Hong et al. 2007; Butler and Joaquin 2002; Chua et al. 2009; Garcia and Tsafack 2011 and references therein) have revealed that asset returns are more correlated when the

B. Kaszuba (✉)
Department of Financial Investments and Risk Management, Wroclaw University of Economics, ul. Komandorska 118/120, 53-345 Wroclaw, Poland
e-mail: bartosz.kaszuba@ue.wroc.pl

M. Spiliopoulou et al. (eds.), *Data Analysis, Machine Learning and Knowledge Discovery*, Studies in Classification, Data Analysis, and Knowledge Organization, DOI 10.1007/978-3-319-01595-8_29,

market goes down and less when it goes up. This occurrence is known as asymmetric dependence between asset returns. Hence, it influences reduction of diversification during market turbulence (when extreme returns occur).
The asymmetric dependence has been revealed by the concept of exceedance correlation. This concept is used to measure the correlation between asset returns when both are either lower or greater than the given level (e.g., one standard deviation). Therefore, this approach measures the correlation in the tails (upper and lower tail separately) for each pair of assets returns. However, Campbell et al. (2008) show that increase in conditional correlation is due to normality assumption for the return distribution.
The aim of this paper is not to propose an alternative model for the exceedance correlation, but to propose an alternative statistic to measure correlation between multivariate extreme returns. Not only it is more adequate to portfolio allocation but also it is easier to compute, more useful for practitioners and less sensitive to model assumption. The main disadvantage of the exceedance correlation is that it is calculated for each pair separately and may be based on different moments of time. Thus, even strong asymmetries for all assets in portfolio may never occur, especially for a large size portfolio. The proposed approach is defined as the correlation for a subsample in which Mahalanobis distance for multivariate observations is greater than a given level. This definition allows proposed correlation to capture both more and less correlated assets while the exceedance correlation fails.
The rest of this paper is organised as follows. Section 2 introduces the concept of the exceedance correlation. Section 3 proposes new approach for correlation between extreme returns and compares this concept to the exceedance correlation. Section 4 includes empirical analysis for proposed statistic. Section 5 concludes this paper.

2 Exceedance Correlation

The conditional correlation, also known as the exceedance correlation, measures a correlation between extreme assets returns. The exceedance correlation, introduced by Longin and Solnik (2001), is a correlation of two returns R_1 and R_2 conditioned by exceedance of some threshold, that is:

$$\rho_E(\alpha) = \begin{cases} Corr(R_1, R_2 | R_i \leq q_{R_i}(\alpha); i = 1, 2; \alpha \leq 0.5) \\ Corr(R_1, R_2 | R_i > q_{R_i}(\alpha); i = 1, 2; \alpha > 0.5) \end{cases}, \tag{1}$$

where $q_{R_i}(\alpha)$ is the α-quantile of the return R_i. Instead of the quantile of the variable R_i, the θ_i threshold can be used, where θ_i means amount of standard deviations exceeded by the return R_i.

The conditional correlation strictly depends on underlying distributions, which is illustrated in Fig. 1. This example shows the shape of exceedance correlation for standard bivariate normal distribution with correlation 0.5 and for the standard

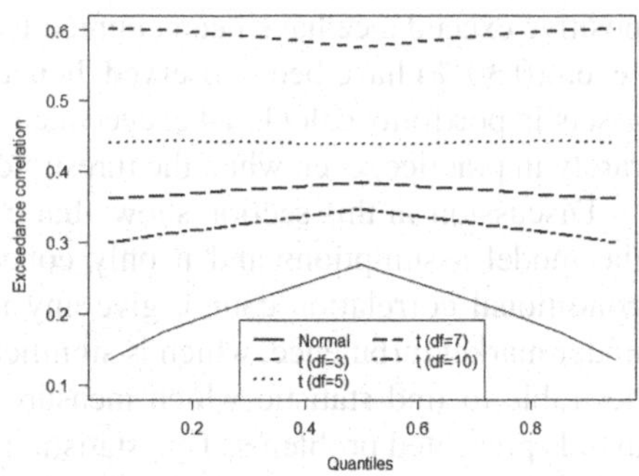

Fig. 1 A shape of the exceedance correlation

bivariate t-distribution with correlation 0.5 with 3, 5, 7 and 10 degrees of freedom. It can be seen that the exceedance correlation for normal distribution is strictly less than 0.5 and it decreases moving toward the tails of distribution. This figure also illustrates the influence of assumed distribution on a shape of the exceedance correlation, e.g. for the t-distribution exceedance correlation increases when decreasing degrees of freedom. It has been shown by many researchers (Longin and Solnik 2001; Ang and Bekaert 2002; Ang and Chen 2002; Hong et al. 2007; Butler and Joaquin 2002; Chua et al. 2009 and others) that asset returns are more correlated when the market goes down and less when it goes up. This result has been obtained by comparing the differences between theoretical exceedance correlation assuming bivariate normal distribution and the empirical exceedance correlation. Unfortunately, for normality assumption the exceedance correlation is strictly less than for the t-distribution assumption, hence the differences between theoretical and empirical conditional correlation are sensitive to distributional assumption. This problem has been analysed by Campbell et al. (2008) who have shown, that assuming the t-distribution, these differences frequently disappear. This assumption is justified in practice due to existence of outliers in real dataset.

Another problem with the exceedance correlation can occur in application to the portfolio theory. Many researchers have argued that higher conditional correlation between asset returns when the market goes down reduces benefits from portfolio diversification (i.e. Ang and Chen (2002), Chua et al. (2009), Ang and Bekaert (2002)). The magnitude of this reduction decreases when the number of assets in a portfolio increases, which results from the definition of the exceedance correlation which is calculated for each pair separately, hence conditional correlation is based on different moments of time. Estimating a conditional correlation matrix with some threshold α, for assets in a portfolio, does not mean that when extreme returns occur all returns will be correlated according to this conditional correlation matrix, but it means that when all returns exceed some threshold α, they will be correlated in accordance with the estimated conditional correlation matrix. Thus, increasing number of assets in portfolio, probability that all returns exceed some threshold α decrease. For example, analysing returns of six stocks from DAX30, between 3 January 2003 and 22 February 2012 (2,332 days) only on 35 days (1.5 %) all returns have been exceeded over one standard deviation (negative exceedance and positive exceedance have been counted together), for 20 stocks from DAX30, only seven (0.30 %) have been observed, hence it can be seen that increasing amount of assets in portfolio, calculated exceedance correlations for all stocks can appear very rarely in practice, even when the threshold is relatively small.

Discussion in this section show, that exceedance correlation is very sensitive to the model assumptions and it only concerns very rare situations. The presented conditional correlation cannot give any information about less correlated returns under market turbulence, which is significant in portfolio allocation. Therefore, it is desirable to find statistic which measures dependence between stock returns and avoids presented problems. This statistic is proposed in next section.

3 Correlation of Multivariate Extreme Returns

The aim of this section is to propose a statistic which measures dependence between extreme returns and to give some reasons for its usefulness in financial data analysis.

The main point of the proposed correlation is to analyse correlation between multivariate extreme returns (also called outliers). A return is called multivariate extreme return when it has a "large" value of an outlyingness measure, e.g. the Mahalanobis distance. The main difference between the exceedance correlation and the proposed correlation of extreme returns is that calculation of the exceedance correlation uses only bivariate extreme returns, while the proposed correlation uses only the multivariate ones (which are not necessarily bivariate).

For a portfolio with N risky assets, proposed correlation of multivariate extreme returns is given as:

$$\rho_M(\alpha) = Corr(R_i, R_j | d^2(\mathbf{R}, \boldsymbol{\mu}, \Sigma) \geq c(\alpha)), \tag{2}$$

where R_i is the return on the i-th asset in the portfolio, $\mathbf{R} = (R_1, \ldots, R_N)$ is the vector of assets returns, $\boldsymbol{\mu}$ and Σ is the mean and covariance matrix of the returns on the N assets, $d^2(\mathbf{R}, \boldsymbol{\mu}, \Sigma)$ is a squared Mahalanobis distance defined as follows:

$$d^2(\mathbf{R}, \boldsymbol{\mu}, \Sigma) = (\mathbf{R} - \boldsymbol{\mu})' \Sigma^{-1} (\mathbf{R} - \boldsymbol{\mu}).$$

The value $c(\alpha)$ is a fixed chosen constant determined by α, a number between 0 and 0.5, which is a fraction of observations taken to calculate $\rho_M(\alpha)$. It is clear that $\rho_M(\alpha)$ estimates correlation between the most extreme returns in the dataset. Hence, the natural choice for $c(\alpha)$ is the value for which $P_{\boldsymbol{\mu},\Sigma}(d^2(\mathbf{R}, \boldsymbol{\mu}, \Sigma) \geq c(\alpha)) = \alpha$ When assuming that returns are multivariate normal, then $c(\alpha) = \chi^2_N(1-\alpha)$ where $\chi^2_N(\alpha)$ is the α-quantile of the chi-square distribution with N degrees of freedom.

The presented approach is based on an elliptical truncation concept described by Tallis (1963) for elliptically truncated multivariate normal distributions. The main advantage of this concept is that the correlation does not change after elliptical truncation when data come from an elliptical distribution. Thus, the proposed conditional correlation can be easily compared to its unconditional counterpart, which makes it more useful in practice than the exceedance correlation is.

The proposed conditional correlation is able to capture both more and less correlated assets under market turbulence, because it is computed for multivariate outliers. It plays an important role in portfolio allocation as it is desirable to increase the allocation to less correlated assets, especially when extreme returns occur.

The presented approach can be easily robustified using robust estimators of the mean and covariance matrix in order to determine robust Mahalanobis distances. Robust estimators are less sensitive to both deviations from the assumed distribution (usually normal distribution) and to outliers. Hence, correlation of multivariate extreme returns is less sensitive to the model assumption. The robustified version of the conditional correlation is applied in empirical analysis and described more widely in the next section.

4 Empirical Analysis

This section includes the methodology for testing the correlation of outliers as well as performs the test and presents the results obtained for the real dataset.

4.1 Methodology

This section considers the problem of testing $H_0 : \rho = \rho_M$, the hypothesis that the correlation of multivariate extreme returns differ from true correlation coefficient. The aforementioned discussion shows that when data come from an elliptical distribution the null hypothesis is valid, since the test procedure consists in measuring two correlation coefficients over two distinct samples. That is, ρ and ρ_M are determined as follows:

$$\begin{aligned} \hat{\rho}_M(\alpha) &= Corr(R_i, R_j | d^2(\mathbf{R}, \hat{\mu}, \hat{\Sigma}) \geq c(\alpha)), \\ \hat{\rho} = \hat{\rho}(\alpha) &= Corr(R_i, R_j | d^2(\mathbf{R}, \hat{\mu}, \hat{\Sigma}) < c(\alpha)), \end{aligned} \tag{3}$$

where $\hat{\mu}$ and $\hat{\Sigma}$ are estimates of multivariate location and dispersion. Assuming normality, the maximum likelihood estimators of mean vector and covariance matrix can be used, albeit stock returns are not normally distributed which makes the MLE estimators inefficient in the case of even minor deviations from normal distribution. Hence, a better approach is determining robust distances using robust estimators, such as Minimum Covariance Determinant (MCD). Correlation coefficients $\hat{\rho}_M(\alpha)$ and $\hat{\rho}$ are estimated using Pearson's correlation due to maintenance impact of the highest returns.

It can be seen, that both correlations $\hat{\rho}_M(\alpha)$ and $\hat{\rho}$ are measured from different groups of returns. The first estimate $\hat{\rho}_M(\alpha)$ measures correlation of extreme returns, the second estimate $\hat{\rho}$ measures correlation within the bulk of the data.

For the estimated values, the test on the difference between two independent correlations is applied. In the first step, both correlation estimates are transformed using Fisher r-to-z transformation (Steiger 1980):

$$Z_i = \frac{1}{2} \ln \frac{1+\rho}{1-\rho},$$

and next, the test statistic $Z = Z_1 - Z_2$ is computed. The test statistic Z is approximately normally distributed with variance $\sigma_Z = (\frac{1}{N_1-3} + \frac{1}{N_2-3})^{1/2}$. The aforementioned procedure is applied for real dataset in the next section. As estimates of multivariate location and dispersion, MCD estimates with 30 % breakdown point are applied. As $c(\alpha)$ the $\chi^2_6(0.95)$ is chosen.

The research utilises historical daily logarithmic rates of return for six largest companies from DAX30: Siemens (SIE), Bayer (BAYN), SAP (SAP), Allianz (ALV), Daimler (DAI), E.ON (EOAN). The empirical analysis focuses on analysing the research hypothesis for two different periods separately: bear market—from 2007-07-01 to 2009-02-28 (422 returns) and bull market—from 2009-02-28 to 2011-07-19 (609 returns).

4.2 Empirical Results

This section focuses on the analysis of the conditional correlations with accordance the aforementioned methodology. Tables 1 and 2 present estimated conditional correlations of extreme returns ($\hat{\rho}_M(\alpha)$) and estimated conditional correlations within the bulk of data ($\hat{\rho}$). An asterisk indicates a significant difference between correlations within the bulk of the data and correlation of extreme returns at the 5 % level.

Table 1 presents estimated conditional correlations under bull market. It can be seen that correlations of extreme returns decrease and this reduction is significant for almost all companies except for Daimler. For Daimler, only one correlation with E.ON fell down from 0.53 to 0.33 while other correlations persist on the same level. These results indicate that under bull market correlations of extreme returns decrease, however there exist stocks for which correlation of outliers is the same as correlation within the bulk of the data.

Table 2 presents the estimated conditional correlations under bear market. It shows that among four companies: Bayer, SAP, Allianz and Daimler each correlation of outliers increases, moreover these correlations are relatively high (greater than 0.6). For other companies (Siemens, E.ON) almost all correlations have not changed significantly except for pairs Siemens and Daimler, Siemens and SAP. The interesting result is that for E.ON all correlations of extreme returns

Table 1 Conditional correlations under bull market

	ALV	BAYN	DAI	EOAN	SIE	SAP	ALV	BAYN	DAI	EOAN	SIE	SAP
	The bulk of the data						*Extreme returns*					
ALV	1	0,64*	0,63	0,70*	0,76*	0,55*	1	0,27*	0,62	0,40*	0,62*	0,39*
BAYN		1	0,55	0,56*	0,66*	0,50*		1	0,42	0,31*	0,40*	0,27*
DAI			1	0,53*	0,67	0,50			1	0,33*	0,66	0,38
EOAN				1	0,66*	0,49*				1	0,40*	0,28*
SIE					1	0,57*					1	0,38*
SAP						1						1

An asterisk indicates a significant difference between correlation within the bulk of the data and correlation of extreme returns at the 5 % level

Table 2 Conditional correlations under bear market

	ALV	BAYN	DAI	EOAN	SIE	SAP	ALV	BAYN	DAI	EOAN	SIE	SAP
	The bulk of the data						*Extreme returns*					
ALV	1	0,47*	0,67*	0,36	0,62	0,48*	1	0,63*	0,77*	0,30	0,70	0,65*
BAYN		1	0,42*	0,45	0,57	0,43*		1	0,70*	0,32	0,67	0,61*
DAI			1	0,32	0,63*	0,46*			1	0,30	0,80*	0,74*
EOAN				1	0,43	0,36				1	0,30	0,27
SIE					1	0,54*					1	0,68*
SAP						1						1

An asterisk indicates a significant difference between correlation within the bulk of the data and correlation of extreme returns at the 5 % level

decreased and were relatively low (between 0.27 and 0.32) in comparison to other correlations of outliers which were greater than 0.61. These results show that under bear market correlations of extreme returns are high and increase significantly. Although, there exist stocks for which correlation of outliers does not change significantly.

The differences between correlations under bear and bull market have been compared in the same manner. Tables of which are not presented here due to limited content of this paper. The results are similar to the presented ones, two most interesting of which are drawn. Comparing the differences between correlations from the bulk of the data, the significant increase has been observed for bull market. For the differences between correlations of extreme returns, the significant increase has been observed for bear market. Nevertheless, correlations did not change significantly for some stocks.

To sum up, it can be seen that correlations of extreme returns significantly increase under bear market and significantly decrease under bull market. However, there exist stocks for which correlation does not change under market turbulence, which shows that it is still possible to keep a diversified portfolio under bear or bull market even when extreme returns occur. Another interesting conclusion is that correlations differ under bear and bull market, since portfolio weights should be adjusted when the market trends change.

5 Conclusions

This paper includes a new proposal for correlation estimate of extreme returns which can be used for practical purposes. It has been shown that proposed conditional correlation is more adequate for portfolio allocation and more useful in practice as against exceedance correlation.

The empirical analysis has revealed that although, correlation of extreme returns significantly change, there exist stocks for which it does not. Thus, it is still possible to keep a diversified portfolio under bear or bull market even when extreme returns occur.

Obtained results can be practically interpreted as follows: under bull market investors make similar decisions for most of companies (higher correlations within the bulk of data) and occurrence of extreme returns results from independent decisions which are based on the specific situation of each company (lower correlations for outliers), e.g. new contract or sound financial results. Under bear market one can observe the opposite situation, investors make independent decision for each companies (lower correlations within the bulk of data), but occurrence of extreme return for one company (high positive or negative return) follow the same decision for other companies.

References

Ang, A., & Bekaert, G. (2002). International asset allocation with regime shifts. *Review of Financial Studies*, *11*, 1137–1187.

Ang, A., & Chen, J. (2002). Asymmetric correlations of equity portfolios. *Journal of Financial Economics*, *63*, 443–494.

Butler, K. C., & Joaquin, D. C. (2002). Are the gains from international portfolio diversification exaggerated? The influence of downside risk in bear markets. *Journal of International Money and Finance*, *21*, 981–1011.

Campbell, R. A. J., Forbes, C. S., Koedijk, K. G., & Kofman, P. (2008). Increasing correlations or just fat tails? *Journal of Empirical Finance*, *15*, 287–309.

Chua, D. B., Kritzman, M., & Page, S. (2009). The myth of diversification. *The Journal of Portfolio Management*, *36*, 26–35.

Garcia, R., & Tsafack, G. (2011). Dependence structure and extreme comovements in international equity and bond markets. *Journal of Banking and Finance*, *35*, 1954–1970.

Hong, Y., Tu, J., & Zhou, G. (2007). Asymmetries in stock returns: statistical tests and economic evaluation. *Review of Financial Studies*, *20*, 1547–1581.

Longin, F., & Solnik, B. (2001). Extreme correlation of international equity markets. *Journal of Finance*, *56*, 649–676.

Steiger, J. H. (1980). Tests for comparing elements of a correlation matrix. *Psychological Bulletin*, *87*, 245–251.

Tallis, G. M. (1963). Elliptical and radial truncation in normal populations. *The Annals of Mathematical Statistics*, *34*, 940–944.

Value-at-Risk Backtesting Procedures Based on Loss Functions: Simulation Analysis of the Power of Tests

Krzysztof Piontek

Abstract The definition of Value at Risk is quite general. There are many approaches that may lead to various VaR values. Backtesting is a necessary statistical procedure to test VaR models and select the best one. There are a lot of techniques for validating VaR models. Usually risk managers are not concerned about their statistical power. The goal of this paper is to compare statistical power of specific backtest procedures but also to examine the problem of limited data sets (observed in practice). A loss function approach is usually used to rank correct VaR models, but it is also possible to evaluate VaR models by using that approach. This paper presents the idea of loss functions and compares the statistical power of backtests based on a various loss functions with the Kupiec and Berkowitz approach. Simulated data representing asset returns are used here. This paper is a continuation of earlier pieces of research done by the author.

1 Introduction: VaR and Backtesting

Value at Risk (VaR) can integrate different types of financial risk and is one of the most popular risk measures used by financial institutions. It is such a loss in market value of a portfolio that the probability of occurring or exceeding this loss over a given time horizon is equal to a prior defined tolerance level q (Jorion 2007; Campbell 2005). VaR is often defined in a relative way, as a conditional or unconditional quantile of the forecasted return distribution (denoted as $VaR_{r,t}(q)$),

K. Piontek (✉)
Department of Financial Investments and Risk Management, Wroclaw University of Economics, ul. Komandorska 118/120, 53-345 Wroclaw, Poland
e-mail: krzysztof.piontek@ue.wroc.pl

M. Spiliopoulou et al. (eds.), *Data Analysis, Machine Learning and Knowledge Discovery*, Studies in Classification, Data Analysis, and Knowledge Organization, DOI 10.1007/978-3-319-01595-8_30,

where r_t is a rate of return and $F_{r,t}^{-1}$ is a quantile of loss distribution related to the probability of $1 - q$ at the instant t (Jorion 2007; Piontek 2010):

$$P\left(r_{t+1} \leq F_{r,t}^{-1}(q)\right) = q, \; VaR_{r,t}(q) = -F_{r,t}^{-1}.$$

This definition does not, however, inform how VaR should be actually estimated. The risk managers never know a priori, which approach or model will be the best or even correct. That is why they should use several models and backtest them. *Backtesting* is an ex-post comparison of a risk measure generated by a risk model against actual changes in portfolio value over a given period of time, both to evaluate a new model and to reassess the accuracy of existing models (Jorion 2007). A lot of backtest approaches have been proposed (Haas 2001; Campbell 2005; Piontek 2010):

1. based on a frequency of failures (Kupiec, Christoffersen),
2. based on a adherence of VaR model to the asset return distribution (on a multiple VaR levels; Crnkovic-Drachman, Berkowitz),
3. based on a various loss functions (Lopez, Sarma, Caporin).

Those approaches have different levels of usage difficulty and they require different sets of information. During the backtesting we should test the frequency of exceptions as well as their independence (Haas 2001; Campbell 2005; Piontek 2010).

In the further part of this paper, the author examines specific methods of testing only the number of exceptions. The null hypothesis for these tests states that the probability of exceptions (approximated by empirically determined frequency of exception) matches the VaR tolerance level. Sections 2 and 3 present methods used only to evaluate models. By using all the methods presented in Sect. 4 it is possible both to rank models and to test them. Unfortunately testing is not so easy as ranking.

The last part compares the power of analyzed backtests and gives conclusions.

2 Test Based on the Frequency of Failures

The simplest and the most popular tool for validation of VaR models (for a length of backtesting time period equal to T units) is the failure (or hit) process $[I_t(q)]_{t=1}^{t=T}$ with the hit function defined as follows (Jorion 2007):

$$I_t(q) = \begin{cases} 1; & r_t \leq -VaR_{r,t}(q) \quad \text{if a violation occurs} \\ 0; & r_t > -VaR_{r,t}(q) \quad \text{if no violation occurs.} \end{cases}$$

The Kupiec test examines how many times VaR is violated over a given span of time (Jorion 2007; Piontek 2010). This statistical test is based on the likelihood ratio:

$$LR_{uc}^{K} = -2\ln\left(\frac{(1-q)^{T_0} q^{T_1}}{(1-\hat{q})^{T_0}\hat{q}^{T_1}}\right) \sim \chi_1^2, \quad (1)$$

$$\hat{q} = \frac{T_1}{T_0 + T_1}, \quad T_1 = \sum_{t=1}^{T} I_t(q), \quad T_0 = T - T_1.$$

3 Tests Based on Multiple VaR Levels

The information contained in the hit sequence is limited. However, the lost information is potentially very useful in backtesting. By using additional information one may construct a test verifying VaR models for any tolerance level. The properties of correct VaR model should hold for any q. Backtests that use multiple VaR levels might be applied. They are based on deviation between the empirical return distribution and the theoretical model distribution, usually across their whole range of values (Haas 2001; Campbell 2005; Piontek 2010).

Usually, the portfolio returns r_t are transformed into a series u_t and then into z_t, where $F_r(\cdot)$ denotes the forecasted return distribution function (see (1)) and $\Phi^{-1}(\cdot)$—is the inverse normal distribution function (Piontek 2010).

$$u_t = F_r(r_t) = \int_{-\infty}^{r_t} f_r(y)dy, \quad z_t = \Phi^{-1}(u_t)$$

If the Value at Risk model is well calibrated it is expected that:

$$u_t \sim i.i.d.U(0,1) \qquad \text{or equivalently} \qquad z_t \sim i.i.d.N(0,1).$$

Under the null hypothesis that the VaR model is correct, the z_t should be independent and identically distributed standard normal random variable. The property that the series z_t is normally distributed is a direct parallel to the unconditional coverage—the proper number of exceptions. The most popular approach of this type is the one proposed by Berkowitz (Haas 2001; Piontek 2010):

$$z_t - \mu = \rho(z_{t-1} - \mu) + \varepsilon_t, \qquad \text{var}(\varepsilon_t) = \sigma^2,$$

$$H_0 : (\mu, \sigma, \rho) = (0, 1, 0).$$

A restricted likelihood can be evaluated and compared to an unrestricted one:

$$LR_{uc}^{B} = 2\left[LLF(\hat{\mu}, \hat{\sigma}, \hat{\rho}) - LLF(0, 1, \hat{\rho})\right] \sim \chi_2^2. \quad (2)$$

4 Tests Based on Loss Function

As it was mentioned before, risk managers are often interested not only in how individual models perform, but also how they compare to each other. By using the backtests presented in Sects. 2 and 3 of this paper, managers cannot rank the correct models (for example by using p-values). So they can only find that some models are good enough to use, but they cannot choose the best one. A group of methods utilizing the concept of the so called loss functions might give us a help here (Lopez 1998; Blanco and Ihle 1999; Campbell 2005). Those techniques could be used to rank models, but also to test them. Unfortunately testing is not so easy as in Kupiec or Berkowitz approaches.

To use the loss function approach one needs (Lopez 1998; Campbell 2005):

- a set of n paired *observations* describing the losses for each period (r_t) and their associated VaR forecasts ($VaR_{r,t}(q)$),
- a *loss function* that gives a score to each observation (the score depends on how the observed loss (or profit) contrasts with the forecasted VaR),
- a *benchmark* which gives us a score that we could expect from a "good" model,
- a *score function* which aggregates information for the whole period.

The idea of the loss function is to measure the accuracy of a VaR model on the basis of the distance between observed returns and VaR values:

$$L\left(VaR_{r,t}(q), r_t\right) = \begin{cases} f\left(VaR_{r,t}(q), r_t\right); & \text{if } r_t \leq -VaR_{r,t}(q) \\ g\left(VaR_{r,t}(q), r_t\right); & \text{if } r_t > -VaR_{r,t}(q) \end{cases} \tag{3}$$

An exception is given a higher score-value than nonexception (backtesting procedures from the supervisory point of view):

$$f\left(VaR_{r,t}(q), r_t\right) \geq g\left(VaR_{r,t}(q), r_t\right),$$

and usually:

$$g\left(VaR_{r,t}(q), r_t\right) = 0.$$

Finally, for each day the score function compares the values of a loss function with a benchmark and aggregates this information for the whole time horizon ($d = 1, 2$):

$$SF = \frac{1}{T} \sum_{t=1}^{T} \left| L\left(VaR_{r,t}(q), r_t\right) - benchmark \right|^d. \tag{4}$$

Benchmark is usually obtained with Monte Carlo simulation by using (3) under H_0, that the assumed model is correct (VaR model the same as data generating process).

The closer the value of a score function value to zero, the better the model. Therefore one can use the score function to rank models.

A very simple size-adjusted loss function was introduced by Lopez (1998). He proposed a testing method which incorporates the magnitude of exceptions in addition to their number:

$$L_t(VaR_{r,t}(q), r_t) = \begin{cases} 1 + (r_t + VaR_{r,t}(q))^2; & \text{if violation} \\ 0; & \text{if no-violation} \end{cases} \tag{5}$$

Many loss functions and their alternatives have been proposed in the literature afterwards (Blanco and Ihle 1999; Angelidis and Degiannakis 2007). A promising candidate is the loss function (6) using Expected Shortfall—a very intuitive and coherent risk measure (Jorion 2007). Usually $k = 1$ or 2.

$$f(VaR_{r,t}(q), r_t) = ||r_t| - ES_{r,t}(q)|^k, \tag{6}$$

where:

$$ES_{r,t}(q) = E[r_t \,|r_t < -VaR_{r,t}(q)]\,.$$

As mentioned before, the loss function approach might be used to rank models but also to evaluate them. The loss function procedure is not a formal statistical test but one can convert this approach into such a test. The null hypothesis (H_0) states that the assumed model is correct.

A test statistic and a critical value are needed to be compared. The way we obtain them depends on adopted testing framework (Campbell 2005).

To obtain the test statistic (SF_0):

1. fit a statistical model to the observed returns,
2. calculate VaR forecasts using the assumed model,
3. compare VaR to the observed returns using the loss function (3),
4. calculate the score function value (SF_0) using (4).

To obtain the test critical value (CV):

1. simulate a large number of return series using the assumed model,
2. calculate corresponding VaR forecasts using the model for each series,
3. calculate the score function (4) values for each series (SF distribution),
4. find the quantile of the empirical distribution of the score function values for each series.

The quantile may be used just as in a standard hypothesis testing framework—as the critical value (CV). Then we can carry out the test by comparing actual score function value to the critical one (if $SF_0 > CV$ we reject H_0 that the model is correct). This approach will be used in the empirical part of this paper.

Backtesting Errors

The tests presented before are, usually, used for evaluating internal VaR models developed by financial institutions. However, one should be aware of a fact that two types of errors may occur: a correct model can be rejected or a wrong one may be not rejected (Jorion 2007; Piontek 2010).

All tests are designed for controlling the probability of rejecting the VaR model when the model is correct. It means that the type I error is known. This type of wrong decisions leads to the necessity of searching for another model, which is just wasting time and money. But the type II error (acceptance of the incorrect model) is a severe misjudgment because it can result in the use of an inadequate VaR model that can lead to substantial negative consequences.

Performance of selected tests needs to be analyzed in regard to the type II error, so the best test can be selected. The analysis is provided for a different (but small) numbers of observations and model misspecifications.

5 Empirical Research: Simulation Approach

Generated returns are independent and follow standardized Student distribution with the number of degrees of freedom between 3 and 25. Also VaR model is based on the standardized Student distribution, but with 6 degrees of freedom always. Thus, it may be incorrect. On this ground we can test the statistical power of backtests. The t-distribution was used as a simplest (but quite realistic) approximation of returns distributions. Results depend on the shape of the distribution. The further research should be continued based on other fatter tailed and skewed distributions (i.e. Pearson type IV or skewed-t distributions). The data series with 100, 250, 500, 750 and one thousand observations were simulated. For each type of inaccuracy of the model and for each length of the data series Monte Carlo simulations with 10.000 draws were done. It allowed for calculating test statistics and for estimation of the frequency at which the null hypotheses were rejected for incorrect models. The last may be treated as an approximation of the test power.

Tables 1 and 2 present the summary of results obtained for different approaches and different tolerance levels ($q = 0.05$ or $q = 0.01$). The central column (for 0.05) represents the type I error (significance level of the test is equal to typical value 0.05), other columns—the power of the test for a given strength of inaccuracy (incorrect frequency of failures).

Interpretation of the results is straightforward. For example, in the case of 250 observations, for a tolerance level of 0.05 and for the Kupiec test, an inaccurate model giving 3 or 7 % of violations (instead of 5 %) was rejected only in about 39 and 30 % of draws. Put differently, in about 61 and 70 % of cases we did not reject the wrong model at 5 % significance level.

Table 1 Power of backtests for VaR tolerance level $q = 0.05, \alpha = 0.05$

Number of obs.	Actual frequency of violations								
	0.030	0.035	0.040	0.045	0.050	0.055	0.060	0.065	0.070
	Power of the Kupiec test (1)								
100	0.191	0.131	0.091	0.072	0.075	0.073	0.084	0.130	0.173
250	0.390	0.235	0.131	0.075	0.061	0.071	0.123	0.211	0.305
500	0.670	0.406	0.211	0.108	0.054	0.074	0.150	0.295	0.443
750	0.802	0.546	0.259	0.105	0.059	0.108	0.257	0.448	0.655
1,000	0.914	0.682	0.349	0.118	0.047	0.100	0.263	0.520	0.737
	Power of the Berkowitz test (2)								
100	0.224	0.149	0.090	0.060	0.047	0.059	0.137	0.264	0.446
250	0.587	0.342	0.173	0.073	0.046	0.091	0.241	0.524	0.771
500	0.919	0.656	0.317	0.109	0.048	0.131	0.416	0.781	0.958
750	0.986	0.858	0.475	0.151	0.048	0.184	0.578	0.915	0.992
1,000	0.999	0.946	0.619	0.179	0.049	0.215	0.701	0.967	0.997
	Power of the Lopez loss function test (5)								
100	0.074	0.068	0.065	0.058	0.051	0.045	0.041	0.039	0.038
250	0.147	0.103	0.076	0.059	0.050	0.046	0.046	0.047	0.051
500	0.204	0.132	0.087	0.063	0.051	0.047	0.052	0.060	0.071
750	0.257	0.159	0.098	0.065	0.050	0.049	0.058	0.073	0.092
1,000	0.312	0.190	0.111	0.067	0.051	0.051	0.064	0.084	0.113
	Power of the expected shortfall loss function test (6)								
100	0.151	0.100	0.069	0.0530	0.051	0.060	0.083	0.117	0.161
250	0.344	0.200	0.110	0.061	0.051	0.070	0.123	0.202	0.310
500	0.629	0.375	0.182	0.080	0.051	0.086	0.187	0.342	0.525
750	0.810	0.530	0.257	0.097	0.050	0.103	0.254	0.474	0.689
1,000	0.913	0.665	0.336	0.115	0.050	0.119	0.316	0.580	0.803

For VaR tolerance level equal to 0.05 the superiority of Berkowitz test over the other approaches is observable for all lengths of return series. For VaR tolerance level of 0.01 the superiority is observable only for 750 and one thousand observations series length. For the series with a shorter length the conclusions are ambiguous.

The statistical power of the tests depends strictly on the type of loss function. Even for the Expected Shortfall (ES) Loss Function (with $d = 2$; higher power than for the Lopez approach) the statistical power of test is lower than for the Berkowitz approach. The superiority of loss function approach over the Berkowitz approach has not been found for VaR models with tolerance level equal to both 5 and 1 %. The Kupiec and the ES Loss Function approaches are comparable with respect to power, however, the Kupiec test is easier to use.

Generally speaking, statistical power of the tests are rather low. Only for the longer series and stronger inaccuracies the power could be acceptable for risk managers. For the tolerance level of 1 % the results indicate that the Kupiec test is not adequate for small samples—even for one thousand observations. This test

Table 2 Power of backtests for VaR tolerance level $q = 0.01$, $\alpha = 0.05$

Number of obs.	Actual frequency of violations								
	0.006	0.007	0.008	0.009	0.010	0.011	0.012	0.013	0.014
	Power of the Kupiec test (1)								
100	0.003	0.006	0.008	0.013	0.018	0.024	0.031	0.046	0.049
250	0.167	0.099	0.066	0.048	0.094	0.081	0.084	0.090	0.095
500	0.197	0.136	0.099	0.076	0.067	0.073	0.101	0.133	0.177
750	0.222	0.177	0.133	0.116	0.041	0.051	0.079	0.124	0.172
1,000	0.285	0.178	0.106	0.063	0.053	0.067	0.112	0.174	0.249
	Power of the Berkowitz test (2)								
100	0.078	0.069	0.057	0.050	0.049	0.055	0.065	0.071	0.083
250	0.125	0.087	0.063	0.051	0.052	0.054	0.071	0.094	0.129
500	0.228	0.138	0.080	0.061	0.049	0.058	0.085	0.139	0.197
750	0.336	0.195	0.098	0.063	0.047	0.068	0.105	0.174	0.278
1,000	0.447	0.246	0.126	0.065	0.050	0.068	0.130	0.224	0.355
	Power of the Lopez loss function test (5)								
100	0.018	0.020	0.022	0.024	0.025	0.026	0.028	0.029	0.030
250	0.014	0.017	0.019	0.022	0.025	0.029	0.031	0.035	0.039
500	0.059	0.056	0.058	0.054	0.050	0.048	0.048	0.051	0.056
750	0.143	0.099	0.068	0.055	0.049	0.048	0.052	0.058	0.066
1,000	0.146	0.105	0.078	0.059	0.051	0.050	0.055	0.062	0.075
	Power of the expected shortfall loss function test (6)								
100	0.005	0.008	0.013	0.018	0.025	0.033	0.043	0.053	0.066
250	0.002	0.005	0.009	0.016	0.026	0.039	0.055	0.076	0.100
500	0.137	0.093	0.066	0.052	0.050	0.060	0.081	0.112	0.153
750	0.197	0.123	0.077	0.055	0.049	0.061	0.091	0.136	0.197
1,000	0.263	0.156	0.092	0.058	0.051	0.066	0.105	0.164	0.242

should not be used for VaR models with the tolerance level of 1 % for typical lengths of the observed series. The Berkowitz test has higher power than the Kupiec test, but the power of this test is probably not sufficient for risk managers using typical lengths of data series. For VaR tolerance level of 1 % the power of the ES Loss Function tests is unacceptable and lower than for the Berkowitz approach.

6 Final Conclusions

One should not rely on one backtest only. No one procedure is perfect. The main differences between the used test procedures arise due to the set of information. The Kupiec approach uses less information than the loss function based one and the loss function based approach less than the Berkowitz one. Tests based on the hit function throw away valuable information about tail losses. It is better to use the tests that are based on the information on both magnitudes and frequencies of tail losses.

The type of loss function and score function are chosen arbitrarily and statistical power of tests depend on them. One huge violation may cause rejection of any model. We have to clean data, which makes automation of the test procedure not so easy. We should try to test models for a different VaR tolerance levels, positions and data sets. It is worth to consider a two stage procedure: (1) find the set of acceptable models by using at least the Berkowitz approach, (2) find the best model by ranking the set with the loss function method.

It is necessary to determine how low may be power of backtests in some typical cases (Piontek 2010) and to discuss the acceptable minimum.

References

Angelidis, T., & Degiannakis, S. (2007). Backtesting VaR models: A two-stage procedure. *Journal of Risk Model Validation, 1*(2), 27–48.

Blanco, C., & Ihle, G. (1999). How good is your VaR? Using backtesting to assess system performance. *Financial Engineering News*, August, 1–2.

Campbell, S. (2005). *A Review of Backtesting and Backtesting Procedures*. Federal Reserve Board, Washington.

Haas, M. (2001). New methods in backtesting. Working paper. Financial Engineering Research Center Caesar, Friedensplatz, Bonn.

Jorion, P. (2007). *Value at risk* (3rd ed.). New York: McGraw-Hill.

Lopez, J. (1998). *Methods for Evaluating Value-at-Risk Estimates*. Federal Reserve Bank of New York Research Paper no. 9802.

Piontek, K. (2010). The analysis of power for some chosen VaR backtesting procedures: Simulation approach. In *Studies in classification, data analysis, and knowledge organization* (pp. 481–490). New York: Springer.

Part V
AREA Data Analysis in Biostatistics and Bioinformatics

Rank Aggregation for Candidate Gene Identification

Andre Burkovski, Ludwig Lausser, Johann M. Kraus, and Hans A. Kestler

Abstract Differences of molecular processes are reflected, among others, by differences in gene expression levels of the involved cells. High-throughput methods such as microarrays and deep sequencing approaches are increasingly used to obtain these expression profiles. Often differences of gene expression across different conditions such as tumor vs inflammation are investigated. Top scoring differential genes are considered as candidates for further analysis. Measured differences may not be related to a biological process as they can also be caused by variation in measurement or by other sources of noise. A method for reducing the influence of noise is to combine the available samples. Here, we analyze different types of combination methods, early and late aggregation and compare these statistical and positional rank aggregation methods in a simulation study and by experiments on real microarray data.

1 Introduction

Molecular high-throughput technologies generate large amounts of data which are usually noisy. Often measurements are taken under slightly different conditions and produce values that in extreme cases may be contradictory and contain outliers.

A. Burkovski
Research Group Bioinformatics and Systems Biology, Institute of Neural Information Processing, Ulm University, 89069 Ulm, Germany

International Graduate School in Molecular Medicine, Ulm University, Ulm, Germany
e-mail: andre.burkovski@uni-ulm.de

L. Lausser · J.M. Kraus · H.A. Kestler (✉)
Research Group Bioinformatics and Systems Biology, Institute of Neural Information Processing, Ulm University, 89069 Ulm, Germany
e-mail: ludwig.lausser@uni-ulm.de; johann.kraus@uni-ulm.de; hans.kestler@uni-ulm.de

M. Spiliopoulou et al. (eds.), *Data Analysis, Machine Learning and Knowledge Discovery*, Studies in Classification, Data Analysis, and Knowledge Organization,
DOI 10.1007/978-3-319-01595-8_31,

One way of establishing more stable relationships between genes is by transforming the data into ordinal scale by ranking their expression values profile-wise. High expression levels are thereby sorted at the top of the ranking. Common patterns can be revealed by combining these rankings via aggregation methods. These methods construct consensus rankings for which all input rankings have least disagreements in some sense. Here, we study the difference between two general combination procedures, namely: (a) early and (b) late aggregation. In early aggregation, gene values are aggregated by methods like mean or median and are ranked based on the aggregated value. In contrast, late aggregation is the process of building a consensus ranking after the data was transformed into ordinal scale individually. To what extent early and late aggregation approaches differ was not reported so far. In this simulation study we observe, that the quality and the results depend strongly on the underlying noise model of the data. If we assume that each sample is affected by slightly different technical noise, e.g. because of differences in the laboratory conditions, late aggregation more accurately reflect the structure of a ground truth ranking.

2 Background

An overview over existing aggregation methods can be found in Schalekamp and Zuylen (2009). Dwork et al. (2001) propose methods based on bipartite graph matching for rank aggregation. Their analysis is focused on the scenario of partial rankings (i.e. rankings including missing values). They introduce different Markov Chain models for this scenario. Based on this work Fagin et al. (2003) showed that there are additional distance functions which can be used to solve rank aggregation in polynomial time using bipartite matching even in the case of partial lists.

In context of candidate gene selection rank aggregation gained increasing interest over the years. DeConde et al. (2006) present an approach using rank aggregation to identify top genes for prostate cancer across five different microarray experiments. They take top-25 lists from each experiment and find a consensus ranking which shows genes that are considered most influential in all studies. Lin (2010) reports additional results with Borda aggregation and Cross-Entropy Monte Carlo methods for the same data. Lin notes that the Borda aggregation method is very competitive regarding Markov Chains aggregation. The main challenge is to handle longer lists, which it is not always possible with Cross-Entropy and other optimization methods in a computationally efficient manner. In this context Pihur et al. (2007, 2008) present Cross-Entropy rank aggregation to identify marker genes in top-k rankings.

3 Methods

Let $X = \{\mathbf{x}_1 \dots \mathbf{x}_k\}$ with $\mathbf{x}_i \in \mathbb{R}^n$ denote a set of samples (profile vectors) of a certain class of objects. A rank aggregation method can be seen as a function $X \mapsto \theta$, where θ is a so-called consensus ranking. A ranking is thereby a permutation of the numbers $1, \ldots, n$. The consensus ranking of a sample should reflect the ordering of the feature levels in a class. In principal two types of aggregation methods exists.

Early aggregation methods initially aggregate the original profiles X into a summary statistic $\tilde{x} \in \mathbb{R}^n$, e.g. the *mean* or the *median*. The consensus ranking corresponds to the ranking of $\tilde{x}$.

$$X \mapsto \tilde{x} \mapsto \theta \tag{1}$$

Late aggregation methods first transform the original samples into ordinal scale $X \mapsto \Pi = \{\pi_1 \ldots \pi_k\}$, where π_i is a ranking of the values of $\mathbf{x}_i$. The aggregation is done on this intermediate representation.

$$X \mapsto \Pi \mapsto \theta \tag{2}$$

Late rank aggregation can be seen as an optimization problem which aims to find a ranking θ that has an overall minimum distance to k given rankings $\pi_1, \ldots, \pi_k$ with regard to some distance function. Finding a optimal aggregate can be computationally expensive. In the case of the Kendall-τ distance, which counts the number of disagreements between rankings, the optimization is NP-hard (Dwork et al. 2001). It can be calculated efficiently, if other distance measures are used (Fagin et al. 2003). For example, an optimal aggregate for Borda count, Spearman footrule (Diaconis and Graham 1977), and Copeland score (Copeland 1951) can be determined in polynomial time. The rest of this section introduces the late aggregation methods used in this study.

3.1 Borda Score

The Borda method (*borda*) calculates a score for each element (gene) of a profile (e.g., Dwork et al. 2001). This score is based on a summary statistic (e.g., mean) of the ranks the gene achieved in the rankings of Π. The consensus ranking is constructed by ranking the scores of all genes. In this work we use the mean rank to calculate the Borda score $\frac{1}{|\Pi|} \sum_{\pi \in \Pi} \pi$. The resulting aggregate is ranked increasing by this score with ties broken randomly.

3.2 Copeland's Score

In Copeland aggregation (*copeland*), the score of an element depends on its relative position in relationship to the other elements (Copeland 1951). A high score is assigned to those elements which are better ranked in the majority of the rankings in Π than the other elements. For an element i the Copeland score is given by the sum of pairwise comparisons with the remaining elements $j \neq i$.

$$C_i = \sum_{j \neq i} \left(\mathbb{I}_{[v_{i,j} > l_{i,j}]} + \frac{1}{2} \mathbb{I}_{[v_{i,j} = l_{i,j}]} \right) \text{ with} \tag{3}$$

$$v_{i,j} = \sum_{\pi \in \Pi} \mathbb{I}_{[\pi(i) > \pi(j)]} \text{ and } l_{i,j} = \sum_{\pi \in \Pi} \mathbb{I}_{[\pi(i) < \pi(j)]}. \tag{4}$$

Here, $v_{i,j}$ denotes the number of ranks for element i that are higher than for element j and $l_{i,j}$ the number of ranks for element i that are lower than for element j. The elements are ranked increasing by their score values.

3.3 Kolde's Robust Rank Aggregation

Kolde et al. (2012) recently presented a rank aggregation method (*robust*) which uses order statistics to compute scores for individual items. Their main idea is to compare input rankings to an uncorrelated ranking and assigning a rank to the items based on their significance score. The null model is that all rank lists are non-informative and the ranks are randomly distributed in the items. The authors improve the method by comparing each item to a random position, calculating the p-value, and deciding on how much better the item has improved its position. The elements in the aggregate are then ordered by the resulting score.

3.4 Pick-a-Perm

Ailon et al. (2008) showed that a consensus ranking can be found by selecting one from the available input rankings $\theta \in \Pi$. The consensus ranking is chosen to have a minimum distance to all other input rankings $\pi_1, \ldots, \pi_k$ according to the Kendall-τ distance. This method will be called *pick* in the following.

3.5 Spearman's Footrule and Canberra Distance

Spearman's footrule (*spearman*) is defined as the sum of rank differences of all elements in two different rankings:

$$\mathcal{F}(\pi, \pi') = \sum_{i=1}^{n} |\pi(i) - \pi'(i)| \tag{5}$$

The Canberra distance (*canberra*) can be seen as modification of Spearman's footrule. It additionally takes into account the relevance of the ranks based on their position:

$$CD(\pi, \pi') = \sum_{i=1}^{n} \frac{|\pi(i) - \pi'(i)|}{\pi(i) + \pi'(i)} \tag{6}$$

The interpretation of this distance is that elements having higher ranks are given less weight in the aggregation due to the denominator $\pi(i) + \pi'(i)$.

4 Experimental Setup

We compared early and late aggregation methods on both artificial and real microarray datasets. For the artificial ones we utilized the microarray generator proposed by Smyth (2004), which is designed to generate two class datasets with differentially expressed genes (we selected the case class for our experiments). This generator assumes the single genes to be normally distributed (mean = 0). Variances are chosen according to an inverse χ^2 distribution with d_0 degrees of freedom for each gene. For differentially expressed genes the mean value of the case class is chosen according to a normal distribution with mean 0 and variance $\sigma \cdot v_0$.

Based on this setup we generated a nested simulation first randomly drawing means and variances for the subsequent normal models of data generation. Our experiments are based on the simulation parameters used by Smyth $d_0 = 4$ and $v_0 = 2$. We defined about 90 % of all genes to be differentially expressed (i.e. mean $\neq 0$). Technical noise for each sample was modeled by a noise vector drawn from $\nu \sim \mathcal{U}^n(1, c)$ and was multiplied feature-wise with the variance vector of the sample. For $c = 1$ our model corresponds to the original model proposed by Smyth. We constructed 100 datasets of 250 genes each. The mean vectors of the corresponding models are used as ground truth. Aim was to estimate the ground truth by aggregating the rankings of 10 samples each drawn according to the previously generated model. The performance is measured by the mean Kendall-τ disagreements.

The aggregation methods are also compared on three real microarray datasets (Table 1). In these experiments the initial ground truth was estimated by applying *mean* on all samples of the selected class. A rank aggregation method was then tested by estimating this ranking on a subset of 10 randomly chosen samples. This experiment was repeated 100 times.

Table 1 Summary of the real microarray datasets

Dataset	Cancer	Class	# sam.	# fea.	Citation
Alon	Colon	Adenocar.	40	2,000	Alon et al. (1999)
West	Breast	ER+	25	7,129	West et al. (2001)
Shipp	Lymphoma	FL	19	7,129	Shipp et al. (2002)

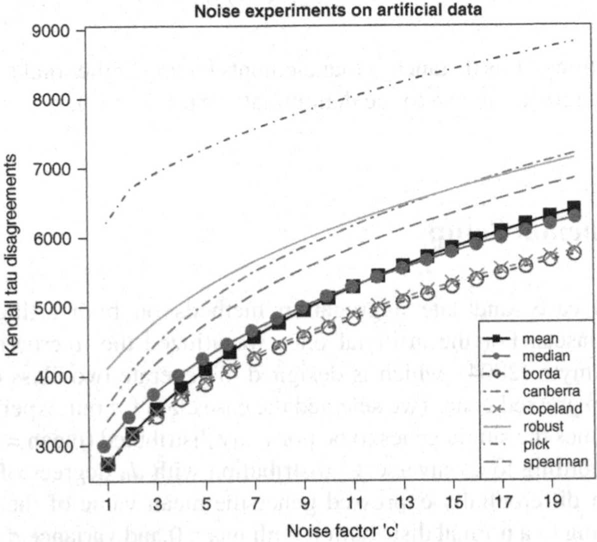

Fig. 1 Summary of the artificial experiments (Sect. 4). The aggregation methods were compared on reconstructing the feature rankings of an artificial model based on a set of 10 samples. This experiment was repeated 100 times. The mean number of Kendall-τ disagreements is shown

5 Results

Figure 1 shows the results for the experiments on the artificial datasets. The figure shows the error curves which measure the Kendall-τ disagreements between the aggregate and the ground truth rank.

For the parameter $c = 1$ the early aggregation method *mean* and late aggregation methods *borda* and *copeland* show almost no difference in their performance. For increasing c, *mean* predicts rankings that disagree more and more with the ground truth ranking of the population. The methods *borda* and *copeland* act more stable than *mean* for larger values of c. For large values of c, *median* predicts the ground truth rank better than *mean* but worse than *borda* and *copeland*. The methods *spearman* and *canberra* predict rankings that are less close to ground truth rank than the previous methods. For large values of c *robust* produces slightly better results than *canberra*. The lowest performance was achieved by *pick*.

The results on the real datasets can be found in Fig. 2. The black color of a cell denotes that the aggregation method indexing the row achieved a significantly lower

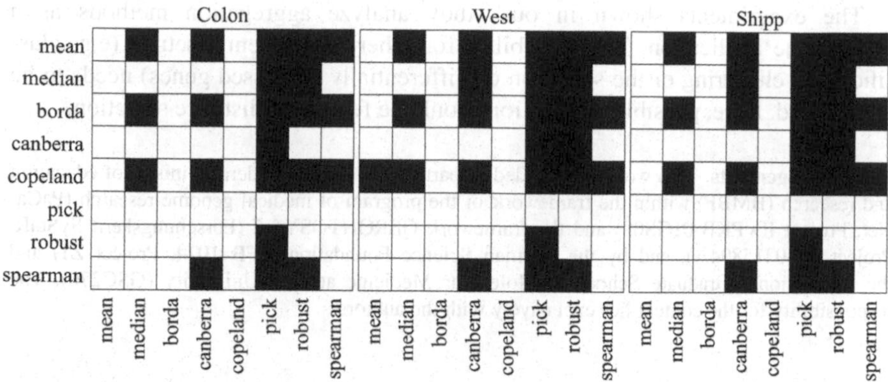

Fig. 2 Results of the microarray experiments: A *black cell* denotes that the aggregation method given in the row significantly outperforms the method in the column according to one-sided Wilcoxon rank tests ($\alpha = 0.05$, Bonferroni, $n = 56$)

mean distance to the ground truth ranking than the method indexing the column. The significance was tested by one-sided Wilcoxon rank tests ($\alpha = 0.05$, Bonferroni, $n = 56$). Two of the methods, the *borda* and *copeland*, are not outperformed by any other method. The early aggregation method *mean* was only beaten by the *copeland* method on the West dataset. The methods *median* and *spearman* showed an identical behavior and are beaten three times. All other methods outperformed *pick*.

6 Discussion and Conclusion

We inspected early and late aggregation approaches in settings where the goal is to find a consensus ranking about the order of elements or genes in the presence of technical noise. In early aggregation, elements are first aggregated and ranked afterwards. In late aggregation, elements are first transformed into ordinal scale and are then aggregated into a consensus ranking using different methods. In most microarray experiments noise is present in the data and the underlying noise model plays a major role for rank aggregation.

On simulated data late aggregation with Copeland or Borda outperforms early aggregation based on mean or median. For the real datasets the Borda and Copeland methods are on par with the mean. Still, these methods constantly outperform the aggregation with median. This suggests that Borda and Copeland methods are more robust predictors than the median.

One can speculate that the reason for this lower performance is due to the small sample sizes as the mean performs well on large datasets (data not shown). The ordinal data used by late aggregation is less prone to differences in noise allowing late aggregation to outperform early aggregation methods.

The experiments shown in our study analyze aggregation methods as a stand-alone application. Their usability for other experimental setups (e.g. classification, clustering or the selection of differentially expressed genes) needs to be determined. Here, possible applications could be feature or distance selection.

Acknowledgements This work was funded in part by the German federal ministry of education and research (BMBF) within the framework of the program of medical genome research (PaCa-Net; Project ID PKB-01GS08) and the framework GERONTOSYS 2 (Forschungskern SyStaR, Project ID 0315894A), and by the German Science Foundation (SFB 1074, Project Z1) and the International Graduate School in Molecular Medicine at Ulm University (GSC270). The responsibility for the content lies exclusively with the authors.

References

Ailon, N., Charikar, M., & Newman, A. (2008). Aggregating inconsistent information: Ranking and clustering. *Journal of the ACM, 55*, 23:1–23:27.

Alon, U., Barkai, N., Notterman, D. A., Gish, K., Ybarra, S., Mack, D., et al. (1999). Broad patterns of gene expression revealed by clustering analysis of tumor and normal colon tissues probed by oligonucleotide arrays. *Proceedings of the National Academy of Sciences of USA, 96*(12), 6745–6750.

Copeland, A. (1951). A 'reasonable' social welfare function. *Seminar on Mathematics in Social Sciences*, University of Michigan.

DeConde, R. P., Hawley, S., Falcon, S., Clegg, N., Knudsen, B., & Etzioni, R. (2006). Combining results of microarray experiments: A rank aggregation approach. *Statistical Applications in Genetics and Molecular Biology, 5*, 1–23.

Diaconis, P., & Graham, R. L. (1977). Spearman's footrule as a measure of disarray. *Journal of the Royal Statistical Society. Series B (Methodological), 39*(2), 262–268.

Dwork, C., Kumar, R., Naor, M., & Sivakumar, D. (2001). Rank aggregation revisited. *Systems Research, 13*(2), 86–93.

Fagin, R., Kumar, R., & Sivakumar, D. (2003). Comparing top k lists. In *Proceedings of the Fourteenth Annual ACM-SIAM Symposium on Discrete Algorithms* (pp. 28–36). Philadelphia: SIAM.

Kolde, R., Laur, S., Adler, P., & Vilo, J. (2012). Robust rank aggregation for gene list integration and meta-analysis. *Bioinformatics, 28*(4), 573–580.

Lin, S. (2010). Rank aggregation methods. *Wiley Interdisciplinary Reviews: Computational Statistics, 2*(5), 555–570.

Pihur, V., Datta, S., & Datta, S. (2007). Weighted rank aggregation of cluster validation measures: A monte carlo cross-entropy approach. *Bioinformatics 23*(13), 1607–1615.

Pihur, V., Datta, S., & Datta, S. (2008). Finding common genes in multiple cancer types through meta-analysis of microarray experiments: A rank aggregation approach. *Genomics, 92*(6), 400–403.

Schalekamp, F., & Zuylen, A. (2009). Rank aggregation: Together we're strong. In *Proceedings of the 11th Workshop on Algorithm Engineering and Experiments* (pp. 38–51). Philadelphia: SIAM.

Shipp, M. A., Ross, K. N., Tamayo, P., Weng, A. P., Kutok, J. L., Aguiar, R. C., et al. (2002). Diffuse large b-cell lymphoma outcome prediction by gene-expression profiling and supervised machine learning. *Nature Medicine, 8*(1), 68–74.

Smyth, G. K. (2004). Linear models and empirical Bayes methods for assessing differential expression in microarray experiments. *Statistical Applications in Genetics and Molecular Biology, 3*(1), 3.

West, M., Blanchette, C., Dressman, H., Huang, E., Ishida, S., Spang, R., et al. (2001) Predicting the clinical status of human breast cancer by using gene expression profiles. *Proceedings of the National Academy of Sciences of USA, 98*(20), 11462–11467.

Unsupervised Dimension Reduction Methods for Protein Sequence Classification

Dominik Heider, Christoph Bartenhagen, J. Nikolaj Dybowski, Sascha Hauke, Martin Pyka, and Daniel Hoffmann

Abstract Feature extraction methods are widely applied in order to reduce the dimensionality of data for subsequent classification, thus decreasing the risk of noise fitting. Principal Component Analysis (PCA) is a popular linear method for transforming high-dimensional data into a low-dimensional representation. Non-linear and non-parametric methods for dimension reduction, such as Isomap, Stochastic Neighbor Embedding (SNE) and Interpol are also used. In this study, we compare the performance of PCA, Isomap, t-SNE and Interpol as preprocessing steps for classification of protein sequences. Using random forests, we compared the classification performance on two artificial and eighteen real-world protein data sets, including HIV drug resistance, HIV-1 co-receptor usage and protein functional class prediction, preprocessed with PCA, Isomap, t-SNE and Interpol. Significant differences between these feature extraction methods were observed. The prediction performance of Interpol converges towards a stable and significantly

D. Heider (✉) · J.N. Dybowski · D. Hoffmann
Department of Bioinformatics, University of Duisburg-Essen, Universitätsstr. 2, 45117 Essen, Germany
e-mail: dominik.heider@uni-due.de; nikolaj.dybowski@uni-due.de; daniel.hoffmann@uni-due.de

C. Bartenhagen
Department of Medical Informatics, University of Münster, Domagkstr. 9, 48149 Münster, Germany
e-mail: christoph.bartenhagen@ukmuenster.de

S. Hauke
CASED, Technische Universität Darmstadt, Mornewegstr. 32, 64293 Darmstadt, Germany
e-mail: sascha.hauke@cased.de

M. Pyka
Department of Psychiatry and Psychotherapy, Philipps-University Marburg, Rudolf-Bultmann-Str. 8, 35039 Marburg, Germany
e-mail: martin.pyka@med.uni-marburg.de

M. Spiliopoulou et al. (eds.), *Data Analysis, Machine Learning and Knowledge Discovery*, Studies in Classification, Data Analysis, and Knowledge Organization, DOI 10.1007/978-3-319-01595-8_32,

higher value compared to PCA, Isomap and t-SNE. This is probably due to the nature of protein sequences, where amino acid are often dependent from and affect each other to achieve, for instance, conformational stability. However, visualization of data reduced with Interpol is rather unintuitive, compared to the other methods. We conclude that Interpol is superior to PCA, Isomap and t-SNE for feature extraction previous to classification, but is of limited use for visualization.

1 Introduction

Machine learning techniques, such as artificial neural networks, support vector machines and random forests are widely applied in biomedical research. They have been used, for instance, to predict drug resistance in HIV-1 (Dybowski et al. 2010) or protein functional classes (Cai et al. 2003). Protein sequences can be encoded with so-called descriptors, mapping each amino acid to real numbers. It has been shown in a recent study that these descriptor encodings lead to a better classification performance compared to the widely used sparse encoding (Nanni and Lumini 2011). Recently, we have developed a method for numerical encoding and linear normalization of protein sequences to uniform length (Heider and Hoffmann 2011) as preprocessing step for machine learning. This normalization procedure has been successfully applied to several classification problems (Dybowski et al. 2010; Heider et al. 2010; Dybowski et al. 2011), and it has been shown that it leads to higher prediction performance compared to other preprocessing schemes involving multiple sequence alignments or pairwise sequence alignments with a reference sequence. However, the longer the protein sequences are, the higher the dimension of the input space for the classifier becomes. This can lead to problems regarding generalization and prediction performance of the resulting classifiers. This is important for protein functional class prediction, as the number of known proteins belonging to a certain group is often limited, and the lengths of the proteins can easily exceed hundreds of amino acids.

Thus, in biomedical research one frequently encounters the "small-n-large-p" problem, where p denotes the number of observed parameters (here number of amino acids) and n the number of samples. As mentioned before, the ratio n/p is an important indicator to avoid overfitting of the machine learning techniques. There are two different approaches to overcome small-n-large-p problems: (1) feature selection methods introduce a ranking of the parameters and pick a subset of features based on the ranking for subsequent analysis, and (2) feature extraction methods transform the high dimensional input data into a lower dimensional target space.

In the current study, we focus on feature extraction methods and demonstrate the use of Interpol (Heider and Hoffmann 2011) to reduce the dimensionality of the input space for subsequent protein classification, and compare the prediction performance of Interpol with PCA, Isomap and t-Distributed Stochastic Neighbor Embedding (t-SNE).

Table 1 The table summarizes the number of sequences within each dataset and the class ratio of positive to negative samples

Dataset	# sequences	Class ratio	Dataset	# sequences	Class ratio
APV	768	1.60:1	d4T	630	1.17:1
ATV	329	0.92:1	ddI	632	0.96:1
IDV	827	1.04:1	TDF	353	2.03:1
LPV	517	0.82:1	DLV	732	1.78:1
NFV	844	0.67:1	EFV	734	1.63:1
RTV	795	0.96:1	NVP	746	1.33:1
SQV	826	1.50:1	V3	1,351	0.17:1
3TC	633	0.45:1	GTP	1,435	0.85:1
ABC	628	0.41:1	AZT	630	1.08:1

2 Methods

2.1 Data

We used 18 real-world (see Table 1) and two synthetic data sets in order to evaluate our procedure. Seventeen data sets are related to HIV-1 drug resistance (Rhee et al. 2006; Heider et al. 2010), one data set is related to HIV-1 co-receptor usage (Dybowski et al. 2010), and one dataset is related to protein functional class prediction (Heider et al. 2009). The cut-offs of the IC_{50} values between susceptible and resistant sequences are in accordance with Rhee et al. and Heider et al. for the HIV-1 drug resistance data sets. The concentration of a drug inhibiting 50 % of viral replication compared to cell culture experiments without a drug is defined as IC_{50} (50 % inhibitor concentration). As we are focussing on binary classification, intermediate resistant and high-level resistant are both defined as resistant. Furthermore, we analyzed the classification of small GTPases (GTP) according to Heider et al.

We also analyzed two synthetic data sets, S1 and S2, consisting of 54 42-mer amino acid sequences and 41-mer sequences, respectively. Each sequence is composed as follows:

$$G^{20}XXG^{20}$$

with $G^{20} = 20$ glycines, $XX = RI$ (arginine and isoleucine) for the positive samples, and $XX = IR$ (isoleucine and arginine) for the negative samples. In S2, the first G is deleted. Furthermore, random point mutations are introduced, where Gs are mutated into one of the other 20 amino acids commonly found in proteins. Arginine and isoleucine were chosen due to their extreme opposed hydropathy values.

2.2 Descriptors

We used the hydropathy scale of Kyte and Doolittle (1982) to encode the protein sequences, as it has been shown in several studies that it leads, in general, to accurate predictions. Moreover, we tested other descriptors, e.g. net charge. However, these alternative descriptors led to worse prediction performance.

2.3 Principal Component Analysis

Principal Component Analysis (PCA) is a statistical approach for decomposing multidimensional data into a set of uncorrelated variables, the so-called principal components (Jolliffe 2002). It is widely-used to reduce the number of variables in highly correlated data and to detect underlying signal sources in noisy data. If a given set of variables is a linear combination of a smaller number of uncorrelated signals, PCA is able to compute these signals, and to provide the linear combination underlying the input data. We used PCA implemented in `R` (http://www.r-project.org/).

2.4 Isomap

Isomap (Tenenbaum et al. 2000) is closely related to Multidimensional Scaling (Cox et al. 2003) and performs dimension reduction on a global scale. It finds a low-dimensional embedding such that the geodesic distances between all samples in a dataset are preserved, i.e. Isomap ensures that the shortest path between each pair of samples (nodes) in a neighborhood graph G is maintained. Therefore, Isomap is able to capture paths along a non-linear manifold allowing application to more complex datasets whose samples are expected to lie on or near non-linear sub-manifolds.

We used Isomap implemented in the `R`-package `RDRToolbox`. The number of neighbors for computing the neighborhood graph was set to 10.

2.5 t-SNE

Although being implemented for visualizing of high-dimensional data, t-SNE (t-Distributed Stochastic Neighbor Embedding) can be easily used for reducing noise (van der Maaten and Hinton 2008). t-SNE is a modified version of Stochastic Neighbor Embedding (SNE). SNE converts the high-dimensional Euclidean distances between samples into conditional probabilities representing similarities.

t-SNE differs from SNE due to its cost function. It used a symmetric version of the SNE cost function and moreover, uses a Student-t distribution rather than a Gaussian to compute the similarities between the samples. We used t-SNE implemented in the `R`-package `tsne`.

2.6 Interpol

The Interpol normalization (Heider and Hoffmann 2011) was also applied in the current study. It transforms the discrete descriptor values into the continuous space. Interpol connects all known data points (here: descriptor values along a given protein sequence), resulting in a continuous curve that represents these data points. The normalization factor k is defined as the number of equally spaced positions at which the curve is sampled. These samples are then used as an input for classification. Although originally developed for normalizing protein sequences of variable length, Interpol can also be used to deliberately *reduce* input dimensionality for classification. This can be done by choosing a normalization factor $k < n$, with n being the length of a given feature vector.

In our study, we used Interpol implemented in `R` with the averaging method. Thus, Interpol divides an encoded sequence of length n in k equally sized intervals $I_1, \ldots, I_k$. All descriptor values lying in interval I_j $(j \in 1, \ldots, k)$ are subsequently averaged. The averaged values represent the reduced sequence of length k. For instance, $k = 2$ lead to two intervals: the first is $I_1 = [1, n/2]$ and the second one is $I_2 = [n/2, n]$. For $k = 3$ the three resulting intervals are $I_1 = [1, n/3]$, $I_2 = [n/3, 2n/3]$ and $I_3 = [2n/3, n]$. For $k = m, m < n$ the resulting intervals are $I_1 = [1, n/m], I_2 = [n/m, 2n/m], \ldots, I_m = [(m-1)n/m, n]$.

2.7 Classification

We used random forest (Breiman 2001) implemented in `R`, as they have been shown to be less prone to overfitting and thus produce highly robust and accurate predictions compared to other machine learning techniques (Caruana and Niculescu-Mizil 2006).

2.8 Performance Measurement

To evaluate the sensitivity, specificity and accuracy for each classifier, we repeated a leave-one-out cross-validation 10 times and averaged the results. Furthermore, we computed the area under the curve (AUC), which is the integral of the Receiver Operating Characteristics (ROC) curve.

Table 2 Hydropathy index of Kyte and Doolittle (1982)

Amino acid	Hydropathy	Amino acid	Hydropathy
R	−4.5	A	1.8
K	−3.9	M	1.9
Q,E,N,D	−3.5	C	2.5
H	−3.2	F	2.8
P	−1.6	L	3.8
Y	−1.3	V	4.2
W	−0.9	I	4.5
S	−0.8		
T	−0.7		
G	−0.4		

3 Results and Discussion

We analyzed several real-world and two synthetic data sets as described in Material and Methods. The dimension of both synthetic data sets (S1 and S2) was successfully reduced by applying Interpol. Low normalization factors (k) lead to a loss of prediction performance, but with increasing dimensionality, the performance converged towards stable AUC values. The prediction accuracy of the synthetic data sets converged towards an AUC of 1 with increasing target dimensionality. The other methods, namely PCA, Isomap and t-SNE, worked as well on the synthetic data.

For the real-world datasets, as mentioned before for the synthetic datasets, low normalization factors (k) lead to a loss of prediction performance. However, with increasing dimensionality, the performance converged towards stable AUC values. The dimension of the input space can be reduced, in general, by a factor of 2 with $n/2$ being typically the minimum to achieve converged prediction performance. However, we recommend to use $k > n/2$ to guarantee high prediction performance.

A reduction to $n/2$ leads to a function f, which maps two successive values y_i and y_{i+1} onto a value $y_{i+0.5}$ between these two feature values:

$$f(x_i, x_{i+1}) = \frac{y_i + y_{i+1}}{2} = y_{i+0.5} \tag{1}$$

The function f is semi-bijective, i.e. one can use the inverse function f^{-1} to calculate the former two data points y_i and y_{i+1} out of the resulting $y_{i+0.5}$ value in most of the cases. However, the order of the two data points is lost.

There are 17 different numerical values for the amino acids with regard to the hydropathy index of Kyte and Doolittle (1982). Combining these values result in $\frac{17^2-17}{2} + 17 = 153$ possible pairs of amino acids. These pairs are mapped onto 109 unique values in the interval $[-4.5, 4.5]$, resulting in the above semi-bijective function. However, non-unique values result from pairs that are highly similar to other pairs. These pairs will lead with high probability to similar prediction results in the classification process.

The results for the proposed $n/2$ border are shown in Table 2. In contrast to Interpol, PCA, Isomap and t-SNE performed significantly worse (with regard to the

Table 3 The performance of the classifiers for the proposed $n/2$ border are shown as $AUC \pm sd$ exemplarily for the datasets APV, D4T, V3 and GTP

Data set	PCA	Isomap	t-SNE	Interpol
APV	0.8686 ± 0.0015	0.8277 ± 0.0015	0.8235 ± 0.0015	0.9374 ± 0.0015
D4T	0.8301 ± 0.0038	0.7246 ± 0.0032	0.7406 ± 0.0033	0.9217 ± 0.0016
V3	0.9561 ± 0.0018	0.9422 ± 0.0019	0.9411 ± 0.0017	0.9690 ± 0.0016
GTP	0.9722 ± 0.0012	0.9553 ± 0.0014	0.9622 ± 0.0013	0.9844 ± 0.0012

Wilcoxon-signed rank test on the AUC distributions from the tenfold leave-one-out cross-validation runs). Interpol performs up to 9 % better than PCA, up to 20 % better than Isomap and up to 11 % better than t-SNE with a reduction to half of the input dimension (Table 3). The coverage of the variance for PCA is about 80–90 % for the different datasets, e.g. 87.50 % (APV), 89.94 % (D4T) and 86.23 % (GTP).

4 Conclusions

We conclude that sequence interpolation by Interpol can be used for reducing the input dimensionality of protein sequences for classification. Using a desired dimensionality $k > n/2$ there is no loss of prediction performance. Compared to PCA, Isomap and t-SNE, sequence interpolation is consistently better for reducing dimensionality. The difference in AUC can be up to 20 % (see http://www.uni-due.de/~hy0546/auc.pdf for comparisons of all datasets). In comparison to PCA, Isomap and t-SNE, sequence interpolation is computationally cheap and processes on sample-by-sample basis. It thus is highly efficient even for large datasets. Interpol performs stably and superior also for smaller dimension ($k < n/2$), and thus, leads to better prediction performance compared to PCA, Isomap and t-SNE for all k, except for a dimensionality of two and three. These results are based on RF classification and other classification methods might lead to different results. Thus, Interpol should be used in classification studies with RFs, whereas PCA, Isomap and t-SNE should be used to visualize the data in the two- or three-dimensional space.

References

Breiman, L. (2001). Random forests. *Machine Learning, 45*, 5–32.

Cai, C. Z., Han, L. Y., Ji, Z. L., Chen, X., & Chen, Y. Z. (2003). SVM-Prot: Web-based support vector machinee software for functional classification of a protein from its primary sequence. *Nucleic Acids Research, 31*, 459–462.

Caruana, R., & Niculescu-Mizil, A. (2006). An empirical comparison of supervised learning algorithms. In *Proceedings of the 23rd International Conference on Machine Learning*, Pittsburgh, PA.

Cox, T. F., Cox, M. A. A., & Raton, B. (2003). Multidimensional scaling. *Technometrics, 45*(2), 182.

Dybowski, J. N., Heider, D., & Hoffmann, D. (2010). Prediction of co-receptor usage of HIV-1 from genotype. *PLOS Computational Biology, 6*(4), e1000743.

Dybowski, J. N., Riemenschneider, M., Hauke, S., Pyka, M., Verheyen, J., Hoffmann, D., et al. (2011). Improved Bevirimat resistance prediction by combination of structural and sequence-based classifiers. *BioData Mining, 4*, 26.

Heider, D., Appelmann, J., Bayro, T., Dreckmann, W., Held, A., Winkler, J., et al. (2009). A computational approach for the identification of small GTPases based on preprocessed amino acid sequences. *Technology in Cancer Research and Treatment, 8*(5), 333–342.

Heider, D., Hauke, S., Pyka, M., & Kessler, D. (2010). Insights into the classification of small GTPases. *Advances and Applications in Bioinformatics and Chemistry, 3*, 15–24.

Heider, D., & Hoffmann, D. (2011). Interpol: An R package for preprocessing of protein sequences. *BioData Mining, 4*, 16.

Jolliffe, I. T. (2002). *Principal component analysis* (2nd ed.). *Springer series in statistics*. New York: Springer.

Kyte, J., & Doolittle, R. (1982). A simple method for displaying the hydropathic character of a protein. *Journal of Molecular Biology, 157*, 105–132.

Nanni, L., & Lumini, A. (2011). A new encoding technique for peptide classification. *Expert Systems with Applications, 38*(4), 3185–3191.

Rhee, S. Y., Taylor, J., Wadhera, G., Ben-Hur, A., Brutlag, D. L., & Shafer, R. W. (2006). Genotypic predictors of human immunodeficiency virus type 1 drug resistance. *Proceedings of the National Academy of Sciences of USA, 103*(46), 17355–17360.

Tenenbaum, J. B., De Silva, V., & Langford, J. C. (2000). A global geometric framework for nonlinear dimensionality reduction. *Science, 290*(5500), 2319–2323.

van der Maaten, L., & Hinton, G. (2008). Visualizing high-dimensional data using t-SNE. *Journal of Machine Learning Research, 9*, 2579–2605.

Three Transductive Set Covering Machines

Florian Schmid, Ludwig Lausser, and Hans A. Kestler

Abstract We propose three transductive versions of the set covering machine with data dependent rays for classification in the molecular high-throughput setting. Utilizing both labeled and unlabeled samples, these transductive classifiers can learn information from both sample types, not only from labeled ones. These transductive set covering machines are based on modified selection criteria for their ensemble members. Via counting arguments we include the unlabeled information into the base classifier selection. One of the three methods we developed, uniformly increased the classification accuracy, the other two showed mixed behaviour for all data sets. Here, we could show that only by observing the order of unlabeled samples, not distances, we were able to increase classification accuracies, making these approaches useful even when very few information is available.

1 Introduction

Classifying tissue samples according to high dimensional gene expression profiles is one of the basic tasks in molecular medicine. These profiles are normally designed for a wide field of applications and rather unspecific in their marker selection for the current classification task. It is assumed that these profiles contain a vast amount of uninformative features. A decision criterion based on such a profile should be based only on a small subset of expression levels. In contrast to other low dimensional embeddings, e.g. PCA (Jolliffe 2002), such a sparse decision criterion allows the direct identification of marker genes or possible drug targets. A classifier suitable in

---

F. Schmid · L. Lausser · H.A. Kestler (✉)
Research Group Bioinformatics and Systems Biology, Institute of Neural Information Processing, Ulm University, 89069 Ulm, Germany
e-mail: florian-1.schmid@uni-ulm.de; ludwig.lausser@uni-ulm.de; hans.kestler@uni-ulm.de

M. Spiliopoulou et al. (eds.), *Data Analysis, Machine Learning and Knowledge Discovery*, Studies in Classification, Data Analysis, and Knowledge Organization, DOI 10.1007/978-3-319-01595-8_33,

this scenario is the set covering machine (SCM) with data dependent rays (Kestler et al. 2011). The decision boundary of this ensemble method can be interpreted as

```
IF gene1 > value1 AND gene2 < value2 ... THEN class1
```

The training of this inductive classifier is solely based on labeled training examples. It is often restrictive in settings in which labeling processes are time consuming or costly. Here, the limitation of labeled samples can lead to overfitting. In this work we present three newly developed transductive versions of this SCM enabling the training algorithm to incorporate unlabeled samples. For all three methods the unlabeled samples are involved in the selection process of the single ensemble members. The selection criteria are based on counting arguments abstracting from scale and density information of the unlabeled samples; only the feature-wise ordering of all samples is utilized. The results indicate that even this ordering information can increase classification accuracies.[1]

2 Methods

The task of classification can be formulated as follows: Given an object by a vector of measurements $\mathbf{x} \in \mathbb{R}^m$ predict the correct category $y \in \{0, 1\}$. The samples of class 1 and 0 are called positive and negative samples. A classification method is allowed to adapt according to a set of labeled training samples $\mathcal{S}_{tr} = \{(\mathbf{x}_i, y_i)\}_{i=1}^n$. The unlabeled samples of this set will also be denoted by $\mathcal{P} = \{\mathbf{x} \mid (\mathbf{x}, 1) \in \mathcal{S}_{tr}\}$ and $\mathcal{N} = \{\mathbf{x} \mid (\mathbf{x}, 0) \in \mathcal{S}_{tr}\}$. The quality of the prediction can be tested on a set of test samples $\mathcal{S}_{te} = \{(\mathbf{x}'_i, y'_i)\}_{i=1}^{n'}$. Omitting the labels, this set will also be called $\mathcal{U} = \{\mathbf{x} \mid (\mathbf{x}, y) \in \mathcal{S}_{te}\}$. Different types of classifiers can be distinguished according to their method of predicting a class label.

An *inductive classification* process can be separated into two major phases. In the induction phase a decision function, a classification model, $h \in \mathcal{H}$ is trained. The applied training algorithm l is a supervised one, i.e. it solely utilizes labeled samples: $l(\mathcal{S}_{tr}, \mathcal{H}) \mapsto h$.In the deduction phase the classifier h is applied on unseen samples: $h(\mathbf{x}) \mapsto \hat{y}$.The aim of an inductive learner is to find a classifier h that is highly predictive over the distribution of all labeled samples. It is often misleading as a general good performance does not imply a good performance on a particular test set $\mathcal{S}_{te}$ of objects of interest. A decision could achieve better results on $\mathcal{S}_{te}$, if it is specialist on this set. In contrast a *transductive classification* (Vapnik 1998) process is only concerned with predicting the class labels of a fixed set of test samples $\mathcal{S}_{te}$. The classification is achieved in a single step by utilizing all samples and not only training a concept class h on the labeled data.

[1]We only utilize the information of orderings of features values, not the information given by a ordinal structure of the class labels (e.g. Herbrich et al. 1999).

2.1 Inductive Set Covering Machine

An example for an inductive classifier is the set covering machine (SCM) proposed by Marchand and Taylor (2003). It can be seen as an ensemble method merging the predictions of a set of selected classifiers $\mathcal{E} \subseteq \mathcal{H}$.

$$h(\mathbf{x}) = \bigwedge_{h_j \in \mathcal{E}} h_j(\mathbf{x}) \tag{1}$$

The fusion strategy of a SCM is a logical conjunction (unanimity vote), which is a strategy with an asymmetric behavior. A conjunction is more likely to misclassify a positive example than a negative one. A sample will only receive the class label 1, if all ensemble members vote for class 1. It will receive class label 0 if at least one base classifier predicts class 0; an example of class 0 will be classified correctly, if it is covered by at least one correctly classifying ensemble member. A trained ensemble should fulfill two properties to counteract this characteristic of a conjunction:

1. The base classifiers should achieve a high sensitivity.
2. The ensemble size $|\mathcal{E}|$ should be relatively small.

The SCM training algorithm tackles this problem according to a greedy strategy. It iteratively extends the current ensemble $\mathcal{E}$ by a classifier $h_j \in \mathcal{H}$ achieving a maximal usefulness U_j.

$$U_j = |\mathcal{Q}_j| - p|\mathcal{R}_j| \tag{2}$$

The usefulness U_j can be seen as a tradeoff of two objectives representing the positive and negative effects the candidate h_j would have on the existing ensemble $\mathcal{E}$. The first objective $|\mathcal{Q}_j|$ denotes the number of uncovered negative examples h_j classifies correctly. The second one $|\mathcal{R}_j|$ is the number of positive examples correctly classified by the current ensemble $\mathcal{E}$ but misclassified by h_j. The two objectives are linked via a tradeoff parameter $p \in [0, \infty)$. It is often set to $p = \infty$ in order to maximize the sensitivity of the ensemble members. The algorithm stops when all negative samples are covered by at least one classifier or a chosen maximal number of ensemble members s was trained.

2.2 Set Covering Machine with Data Dependent Rays

A special version of the SCM is the set covering machine operating on data dependent rays (Kestler et al. 2011). In contrast to other versions of the SCM this classifier allows variable selection and constructs an interpretable decision boundary. A single threshold classifier determines the class label of a sample $\mathbf{x}$

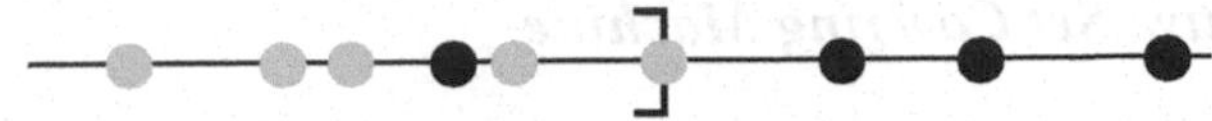

Fig. 1 A ray optimized according to (4). Symbols *gray circle/black circle* denote samples of class 1/0

according to a single threshold $t \in \mathbb{R}$ on a selected feature dimension $j \in \{1, \ldots, n\}$.

$$r(\mathbf{x}) = \mathbb{I}_{[d(x^{(j)}-t)\geq 0]} \tag{3}$$

Here $d \in \{-1, 1\}$ denotes the direction of the classifier. A ray is called data dependent if $t \in \{x^{(j)} : (\mathbf{x}, y) \in \mathcal{S}_{tr}\}$. In order to increase the sensitivity of the final ensemble, the rays are trained to avoid errors on the positive training samples. At the same time the error on the negative ones is minimized.

$$\min_{r' \in \mathcal{H}(\mathcal{S}_{tr})} \left(R_{emp}(r', \mathcal{N}) + |\mathcal{N}| R_{emp}(r', \mathcal{P}) \right) \tag{4}$$

Figure 1 shows an example of a data dependent ray optimized according to (4). The threshold t must be equal to the largest or smallest value of a positive example. The closest point on the negative side of the threshold must be a negative one. The open interval between this point and the threshold will be of interest for the following. It will be denoted as $\delta(r)$.

2.3 Transductive Set Covering Machines

The inductive set covering machine chooses the current ray r_k among those candidates having a maximal usefulness.

$$r_k \in \mathcal{K} = \left\{ r_i \in \mathcal{H} \mid U_i = \max_{r_j \in \mathcal{H}} U_j \right\} \tag{5}$$

In the context of small sample sizes, the number of possible candidates can be relatively large as the number of ray classifiers is very large ($\geq m$) in contrast to a relatively small number of values the utility can take ($\leq |\mathcal{P}||\mathcal{N}|$) on the set of feasible rays. The usefulness is a rather unspecific selection criteria in this scenario. A wrong choice can lead to a diminished predictive performance.

To improve the selection process, we include a post processing step on $\mathcal{K}$ based on the unlabeled samples $\mathcal{U}$ available in the transductive training. The unlabeled samples are used to rank the candidates in $\mathcal{K}$ in up to $|\mathcal{U}|$ additional categories. We have tested three different post processing strategies resulting in three transductive set covering machines (SCM_{min}, SCM_{max}, SCM_{prev}).

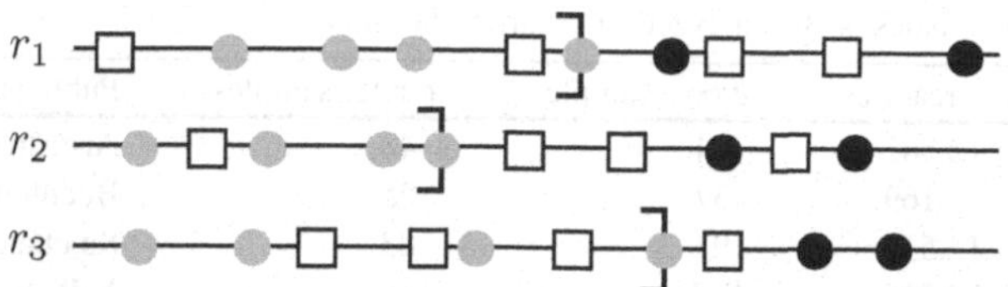

Fig. 2 Three rays of equal usefulness. Symbols *gray circle/black circle* indicate samples of class 1/0. The unlabeled samples are denoted by *square*. Each of the rays is preferred by a different transductive SCM. Ray r_1 is optimal for SCM_{min}. Here, no unlabeled samples occur in $\delta(r_1)$. Ray r_2 has the maximal number of unlabeled samples in $\delta(r_2)$. It is optimal for SCM_{max}. SCM_{prev} would select r_3. Here, the most unlabeled samples are on the positive side of the threshold corresponding to a higher prevalence of the positive class

The SCM_{min} selects a ray r in $\mathcal{K}$ according to the number of unlabeled samples in the interval $\delta(r)$. It chooses those rays which have a minimal number of unlabeled samples in this interval. The SCM_{max} can be seen as an opposite version to the SCM_{min}. Here the set of candidates is restricted to those having a maximal number of unlabeled samples in $\delta(r)$. The SCM_{prev} incorporates the prevalence of the classes in the candidate selection process. Here, rays are preferred for which more unlabeled samples can be found on the side of the decision boundary assigned to the class with the higher prevalence.

An example, describing the differences of the reordering steps of all three versions is shown in Fig. 2. It shows three rays of equal usefulness. While the SCM_{ind} would randomly select one of the given rays, each of the transductive SCMs would add a different ray to the classification model. The SCM_{min} would select r_1 which has no unlabeled samples in $\delta(r_1)$. SCM_{max} as opposite version would add r_2 because of the two unlabeled points in $\delta(r_2)$. The SCM_{prev} would select r_3. Here are most unlabeled samples on the positive side of the threshold which is the one having higher prevalence.

3 Experimental Setup

The transductive versions of the set covering machine are tested and compared to the inductive one in sequences of k-fold cross-validation experiments (e.g. Bishop 2006), where $k \in \{10, \dots, 2\}$. The cross-validation experiments are adapted to the transductive learning scheme by utilizing the single folds as proposed in Lausser et al. (2011). A transductive learner does not only get access to the labeled training folds but also to the unlabeled test fold. Decreasing the number of folds has slightly different consequences for the training set of inductive and transductive learners. While an inductive learner has to cope with a smaller set of (labeled) samples the number of available samples remains constant for the transductive one. Here, the tradeoff between labeled and unlabeled samples is varied. While 90 % of all

Table 1 Characteristics of the analyzed microarray datasets

Dataset	# features	# pos. samples	# neg. samples	Publication
Armstrong	12,582	24	48	Armstrong et al. (2002)
Buchholz	169	37	25	Buchholz et al. (2005)
Su	12,533	91	83	Su et al. (2001)
Valk	22,215	177	116	Valk et al. (2004)

training samples will be labeled in a tenfold cross-validation only 50 % will be labeled in the twofold experiment. Transductive learners are likely to be applied in settings, in which much more unlabeled than labeled samples exist. The setting can be mimicked by switching the role of labeled and unlabeled folds. The classifiers are tested on real microarray data and artificial data. A list of real datasets is given in Table 1.

For the real datasets a classification and regression tree (CART) was chosen as reference classifier (Breiman et al. 1984). It was trained according to the semi-supervised self-training scheme (Weston et al. 2003). In this iterative process the unlabeled samples and class labels predicted by the current model are used as additional training data for the following classifier. The training ends if the label vector does not change anymore or a maximal number of iterations ($t = 20$) was achieved.

The artificial datasets contain $n = 100$ samples of a dimensionality of $m = 1{,}000$. The single samples are drawn from a uniform distribution $\mathbf{x}_j \sim \mathcal{U}^m(-1, 1)$. The labels of the samples are determined according to a conjunction of $|\mathbf{j}| = 5$ single threshold classifiers.

$$y_i = \begin{cases} 0 & \text{if } \bigwedge_{j \in \mathbf{j}} \mathbb{I}_{\left[x_i^{(j)} > 0\right]} \\ 1 & \text{otherwise} \end{cases} \tag{6}$$

Here $\mathbf{j} \subset \{1, \ldots, m\}$ denotes an index vector for the selected feature dimensions. The datasets are collected with different tradeoffs of positive and negative labeled samples. The number of positive samples varies in $p \in \{10, 20, \ldots, 90\}$.

4 Results

A summary of the results on the artificial datasets is shown in Fig. 3. They show that the transductive versions work better than the inductive SCM when more unlabeled than labeled samples occur. While the SCM_{min} is better than the SCM_{ind} in the presence of more positive samples, the SCM_{max}, as opposite version of the SCM_{min}, is better for more negative samples. The SCM_{prev} shows a better performance than the SCM_{ind} over all class proportions.

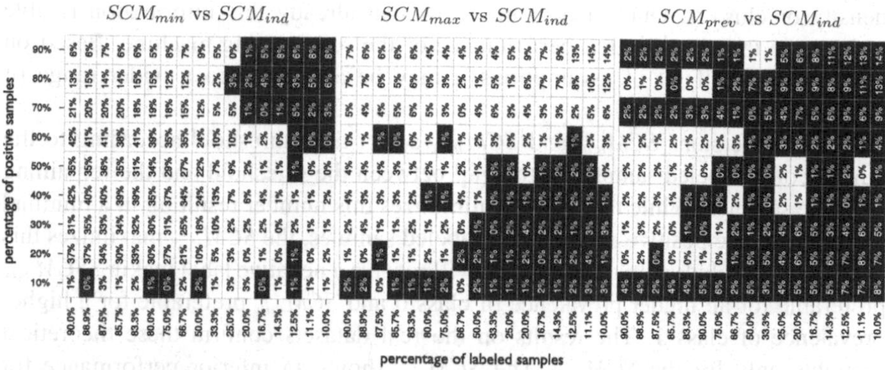

Fig. 3 The cross-validation results on the artificial datasets. The different datasets with different proportions of positive and negative samples are shown on the y-axis. A *black square* denotes an increased accuracy of a transductive SCM. Values given in the *squares* show the differences of the cross-validation accuracies between the variants

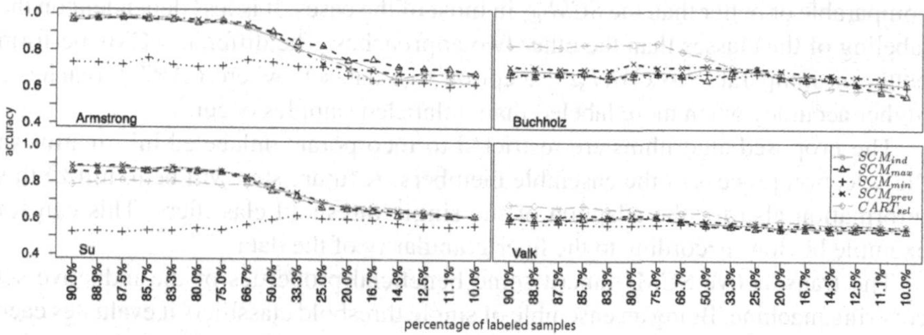

Fig. 4 Cross-validation accuracies on the microarray datasets

The results on the real datasets are shown in Fig. 4. SCM_{max} and SCM_{prev} show an equal or better performance than SCM_{ind} on all datasets. On the Buchholz and on the Valk dataset $CART_{sel}$ performs better than the SCMs when there are more labeled than unlabeled samples. While SCM_{max} works best on the Armstrong dataset, the SCM_{prev} dominates all other versions of the SCM over all experiments on the Buchholz and on the Valk dataset. The results of the SCM_{min} are comparable for the Buchholz and the Valk dataset. It is outperformed by all other SCMs on the remaining datasets.

5 Discussion and Conclusion

The transductive versions of the SCM discussed in this work incorporate information of unlabeled samples by solely evaluating their feature-wise ordinal position. They do not rely on structural properties of a data collection like regions of high density or clusters. Our experiments show that already this information is able to improve the classification performance of the set covering machine. Based on a relatively small set of assumptions it may be applicable to a wider range of classification algorithms.

The two versions SCM_{min} and SCM_{max} select their threshold according to the number of unlabeled samples in the gap between the threshold and the remaining negative samples. While the SCM_{min} minimizes this number resulting in a minimal gap between the positive and negative labeled samples, the SCM_{max} maximizes this gap. The unlabeled samples in this gap will receive a negative label making SCM_{max} preferable for an higher prevalence of class 0 and SCM_{min} preferable for a higher prevalence of class 1. The results on the real datasets confirm these theoretical thoughts only for the SCM_{max}. The SCM_{min} shows an inferior performance for datasets with a larger positive class. This may be due to the greedy sensitivity maximization of the SCM coupled to the new selection strategy. In contrast the specificity maximization of the SCM_{max} seems to be beneficial in this scenario. The SCM_{prev}, which estimates the class prevalence from the training data, performs comparable or better than the SCM_{ind} in most of the cases. It is less dependent on the labeling of the classes than the other two approaches. The different SCMs perform better or comparable to $CART_{sel}$ except on two datasets where $CART_{sel}$ reaches a higher accuracy when more labeled than unlabeled samples occur.

The proposed algorithms are restricted to incorporate unlabeled information in the selection process of the ensemble members. A future step will be to utilize this information also for the adaption of the single threshold classifiers. This can for example be done according to the finer granularity of the data.

The transductive SCMs inherit some beneficial properties of the inductive set covering machine. Being an ensemble of single threshold classifiers it evaluates each feature dimension independently. It is not affected by different scales or varying feature types. In the context of bimolecular data it may be a possible candidate for a transductive classification process suitable for combining data from different sources as transcriptomics and proteomics.

Acknowledgements This work was funded in part by a Karl-Steinbuch grant to Florian Schmid, the German federal ministry of education and research (BMBF) within the framework of the program of medical genome research (PaCa-Net; Project ID PKB-01GS08) and the framework GERONTOSYS 2 (Forschungskern SyStaR, Project ID 0315894A), and by the German Science Foundation (SFB 1074, Project Z1) to Hans A. Kestler. The responsibility for the content lies exclusively with the authors.

References

Armstrong, S. A., Staunton, J. E., Silverman, L. B., Pieters, R., den Boer, M. L., Minden, M. D., et al. (2002). Mll translocations specify a distinct gene expression profile that distinguishes a unique leukemia. *Nature Genetics, 30*(1), 41–47.

Bishop, C. M. (2006). *Pattern recognition and machine learning*. Secaucus: Springer.

Breiman, L., Friedman, J. H., Olshen, R. A., & Stone, C. I. (1984). *Classification and regression trees*. Belmont: Wadsworth.

Buchholz, M., Kestler, H. A., Bauer, A., Böck, W., Rau, B., Leder, G., et al. (2005). Specialized DNA arrays for the differentiation of pancreatic tumors. *Clinical Cancer Research, 11*(22), 8048–8054.

Herbrich, R., Graepel, T., & Obermayer, K. (1999). *Regression Models for Ordinal Data: A Machine Learning Approach*. Technical report, TU Berlin.

Jolliffe, I. T. (2002). *Principal component analysis*. New York: Springer.

Kestler, H. A., Lausser, L., Lindner, W., & Palm, G. (2011). On the fusion of threshold classifiers for categorization and dimensionality reduction. *Computational Statistics, 26*, 321–340.

Lausser, L., Schmid, F., & Kestler, H. A. (2011). On the utility of partially labeled data for classification of microarray data. In F. Schwenker & E. Trentin (Eds.), *Partially supervised learning* (pp. 96–109). Berlin: Springer.

Marchand, M., & Taylor, J. S. (2003). The set covering machine. *Journal of Machine Learning Research, 3*, 723–746.

Su, A. I., Welsh, J. B., Sapinoso, L. M., Kern, S. G., Dimitrov, P., Lapp, H., et al. (2001). Molecular classification of human carcinomas by use of gene expression signatures. *Cancer Research, 61*(20), 7388–7393.

Valk, P. J., Verhaak, R. G., Beijen, M. A., Erpelinck, C. A., Barjesteh van Waalwijk van Doorn-Khosrovani, S., Boer, J. M., et al. (2004). Prognostically useful gene-expression profiles in acute myeloid leukemia. *New England Journal of Medicine, 16*(350), 1617–1628.

Vapnik, V. N. (1998). *Statistical learning theory*. New York: Wiley.

Weston, J., Pérez-Cruz, F., Bousquet, O., Chapelle, O., Elisseeff, A., Schölkopf, B., et al. (2003). Feature selection and transduction for prediction of molecular bioactivity for drug design. *Bioinformatics, 19*(6), 764–771.

Part VI
AREA Interdisciplinary Domains: Data Analysis in Music, Education and Psychology

Tone Onset Detection Using an Auditory Model

Nadja Bauer, Klaus Friedrichs, Dominik Kirchhoff, Julia Schiffner, and Claus Weihs

Abstract Onset detection is an important step for music transcription and other tasks frequently encountered in music processing. Although several approaches have been developed for this task, neither of them works well under all circumstances. In Bauer et al. (Einfluss der Musikinstrumente auf die Güte der Einsatzzeiterkennung, 2012) we investigated the influence of several factors like instrumentation on the accuracy of onset detection. In this work, this investigation is extended by a computational model of the human auditory periphery. Instead of the original signal the output of the simulated auditory nerve fibers is used. The main challenge here is combining the outputs of all auditory nerve fibers to one feature for onset detection. Different approaches are presented and compared. Our investigation shows that using the auditory model output leads to essential improvements of the onset detection rate for some instruments compared to previous results.

1 Introduction

A tone onset is the time point of the beginning of a musical note or other sound. A tutorial on basic onset detection approaches is given by Bello et al. (2005). The algorithm we will use here is based on two approaches proposed in Bauer et al. (2012): in the first approach the amplitude slope and in the second approach the change of the spectral structure of an audio signal are considered as indicators for tone onsets. In Sect. 2 we briefly describe these two approaches.

N. Bauer (✉) · K. Friedrichs · D. Kirchhoff · J. Schiffner · C. Weihs
Chair of Computational Statistics, Faculty of Statistics, TU Dortmund, Dortmund, Germany
e-mail: bauer@statistik.tu-dortmund.de; friedrichs@statistik.tu-dortmund.de; kirchhoff@statistik.tu-dortmund.de; schiffner@statistik.tu-dortmund.de; weihs@statistik.tu-dortmund.de

M. Spiliopoulou et al. (eds.), *Data Analysis, Machine Learning and Knowledge Discovery*, Studies in Classification, Data Analysis, and Knowledge Organization, DOI 10.1007/978-3-319-01595-8_34,

Since the human ear still is the best system for music perception the onset detection approach is extended by using the auditory model output instead of the original signal. A similar idea was proposed in Benetos (2009) where the extension led to improved results. In contrast to our work, in their study two other features are used for onset detection (Spectral Flux and Group Delay). Furthermore, they use fixed parameter settings for the detection algorithm while the goal of our study is to explore different settings by an experimental design.

Section 3 describes the auditory model utilized for this study. As the output is a multidimensional time series the onset detection indicators of all dimensions have to be combined to one indicator. Therefore, in Sect. 4 we propose several combination strategies. To test our approach we use randomly generated tone sequences of six music instruments with two tempo settings. Section 5 provides the detailed generation procedure. Section 6 presents the results in respect of the music instruments, the tempo and the chosen combination strategy. Additionally, the results of the extended approach are compared with the results of the approach based on the original signal proposed in Bauer et al. (2012). Finally, Sect. 7 summarizes the work and provides ideas for future research.

2 Onset Detection Approach

The ongoing audio signal is split up into T windows of length $\boldsymbol{L}$ with an overlap of $\boldsymbol{O}$ per cent of $\boldsymbol{L}$. Features are evaluated in each window to obtain onset detection functions. Several features have been proposed for this task like High Frequency Content, Spectral Flux, Group Delay, Weighted Phase Deviation, Modified Kullback–Leibler Distance or Zero Crossing Rate (Böck et al. 2012; Krasser et al. 2012). Here, we use the two features proposed in Bauer et al. (2012) in order to compare with those results: the difference between amplitude maxima ($\boldsymbol{F_A}$) and the correlation coefficient between the spectra ($\boldsymbol{F_S}$) of the current and the previous window. Each of the vectors $\boldsymbol{F_A}$ and $\boldsymbol{F_S}$ is then rescaled into the interval [0,1].

For each window a combined feature $\boldsymbol{CombF}$ can be calculated as $\boldsymbol{CombF} = \boldsymbol{W_A} \cdot \boldsymbol{F_A} + (1 - \boldsymbol{W_A}) \cdot \boldsymbol{F_S}$, where the weight $\boldsymbol{W_A} \in [0, 1]$ is a further parameter, which specifies the influence of the amplitude based feature on the weighted sum. In Bauer et al. (2012) we investigated further combination approaches and this approach provided the best results. Let $detect(s, \boldsymbol{L}, \boldsymbol{O}, \boldsymbol{W_A})$ denote the onset detection function which returns the $\boldsymbol{CombF}$-vector for the audio signal s depending on $\boldsymbol{L}$, $\boldsymbol{O}$ and $\boldsymbol{W_A}$.

The peak-picking procedure consists of the following two steps: thresholding of the onset detection function and localization of tone onsets. In order to assess if a window contains a tone onset, based on $\boldsymbol{CombF}$, a threshold is required. We will use here a $\boldsymbol{Q}$%-quantile of the $\boldsymbol{CombF}$-vector as such threshold. If the $\boldsymbol{CombF}$-value for the current window, but neither for the preceding nor for the succeeding window, exceeds the threshold, an onset is detected in this window. If the threshold is exceeded in multiple, consecutive windows, we assume that there is only one onset, located in that window with the maximal ***CombF***-value in this sequence. This is a simple fixed thresholding approach, but as we consider only music signals with small changes of sound intensities (see Sect. 5), it is assumed to be appropriate here (see Bello et al. 2005). For music signals with higher sound volume changes adaptive thresholds are more reasonable (Rosão et al. 2012). For each window with an onset detected its beginning and ending time points are calculated and the onset time is then estimated by the centre of this time interval. Let $onsets(CombF, \boldsymbol{Q})$ denote the onset estimation function which returns for the onset detection feature $CombF = detect(s, \boldsymbol{L}, \boldsymbol{O}, \boldsymbol{W_A})$ and quantile $\boldsymbol{Q}$ the estimated time points of the tone onsets.

In this work we assume that a tone onset is correctly detected, if the absolute difference between the true and the estimated onset time is less than 50 ms (see Dixon 2006). As quality criterion for the goodness of the onset detection the so called F-value is used: $F = \frac{2c}{2c+f^{+}+f^{-}}$, where c is the number of correctly detected onsets, f^{+} is the number of false detections and f^{-} denotes the number of undetected onsets. Note that the F-value always lies between 0 and 1. The optimal F-value is 1. For the given vector of the true onset time points *true*, the function $f\_value(est, true)$ returns the F-value for the vector of the estimated onset times $est = onsets(CombF, \boldsymbol{Q})$.

3 Auditory Model

An auditory model is a computer model of the human auditory system. It requires an acoustic signal as input and outputs the spike firing rates of the auditory nerve fibers. The human auditory system consists of roughly 3,000 auditory nerve fibers. Auditory models contain usually a much smaller number of fibers. For this study the popular model of Meddis (2006) is employed. To put it simply, in this model the auditory system is coded by a multichannel bandpass filter where each channel represents one specific nerve fiber. Each channel has its specific best frequency by which its perceptible frequency range is defined. In this work we use 40 channels with best frequencies between 250 Hz (channel 1) and 7,500 Hz (channel 40). An exemplary output of the model, called auditory image, can be seen in Fig. 1. While the 40 channels are located on the vertical axis and the time response on the horizontal axis, the grey level indicates the channel activity per second.

4 Auditory Model Based Onset Detection Approach

The output of the auditory model is, according to Sect. 3, a set of 40 audio signals $(chan_1, chan_2, \ldots, chan_{40})$, where each signal corresponds to one of 40 hearing channels. For all channels the combined features ***CombF*** are computed as described in Sect. 2 and then combined to one onset indicator. While there are

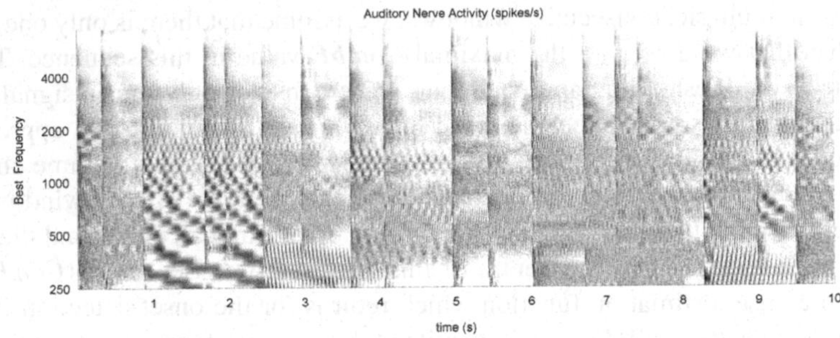

Fig. 1 Auditory image

more comprehensive ways, we tried a simple method first. The detailed procedure is described in the following[1]:

1. For each channel $chan_k$, $k = 1, \dots, 40$, calculate the onset detection function $\boldsymbol{CombF}_k = detect(chan_k, \boldsymbol{L}, \boldsymbol{O}, \boldsymbol{W_A})$.
2. For each window i, $i = 1, \dots, T$, do:
   - $v[i] = (\boldsymbol{CombF}_1[i], \dots, \boldsymbol{CombF}_{40}[i])'$,
   - $output[i] = quantile(v[i], \boldsymbol{C})$, where the function *quantile* returns the $\boldsymbol{C}$%-quantile of the vector $v[i]$.
3. $output = (output[1], \dots, output[T])'$.
4. $est = onsets(output, \boldsymbol{Q})$.
5. $error = f\_value(est, true)$.

In the following we will investigate the influence on the onset detection accuracy of the parameters $\boldsymbol{L}$, $\boldsymbol{O}$, $\boldsymbol{W_A}$ and $\boldsymbol{C}$. In order to reduce the complexity, we estimate the parameter $\boldsymbol{Q}$ by means of grid search in the interval from 0 to 1 with step size 0.05 (the same approach as in Bauer et al. 2012).

The following settings of the four parameters are tested: $\boldsymbol{L}$ with two levels (512 and 2,024 samples), $\boldsymbol{O}$ with two levels (0 and 50 %), $\boldsymbol{W_A}$ with three levels (0, 0.5 and 1) and $\boldsymbol{C}$ with three levels (5, 50 and 95 %). Table 1 lists the 36 parameter settings.

5 Music Data Set

In order to compare the extended onset detection approach proposed in this paper with the original one the same data set is used as in Bauer et al. (2012), which consists of 24 randomly generated tone sequences. In the following we briefly describe the construction principles of this data set.

[1] $v[i]$ is the notation for the i-th element of vector v.

Table 1 Parameter settings for auditory model based onset detection

ID	L	O	W_A	C	ID	L	O	W_A	C	ID	L	O	W_A	C	ID	L	O	W_A	C
1	512	0	0	5	10	512	50	0	5	19	2,048	0	0	5	28	2,048	50	0	5
2	512	0	0	50	11	512	50	0	50	20	2,048	0	0	50	29	2,048	50	0	50
3	512	0	0	95	12	512	50	0	95	21	2,048	0	0	95	30	2,048	50	0	95
4	512	0	0.5	5	13	512	50	0.5	5	22	2,048	0	0.5	5	31	2,048	50	0.5	5
5	512	0	0.5	50	14	512	50	0.5	50	23	2,048	0	0.5	50	32	2,048	50	0.5	50
6	512	0	0.5	95	15	512	50	0.5	95	24	2,048	0	0.5	95	33	2,048	50	0.5	95
7	512	0	1	5	16	512	50	1	5	25	2,048	0	1	5	34	2,048	50	1	5
8	512	0	1	50	17	512	50	1	50	25	2,048	0	1	50	35	2,048	50	1	50
9	512	0	1	95	18	512	50	1	95	27	2,048	0	1	95	36	2,048	50	1	95

There are many characteristics that describe a music signal like tempo, genre, instrumentation or sound volume. We consider merely the instrumentation and the tempo as control variables when designing the data set: the same tone sequences are recorded by different music instruments with different tempo settings. This allows to explicitly measure the influence of these two control variables on the accuracy of the onset detection.

Six music instruments are considered: piano, guitar, flute, clarinet, trumpet and violin. The tone sequences are generated randomly by considering the following settings:

1. *Sound intensities* follow the uniform distribution in the interval [70,90] (in MIDI-coding).
2. *Notes* follow the uniform distribution in the interval [60,76] (in MIDI-coding). This interval corresponds to the common pitch range (from C4 to E5) of the instruments under consideration.
3. *Tone durations* (in seconds) are
   (a) absolute values from the normal distribution with parameters $\mu = 0.5$ and $\sigma = 0.2$,
   (b) absolute values from the normal distribution with parameters $\mu = 0.2$ and $\sigma = 0.1$.

The first tone duration setting generates slow music pieces with two beats per second on average (or 120 BPM[2]), whereas the second setting generates fast music pieces with five beats per second on average (300 BPM). For each of the six music instruments and for each of the two tone duration settings five tone sequences (with the duration of 10 s) are generated. The fast tone sequences hence include ca. 50 tone onsets and the slow sequences ca. 20 onsets.

The generated MIDI-files are converted to WAVE-files using real tone samples from the RWC data base (Goto et al. 2003).

[2]BPM: beats per minute.

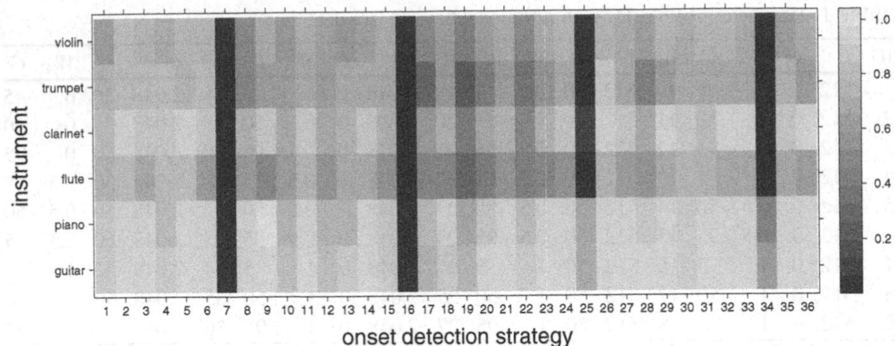

Fig. 2 Comparison of the 36 onset strategies (see Table 1): for each instrument and each strategy the average F-value over five slow and five fast sequences is presented

6 Results

In this section we will compare the 36 proposed parameter settings of the auditory model based onset detection in order to investigate the influence of these settings on the goodness of onset detection for each instrument. Further, we will compare the results of original signal and auditory model based detection for the same algorithm's parameter settings. This should reveal whether an extension of the onset detection algorithm (proposed in our previous work) with an auditory model leads to better detection accuracy.

Comparison of Auditory Image Based Onset Detection Approaches

In order to compare the goodness of the proposed 36 onset detection strategies (see Sect. 4) we calculate the F-measure for each music instrument and for each of the five slow and five fast sequences (see Sect. 5). The resulting average F-measures of the six music instruments and 36 strategies are presented in Fig. 2. Most strategies provide acceptable results but few strategies obviously fail. Furthermore, it can be seen, that for the music instruments piano, guitar and clarinet—in contrast to the others—systematically better detection accuracies are achieved.

Table 2 presents the best parameter settings for each instrument and on average. We can observe an interaction between parameters $\boldsymbol{C}$ and $\boldsymbol{W_A}$: if $\boldsymbol{W_A} = 1$, $\boldsymbol{C}$ should be chosen high ($\boldsymbol{C} = 95$). Whereas the best setting on average is the one with the small window length ($\boldsymbol{L} = 512$ samples), without overlap ($\boldsymbol{O} = 0\,\%$), considering merely the spectral based feature ($\boldsymbol{W_A} = 0$) and middle quantile (median, $\boldsymbol{C} = 50\,\%$). Further, in Table 3 the worst strategies are listed: considering

Table 2 Best parameter settings for each instrument and on average

Instrument	ID	L	O	W_A	C
Piano	36	2,048	50	1	95
Guitar	23	2,048	0	0.5	50
Flute	9	512	0	1	95
Clarinet	36	2,048	50	1	95
Trumpet	2	512	0	0	50
Violin	2	512	0	0	50
On average	2	512	0	0	50

Table 3 Worst parameter settings

ID	L	O	W_A	C
7	512	0	1	5
16	512	50	1	5
34	2,048	50	1	5
25	2,048	0	1	5
19	2,048	0	0	5

only the amplitude based feature ($W_A = 1$) in combination with small C-values does not seem to lead to acceptable results.

Comparison of Auditory Image Based and Original Signal Based Onset Detection

In Bauer et al. (2012) the original signal based tone onset detection algorithm was tested on the data set used in this work for eight possible combinations of the following parameter settings: L with two levels (512 and 2,048 samples), O with two levels (0 and 50 %) and W_A with two levels (0 and 1). As the parameter C is only given in the extended approach, it was set to the average best setting $C = 50\,\%$ (see Table 2).

Table 4 shows the average F-values of auditory model based and original signal based onset detection over the eight parameter settings mentioned above. It also gives the median of the percentage change of the results when using the extended approach. Further, Fig. 3 compares the detection accuracy (F-measure) of both approaches for an example of slow trumpet tone sequences.

According to Table 4 the auditory image based onset detection is on average (over all parameter settings) better than the original signal based. However, when considering the median of the vector of 40 improvement-rates—for each setting and each tone sequence—we observe for flute and trumpet, in contrast to the other music instruments, a decline of the onset detection accuracy when using the auditory model based approach (especially for the slow sequences). This is basically caused by the fact that many onsets for these two instruments were detected with a not accepted delay of more than 50 ms. Figure 4 shows a fragment of the auditory image for a slow trumpet sequence. Obviously the channels with

Table 4 Averaged F-values of the auditory model and original signal based onset detection over eight different parameter settings and the median of the percentage change of the results when using the extended approach

	Sequence					
Approach	Piano slow	Piano fast	Guitar slow	Guitar fast	Flute slow	Flute fast
Auditory model	0.961	0.934	0.950	0.928	0.576	0.662
Original signal	0.918	0.854	0.949	0.895	0.553	0.606
Improvement (in %)	1.33	5.75	0.00	−0.98	−6.73	−1.64
Approach	Clarinet slow	Clarinet fast	Trumpet slow	Trumpet fast	Violin slow	Violin fast
Auditory model	0.849	0.875	0.564	0.708	0.597	0.773
Original signal	0.775	0.801	0.514	0.658	0.466	0.638
Improvement (in %)	4.53	3.95	−6.78	−0.46	32.56	23.92

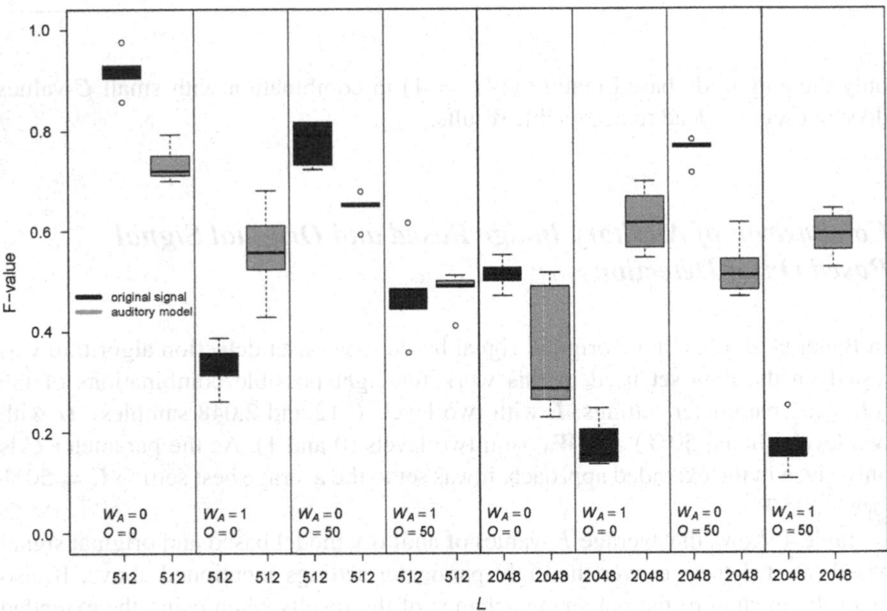

Fig. 3 Comparison of F-values of the original signal (*dark*) and auditory image based (*light*) onset detection for slow trumpet tone sequences. The window length $\boldsymbol{L}$ is given on the x-axis, while the values of the two other parameters ($\boldsymbol{W_A}$ and $\boldsymbol{O}$) are listed within the figure

high best frequencies (over 3,000 Hz) respond later than the channels with low best frequencies. Interestingly such delays do not occur very often for short tones from staccato tone libraries (for trumpet and flute). This may be due to the fact that short and especially staccato tones in contrast to long tones usually have stronger attacks. Thus, in order to improve the onset detection a systematic investigation of these effects is required.

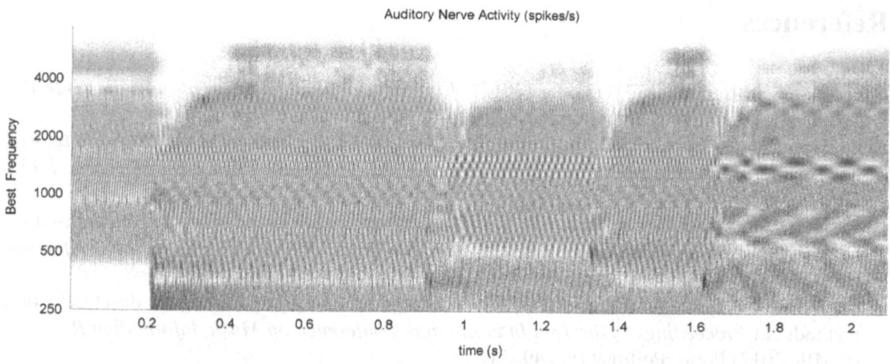

Fig. 4 Cut-out of the audiogram for a slow trumpet sequence

7 Summary

In this work we proposed a strategy for using the multiple channel output of an auditory model for the tone onset detection problem. We compared auditory model and original signal based approaches on the same data set for several algorithm parameter settings. The data set consists of random sequences which were generated for six music instruments and two tempi (fast and slow) using real tone recordings.

An essential improvement of the onset detection when using the auditory model can be noted for the violin (over 20 %). However, for the slow trumpet and flute tone sequences a decline of the detection accuracy can be observed using the auditory model based approach. This can be explained through the delayed response of high frequency channels of the auditory model. In order to improve the auditory model based onset detection the detailed study of auditory model behaviour depending on musical characteristics of audio signals is required and is the aim for our further research. Additionally, we plan to compare our two features with others mentioned in Sect. 2, especially by considering polyphonic music pieces. Further, other approaches for processing the channel outputs of the auditory model can be considered, for example by taking the channel order into account.

Acknowledgements This work was supported by the Collaborative Research Centre "Statistical Modelling of Nonlinear Dynamic Processes" (SFB 823) of the German Research Foundation (DFG), within the framework of Project C2 "Experimental Designs for Dynamic Processes" and Project B3 "Statistical Modelling of Highly-Resolved Spectro-Temporal Audio Data in Hearing Aids".

References

Bauer, N., Schiffner, J., & Weihs, C. (2012). *Einfluss der Musikinstrumente auf die Güte der Einsatzzeiterkennung*. Discussion Paper 10/2012. SFB 823, TU Dortmund.

Bello, J. P., Daudet, L., Abdallah, S., Duxbury, C., Davies, M., & Sandler, M. B. (2005). A tutorial on onset detection in music signals. *IEEE Transactions on Speech and Audio Processing, 13*(5), 1035–1047.

Benetos, E., Holzapfel, A., & Stylianou Y. (2009). Pitched instrument onset detection based on auditory spectra. In *10th International Society for Music Information Retrieval Conference* (ISMIR 2009) Kobe, Japan (pp. 105–110).

Böck, S., Krebs, F., & Schedl, M. (2012). Evaluating the online capabilities of onset detection methods. In *Proceedings of the 13th International Conference on Music Information Retrieval* (ISMIR 2012) Porto, Portugal (pp. 49–54).

Dixon, S. (2006). Onset detection revisited. In *Proceedings of the 9th International Conference on Digital Audio Effects* (DAFx-06) Montreal, Canada (pp. 133–137).

Goto, M., Hashiguchi, H., Nishimura, T., & Oka, R. (2003) RWC music database: Music genre database and musical instrument sound database. In *Proceedings of the 4th International Conference on Music Information Retrieval* (ISMIR 2003) Baltimore, USA (pp. 229–230).

Krasser, J., Abeßer, J., & Großmann, H. (2012). Improved music similarity computation based on tone objects. In *Proceedings of the 7th Audio Mostly Conference* Corfu, Greece (pp. 47–54).

Meddis, R. (2006). Auditory-nerve first-spike latency and auditory absolute threshold: A computer model. *Journal of the Acoustical Society of America, 116*, 406–417.

Rosão, C., Ribeiro, R., & Martins de matoset, D. (2012). Influence of peak selection methods on onset detection. In *Proceedings of the 13th International Conference on Music Information Retrieval* (ISMIR 2012) Porto, Portugal (pp. 517–522).

A Unifying Framework for GPR Image Reconstruction

Andre Busche, Ruth Janning, Tomáš Horváth, and Lars Schmidt-Thieme

Abstract Ground Penetrating Radar (GPR) is a widely used technique for detecting buried objects in subsoil. Exact localization of buried objects is required, e.g. during environmental reconstruction works to both accelerate the overall process and to reduce overall costs. Radar measurements are usually visualized as images, so-called radargrams, that contain certain geometric shapes to be identified.

This paper introduces a component-based image reconstruction framework to recognize overlapping shapes spanning over a convex set of pixels. We assume some image to be generated by interaction of several base component models, e.g., hand-made components or numerical simulations, distorted by multiple different noise components, each representing different physical interaction effects.

We present initial experimental results on simulated and real-world GPR data representing a first step towards a pluggable image reconstruction framework.

1 Introduction

This paper introduces a component-wise decomposition framework for images which is applied in the context of Ground Penetrating Radar data analysis. An image is assumed to be decomposable into several information-bearing components and noise whose areas within an image may be overlapping. The latter assumption is a characteristic property of GPR data analysis and requires us to enable our model to capture interaction effects. In relation to common image analysis, our problem is best described as dissecting spotlights in a theater scene from some whole picture.

A. Busche (✉) · R. Janning · T. Horváth · L. Schmidt-Thieme
ISMLL—Information Systems and Machine Learning Lab, University of Hildesheim, Hildesheim, Germany
e-mail: busche@ismll.uni-hildesheim.de; janning@ismll.uni-hildesheim.de; horvath@ismll.uni-hildesheim.de; schmidt-thieme@ismll.uni-hildesheim.de

M. Spiliopoulou et al. (eds.), *Data Analysis, Machine Learning and Knowledge Discovery*, Studies in Classification, Data Analysis, and Knowledge Organization, DOI 10.1007/978-3-319-01595-8_35,

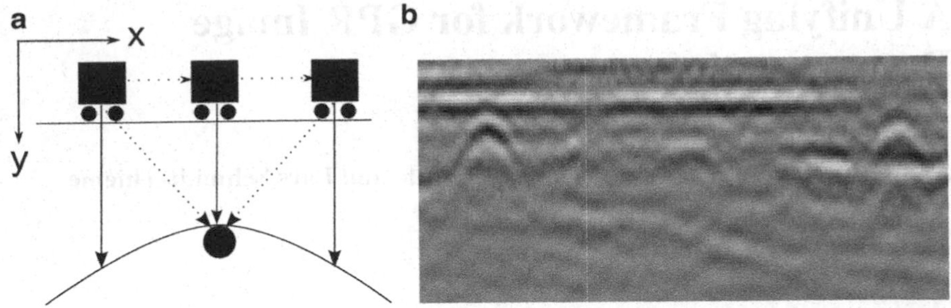

Fig. 1 Hyperbolas are caused by steady movement of a radar vehicle across a buried pipe (**a**). At each horizontal position, an A-Scan (column) is measured, whose stacked visualization results in so-called radargrams, or B-Scans (**b**)

Our framework allows to define individual base image components f_k along with multiple noise-components g_l whose exact semantics are task-dependent. Components contribute pixelwise at position (x, y) to a final image $\hat{I}$:

$$\hat{I}(x, y) = \bigodot_k f_k(x, y; \theta) + \bigodot_l g_l(x, y; \theta) \tag{1}$$

Let $\odot$ denote a component mixing operation which combines individual base components. In the most simple, yet uncommon scenario, aggregating base components can be realized by component-wise summation at pixel level. More realistic real-world application scenarios require more complex operations, e.g. if one thinks of RGB color spaces or convolution (Daniels 2004).

The analysis of Ground Penetrating Radar (GPR) data aims at subsoil imaging, for which many applications exists, e.g., landmine detection and pipe location recognition (cf. Daniels 2004, Chaps. 12, resp. 9). In this paper deals with the pipe recognition problem from B-Scans as depicted in Fig. 1.

Within our framework, we define valuable information as being hyperbola-shaped image components containing a certain reflection pattern, whereas clutter is defined as all other elements within that image.

Since we are dealing with real-world data, several artifacts exist in our radargrams: (a) varying soil conditions cause hyperbolas to be jittered, (b) shadowing or masking effects occur, e.g., if obstacles are buried beneath the pipes, (c) multiple vertical hyperbolas may represent the same pipe, as its inner filling material directly influences the propagation of the wave.

To motivate our approach presented here, we are having a closer look at the radargram image in Fig. 1b and identify the following characteristics: along the depth (y) axis, a clear reflection pattern exists along the hyperbola (Sect. 3.1). For our real-world data measured on a test-site, we can define synthetic hyperbola components (Sect. 3.2) being used within our framework to derive the actual curvature of a hyperbola (Sect. 4).

2 Related Work

Relevant related work is summarized from two areas: (a) the GPR data analysis focusing on pipe recognition and (b) object/shape identification in images.

Hyperbola Detection for GPR Data Interpretation is concerned with hyperbola recognition and can generally be divided in three different approaches: (a) hyperbola estimation from sparse data (Chen and Cohn 2010; Janning et al. 2012; Ahn et al. 2001), (b) utilizing brute-force methods such as the Hough transform to estimate model parameters (Simi et al. 2008), (c) supervised machine-learning, e.g., Neural Networks, for which training data needs to be collected beforehand (Al-Nuaimy 2000).

Since our framework does not define any learning strategy, existing approaches may be well formalized therein, such as:

- *E/M algorithms* may be trained on the pixel probability belonging to one of the base components (done in a sparse way in Chen and Cohn 2010),
- The *Hough Transform* may be phrased within our framework if components are derived in an iterative fashion, see e.g. Buza et al. (2009).

Image Reconstruction techniques are presented in Tu et al. (2006). The author also presents a unifying framework which partitions the source image into crisp subsets. Their generative approach closely relates to ours, but lacks the ability to model overlapping components which we need here.

Forsyth and Ponce (2004) present a technique called "hypothesize and test" which tries to identify previously known shapes from a template database in an image while being robust to affine transformations. Their closely related approach bears two key differences: (a) we are unable to build up an exact template database so that we would be required to infer parameters for a method emitting such templates, and (b) convolution of radar waves causes effects on the radargram images which might completely alter the shape of a template.

Zhao et al. (2012) utilize a numerical simulator to approximate radargram images, but neither evaluate their approach on real-world data nor promise applicability to any real-world data, since their approach is "seriously limited", so that their applicability is "impossible" for the general case.

3 Component-Wise Decomposition of GPR Images

Our framework is utilized for GPR data analysis as follows. Components are assumed to contribute to the final radargram in an additive way, which is known as the first-order Born approximation (Daniels 2004, Chap. 3), so that $\odot \equiv \sum$. This assumption is preliminarily made since our current study focuses on the suitability of our approach. Modeling more complex interaction effects such as convolutional

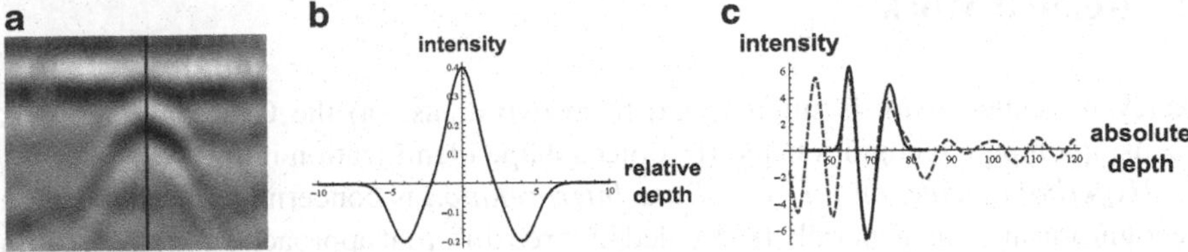

Fig. 2 Characteristic Reflection Patterns are sought in a B-Scan (**a**). These are simulated by three Gaussians (**b**) which are identified in an A-Scan (**c**)

methods is justified (Abubakar and Berg 2002), as e.g., pipes buried close to each other alter their pattern non-linearly. Our framework comprises both

- *Base Components* containing hyperbola-shaped patches along the x axis and a characteristic reflective pattern along the depth axes. Soil conditions and antenna characteristics cause hyperbolas to be distinguishable from the background only to a certain extend, the "fading window".
- *Noise* is regarded as being all other artifacts within an image, e.g., linear patches which are caused by buried walls or underground water.

To learn model parameters, we define a closed-form loss function that allows us to use direct optimization algorithms, e.g. Gradient Descent. Given an image $I(x, y)$, it is defined pixelwise as the summed squared error on intensities:

$$l(I, \hat{I}) = \sum_{x,y} (I(x, y) - \hat{I}(x, y))^2 \tag{2}$$

3.1 Pipe Reflection Patterns

We investigated reflection patterns along the depth axes shown in Fig. 2. The black vertical line represents an A-Scan being visualized in subfigure (c).

Clearly visible are either black–white–black or white–black–white patterns, whose actual excitation (intensity) differs across various real-world radargrams. Three displaced Gaussians $G(\mu, \sigma)$ define our synthetic

$$G(\mu, \sigma_h, \sigma_u, \sigma_d, g_u, g_d) = v_h \cdot G(\mu, \sigma_h) + v_u \cdot G(\mu - g_u, \sigma_u) + v_d \cdot G(\mu + g_d, \sigma_d) \tag{3}$$

which is centered at position μ with standard deviation σ_h "here" (h), having potentially varying extends g_u "upwards" (u) and g_d "downwards" (d) at different magnitudes (v) and standard deviations σ (Fig. 2b). Figure 2c contains an excerpt of an A-scan along with the reflective pattern identified.

3.2 Components for GPR Image Reconstruction

We define our synthetic hyperbola base component by combination of the reflection pattern with a hyperbola equation as:

$$\begin{aligned}\tilde{f}_k(x,y;\theta) &= \frac{v_h}{\sqrt{2\pi}\sigma_h} exp(-\frac{1}{2\sigma_h^2}(h(a,b,x_0,x)-y+y_0-a)^2)\\ &+ \frac{v_d}{\sqrt{2\pi}\sigma_d} exp(-\frac{1}{2\sigma_d^2}(h(a,b,x_0,x)-y-g_d+y_0-a)^2)\\ &+ \frac{v_u}{\sqrt{2\pi}\sigma_u} exp(-\frac{1}{2\sigma_u^2}(h(a,b,x_0,x)-y+g_u+y_0-a)^2) \qquad (4)\end{aligned}$$

with $\theta := \{a, b, x_0, y_0, v_h, v_d, v_u, g_d, g_u\}$. The component yet does not account for the above mentioned "fading window" which will be added later on. The equation combines the three-parametric hyperbola equation

$$h(a,b,x_0,x) = \sqrt{a^2 + b(x-x_0)^2} \qquad (5)$$

from Capineri et al. (1998) with the Gaussian excitation of (3), while the term $(y_0 - a)$ ensures a fixed apex location (x_0, y_0) of the hyperbola when varying parameter a. Consequently, parameters a, b only affect the curvature of the hyperbola, whose values can be constrained for a real-world scenario:

$$a > 0 \qquad b > 0 \qquad (6)$$

$$v_h \cdot v_d < 0 \qquad v_h \cdot v_u < 0 \qquad (7)$$

$$\tfrac{a}{b} < 10 \qquad (8)$$

These conditions are not required in general for the estimation process, but have experimentally shown to guide the convergence of the algorithm to a desired result, since unrealistic local minima are avoided. Condition (6) ensures the correct opening of the hyperbola, while condition (7) takes care of maintaining the desired reflection shape as in Fig. 2b. It has been observed that the optimization tends to converge to straight-line estimates, so that condition (8) ensures the hyperbolas to be shaped like those occurring in real-world scenarios. Large a and b values result in more smooth hyperbolas at the apex.

The choice of a good "fading window" to account for antenna and soil influences is both task- and data-dependent. For our experiments, we got satisfactory results using a triangle weighting function,

$$w(x;\theta) := \begin{cases} 1 - \frac{|x-x_0|}{z}, & \text{if } |x-x_0| < z \\ 0, & \text{else} \end{cases} \qquad (9)$$

parameterized by $\theta := \{x_0, z\}$ so that the base component is parameterized on the union of both individual parameters sets θ.

$$f_k(x, y; \theta) = w(x; \theta) \cdot \tilde{f}_k(x, y; \theta) \quad (10)$$

For our real-world data, setting $z = 14$ units (approx. 26 cm) in (9) resulted in satisfying results as shown below.

4 Experiments

To evaluate our framework, we ran two experiments which were implemented using Mathematica 8 using its built-in optimization routines: (a) we confirmed correct estimation of parameters using loss function (2) for simulated data derived from (4) for increasing noise and jittering, and thereafter (b) applied the framework to real-world data using the same loss but (10). In both experiments, we restricted ourselves to identify a single hyperbola.

4.1 Robustness to Noise

We simulated two kinds of noise on top of synthetic radargrams as follows:

1. Global pixelwise random uniform noise of magnitude relative to the maximum excitation (v_h) of the hyperbola. Within our framework, let noise $g(x, y) := rand(0, \epsilon)$ be some random number in the interval $(0, \epsilon)$. Clearly, if $\epsilon = v_h$, the noise amount equals the excitation of the hyperbola. Let $n_\epsilon = \frac{\epsilon}{v_h}$ denote the normalized fraction.
2. Random Gaussian jittering which we used to simulate heterogeneous soil-conditions. For this kind of noise, we added a y-offset to (4) so that $y_0 := y_0 + G(0, \sigma)$ with G denoting a random value sampled from a Gaussian distribution having variance σ.

Since we are aiming at exact determination of the parameter values on simulated data, we are measuring the error in terms of accuracy regarding the actual parameter values using RMSE. For both results shown in Fig. 3, uniform noise increases up to twice the excitation amount of the hyperbola ($n_\epsilon = 2$). Subfigure (a) contains results for the non-jittered case, whereas the target hyperbola in subfigure (b) was jittered along the y-axis. All results were obtained by averaging over 10 runs.

As stated in Sect. 3.2, the fraction a/b may be used as an indicator for the smoothness at the apex location, so that absolute values for a and b only give limited insight to the actual performance. Just focusing on this fraction and its excitation v_h, we conclude that our approach is suitable for both the jittered and non-jittered case until the noise amount even exceeds the actual excitation of the hyperbola ($n_\epsilon = 1.5$, $\sigma = 0.4$).

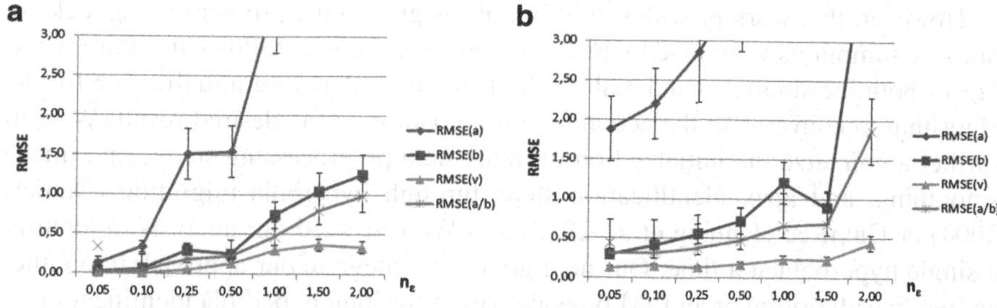

Fig. 3 RMSE on hyperbola parameters (4) for increasing noise and jitter. (**a**) Non-distorted hyperbola. (**b**) Jittered Hyperbola ($\sigma = 0.4$)

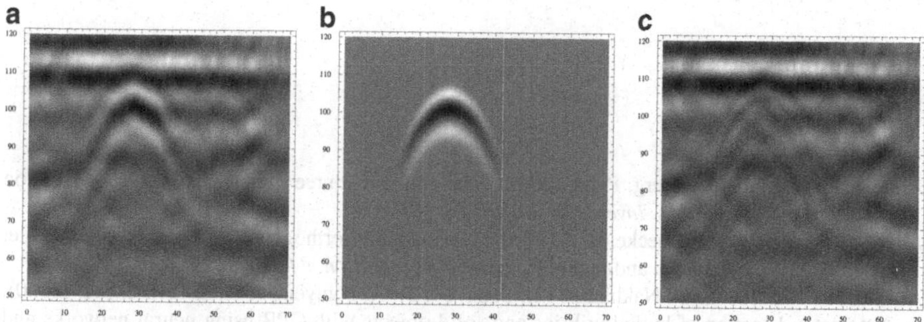

Fig. 4 Our approach successfully extracts hyperbolas (**b**) from B-Scans (**a**)

4.2 *Application to Real-World Data*

We visually evaluated our approach on real-world data on some sample radargrams. Since we are having no exact parameter values to measure the quality quantitatively, Fig. 4 shows an example setting within which subfigure (a) contains an excerpt of a real-world radargram. Initializing and fixing the component to the actual apex position (our only assumption here) and running a parameter estimation for (10) results in an image component as shown in subfigure (b) whose estimated reflection pattern at its apex location is overlayed in Fig. 2c. It can clearly be seen that the remaining noise part in subfigure (c) contains all information but the hyperbola, so that our approach may be regarded as being suitable for identifying the exact shape of a hyperbola.

5 Conclusion and Future Work

This paper presented a general formalization for component-based image decomposition and showed its application to both simulated and real-world GPR data. Based on a thorough data analysis, a two-dimensional hyperbola components was derived and empirically shown to be suitable in both settings

However, this work provides only initial insights to the problem being tackled. Some assumptions were made which we aim to address as follows in future work: (a) for both the simulated and real-world data we need a close initialization for the algorithm to converge to the actual solution/position to the desired result. We will further automatize the initialization utilizing data preprocessing steps, e.g., image smoothing, and apex identification steps through hyperbola migration (Daniels 2004) or GamRec (Janning et al. 2012). (b) We restricted our analysis to identify a single hyperbola at a time. Our next goal is to integrate our approach into either an iterative algorithm or an E/M type algorithm for joint hyperbola identification.

Acknowledgements This work is co-funded by the European Regional Development Fund project AcoGPR (http://acogpr.ismll.de) under grant agreement no. WA3 80122470.

References

Abubakar, A., & van den Berg, P. M. (2002). The contrast source inversion method for location and shape reconstructions. *Inverse Problems, 18*, 495–510.

Ahn, S. J., Rauh, W., & Warnecke, H.-J. (2001). Least-squares orthogonal distances fitting of circle, sphere, ellipse, hyperbola, and parabola. *Pattern Recognition, 34*(12), 2283–2303.

Al-Nuaimy, W., Huang, Y., Nakhkash, M., Fang, M. T. C., Nguyen, V. T., & Eriksen, A. (2000). Automatic detection of buried utilities and solid objects with GPR using neural networks and pattern recognition. *Journal of Applied Geophysics, 43*(2–4), 157–165.

Buza, K., Preisach, C., Busche, A., Schmidt-Thieme, L., Leong, W. H., & Walters, M. (2009). Eigenmode identification in Campbell diagrams. In *International Workshop on Machine Learning for Aerospace*.

Capineri, L., Grande, P., & Temple, J. A. G. (1998). Advanced image-processing technique for real-time interpretation of ground-penetrating radar images. *International Journal of Imaging Systems and Technology, 9*(1), 51–59.

Chen, H., & Cohn, A. G. (2010). Probabilistic robust hyperbola mixture model for interpreting ground penetrating radar data. In *The 2010 International Joint Conference on Neural Networks (IJCNN)* (pp. 1–8).

Daniels, D. J. (2004). *Ground penetrating radar*. London: Institute of Engineering and Technology.

Forsyth, D. A., & Ponce, J. (2004) *Computer vision: A modern approach* (1st ed.). *Prentice Hall series in artificial intelligence*. Upper Saddle River: Pearson Education.

Janning, R., Horvath, T., Busche, A., & Schmidt-Thieme, L. (2012). GamRec: A clustering method using geometrical background knowledge for GPR data preprocessing. In *8th Artificial Intelligence Applications and Innovations*. Berlin: Springer.

Simi, A., Bracciali, S., & Manacorda, G. (2008). Hough transform based automatic pipe detection for array GPR: Algorithm development and on-site tests. In *RADAR 2008* (pp. 1–6).

Tu, Z., Chen, X., Yuille, A., & Zhu, A. C. (2006). *Image parsing: Unifying segmentation, detection, and recognition* (Vol. 4170). *Lecture notes in computer science* (pp. 545–576). Dordrecht: Kluwer Academic.

Zhao, Y., Wu, J., Xie, X., Chen, J., & Ge, S. (2012) Multiple suppression in GPR image for testing backfilled grouting within shield tunnel. In *13th International Conference on Ground Penetrating Radar (GPR)* (pp. 271–276).

Recognition of Musical Instruments in Intervals and Chords

Markus Eichhoff and Claus Weihs

Abstract Recognition of musical instruments in pieces of polyphonic music given as mp3- or wav-files is a difficult task because the onsets are unknown. Using source-filter models for sound separation is one approach. In this study, intervals and chords played by instruments of four families of musical instruments (strings, wind, piano, plucked strings) are used to build statistical models for the recognition of the musical instruments playing them by using the four high-level audio feature groups Absolute Amplitude Envelope (AAE), Mel-Frequency Cepstral Coefficients (MFCC) windowed and not-windowed as well as Linear Predictor Coding (LPC) to take also physical properties of the instruments into account (Fletcher, The physics of musical instruments, 2008). These feature groups are calculated for consecutive time blocks. Statistical supervised classification methods such as LDA, MDA, Support Vector Machines, Random Forest, and Boosting are used for classification together with variable selection (sequential forward selection).

1 Introduction

What characterizes the sound of a musical instrument? Because pitch and loudness are not specific enough to discriminate between single tones of different instruments it is important to look at timbre represented by the distribution of overtones in periodograms. This distribution depends on the physical structure of the musical instrument (see Fletcher 2008; Hall 2001). On the other side the energy of every single signal has got a temporal envelope that differs from one musical

M. Eichhoff (✉) · C. Weihs
Chair of Computational Statistics, Faculty of Statistics, TU Dortmund, Dortmund, Germany
e-mail: eichhoff@statistik.tu-dortmund.de; weihs@statistik.tu-dortmund.de

M. Spiliopoulou et al. (eds.), *Data Analysis, Machine Learning and Knowledge Discovery*, Studies in Classification, Data Analysis, and Knowledge Organization, DOI 10.1007/978-3-319-01595-8_36,

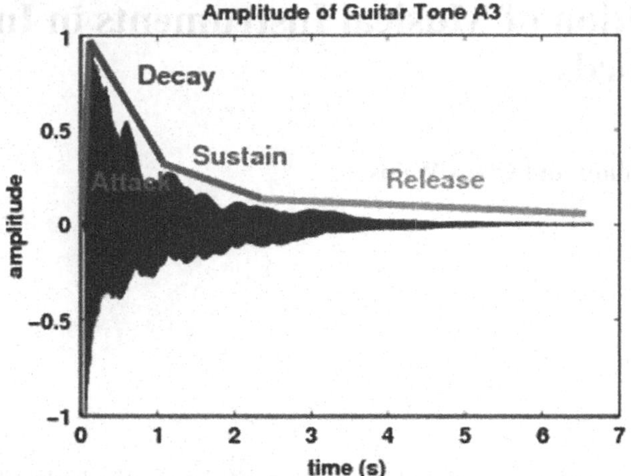

Fig. 1 ADSR-curve of a musical signal

instrument to the other and is therefore also considered in this work. In total, four groups of features are taken into account: a spectral envelope derived from linear predictor coefficients (LPC, 125 var.), Mel-frequency Cepstral Coefficients (MFCC, windowed: 80) and MFCC not-windowed (16 var.) as well as an Absolute Amplitude Envelope (AAE, 132 var.). MFCCs have already shown to be useful for classification tasks in speech processing (see Rabiner and Juang 1993; Zheng et al. 2001) as well as in musical instrument recognition (Krey and Ligges 2009; Wold et al. 1999; Brown 1999).

On the one hand, each musical signal is windowed by half-overlapping segments w_s, $s \in \{1, \ldots, 25\}$, of a size of 4,096 samples. On the other hand, each musical signal can be divided into four phases: attack, decay, sustain and release (ADSR) (see Fig. 1). In order to model these phases, blocks of five consecutive windows each are constructed (see Fig. 2) and so-called block-features are calculated for each block. The number five is chosen arbitrarily.

In this paper, musical signals are intervals and chords. All intervals are built by two different tones played simultaneously by the same or different instruments. All chords consist of three or four tones played simultaneously by the same or different instruments. See Sect. 3.1 for more information.

The four instrument classes consist of the following instruments:

Strings: violin, viola, cello
Wind: trumpet, Flute
Piano
Plucked strings: E-guitar, acoustic guitar

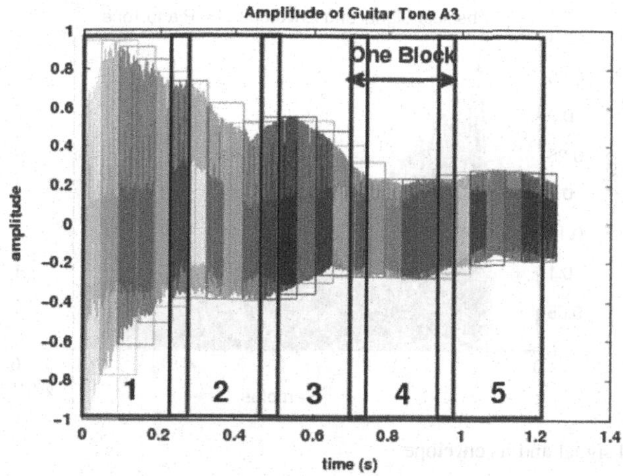

Fig. 2 Blocks of a tone

2 Groups of Features

Each interval or chord is an audio signal $x[n]$, $n \in \{1, \ldots 52920\}$, of length 1.2 s with the sampling rate $sr = 44{,}100$ Hz. On these signals the features of each feature group (LPC, MFCC windowed, MFCC not-windowed, AAE) are calculated.

As described in Eichhoff and Weihs (2010) all features but the Absolute Amplitude Envelope are calculated on each of the five consecutive time windows using the component-by-component median for the blocks. In case of the Absolute Amplitude Envelope calculation was done by using non-overlapping windows of size 400.

The MFCC features are also calculated on the whole 1.2 s ("unwindowed").

Visualizations of these features can be found in Figs. 3–6. Further details concerning the four mentioned feature groups may be found in Eichhoff and Weihs (2010).

3 Classification Tasks and Test Results

For each instrument class piano, plucked strings, strings and wind, classification is carried out separately with class variables with values $\{1, 0\}$, meaning that instrument class is involved in the interval/chord or not. As classification methods Linear Discriminant Analysis (LDA), Multivariate Discriminant Analysis (MDA), Random Forest (RaFo), the boosting method Adaboost (Adaboost) and Decision Trees (Rpart), k-nearest neighbours (kNN) and Support Vector Machines (SVM) are used. The last five methods use hyper-parameters tuned on a grid.

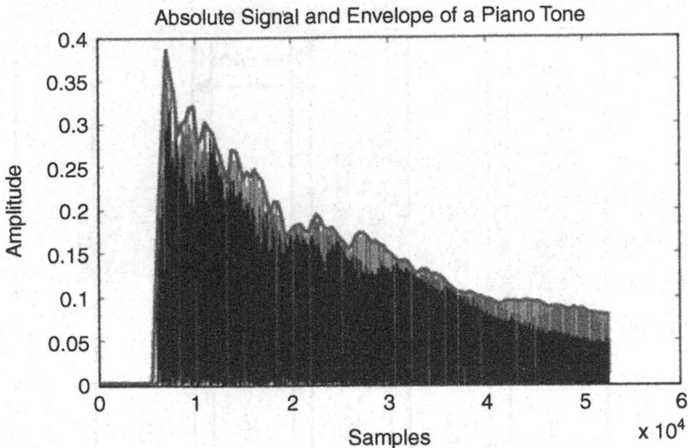

Fig. 3 Musical signal and its envelope

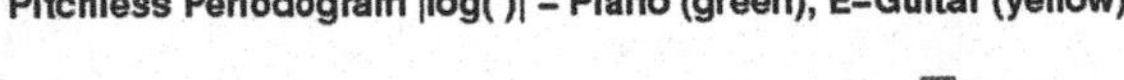

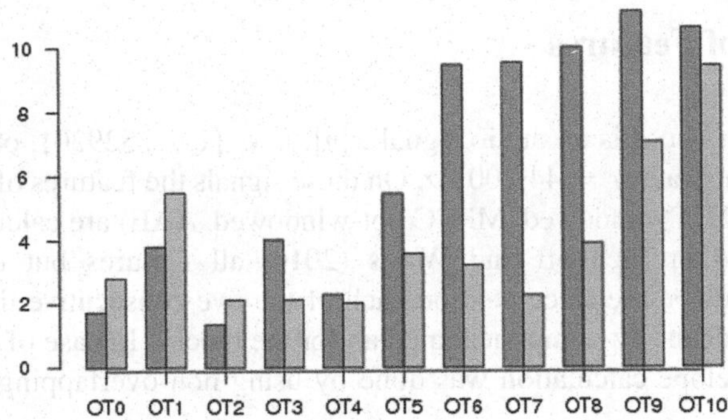

Fig. 4 Pitchless periodogram

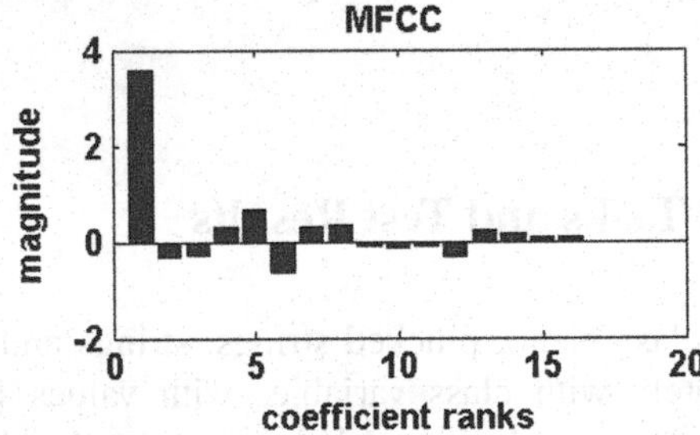

Fig. 5 16 MFCC of a musical signal

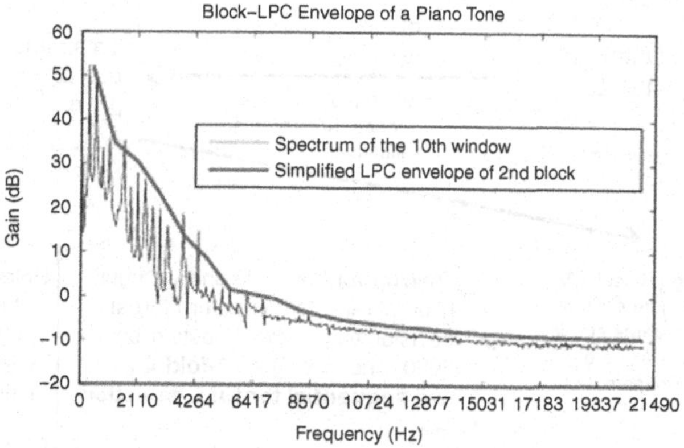

Fig. 6 LPC simplified spectral envelope

3.1 *Database of Intervals and Chords*

Only those intervals and chords are used for classification that often appear in pop music. For this reason the small/large third, fourth, quint, large sixth and small seventh are used in case of the intervals and steps I, IV, V, and VI in case of the chords. The latter consists of four, all other chords of three tones.

The tones of the intervals as well as the chords are i.i.d. chosen out of 12 possible tones as well as the instruments out of the possible eight instruments, the key from major/minor, the interval size, and the steps.

The data set consists of 2,000 tones as training and 1,000 tones as a hold out set. All these tones are built by using 1,614 guitar, 1,202 string, 313 piano and 418 wind tones from the McGill University Master Samples (see McGill University 2010), RWC Database (see Goto et al. 2003) and music samples of the Electronic Music Studios, Iowa (see University of Iowa 2011).

The sound intensity of all tones is mezzo-forte (mf) as stated at the labels of the tones. Of course, this has to be regarded with caution because even if the tones are labeled with "mf", the instrumentalists may have played differently and the recording conditions could have been different in all three cases.

The pitch range of the involved individual tones is E2 up to Gb6 (approx. 82–1,479 Hz).

3.2 *Classification Steps*

Each data frame $A \in \mathbb{R}^{m \times n}$ (rows refer to objects, columns to features) is pre-filtered. For each iteration of a tenfold cross-validation a logistic model is calculated and by stepwise forward selection those variables are selected that minimize the

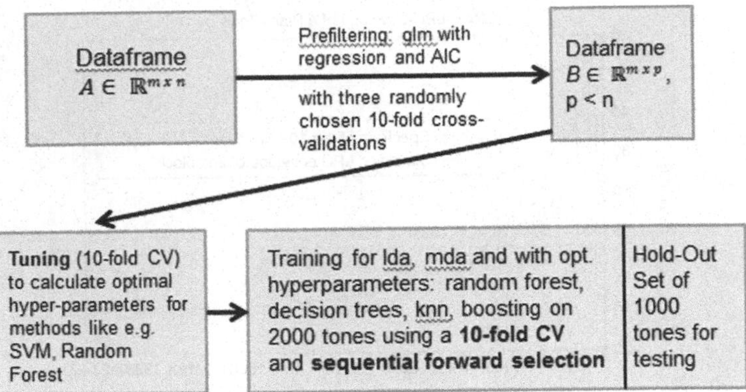

Fig. 7 Classification process

Table 1 Intervals—evaluation result and number of selected variables

	LPC, MFCC win. + not-win., AAE (353 var.)		MFCC win. + not-win. (96 var.)	
	Misclass. error (best tuning result after prefiltering)	Number of selected var. (only prefiltering)	Misclass. error (best tuning result after prefiltering)	Number of selected var. (only prefiltering)
Guitar	0.095 (0.085)	9 (34)	0.155 (0.137)	14 (36)
Piano	0.135 (0.105)	11 (45)	0.143 (0.130)	9 (16)
Strings	0.100 (0.095)	11 (56)	0.140 (0.158)	18 (23)
Wind	0.089 (0.069)	17 (38)	0.121 (0.097)	11 (24)

(AIC)-criterion (Akaike 1974). This procedure is done for three randomly built tenfold cross-validations. Those variables are selected that have been chosen at least nine of ten times.

This procedure leads to a data frame $B \in \mathbb{R}^{m \times p}$ with $p < n$. Then, hyper-parameter tuning is carried out by a tenfold cross-validation, if necessary. The "optimal" hyper-parameters are then used to carry out the training procedure on the above mentioned 2,000 tones using a tenfold cross-validation and sequential forward selection. The methods without hyper-parameters start directly at this point.

At the end, evaluation on the hold-out set is being done to get the misclassification error on these "new" tones.

Figure 7 shows the process of classification. For each instrument and interval and chord classification the best results are shown in Tables 1 and 2. The corresponding classification method always was SVM except for Strings where RaFo shows best results.

Table 3 shows the features jointly selected for two instrument classes (G = Guitar, P = Piano, S = Strings, W = Wind). From this table one can calculate the percentage of features for guitar, piano, strings, wind (left to right) shared with the other instrument classes: Intervals: 0.67 0.45 0.27 0.35, Chords: 0.18 0.31 0.10 0.14.

Table 2 Chords—evaluation result and number of selected variables, based on all 353 variables

	Misclass. error (best tuning result after prefiltering)	Number of selected var. (only prefiltering)
Guitar	0.130 (0.140)	14 (31)
Piano	0.190 (0.204)	13 (42)
Strings	0.133 (0.138)	10 (33)
Wind	0.208 (0.197)	7 (30)

Table 3 Number of "shared" features

	Intervals				Chords			
	G	P	S	W	G	P	S	W
G	9	3	1	2	14	2	0	0
P	3	11	0	2	2	13	1	1
S	1	0	11	2	0	1	10	0
W	2	2	2	17	0	1	0	7

One can observe that in case of chords the number of shared features is smaller than for intervals. This may be explained by the more complex situation of the chord classification. Due to this fact more specific information is necessary to detect a certain instrument class. Due to the same reason, the misclassification errors in the interval classification Table 1 are higher than in the chord classification Table 2. From Table 1 one can recognize that additionally to windowed and not-windowed MFCC it makes sense to consider also the other feature groups AAE and LPC.

3.3 *Classification Results: Common Features Blockwise*

Furthermore, classification results show that not every block seems to be important and necessary for classification. Although one may note critically that only one replication of the procedure has been done. Naturally, the statement could be more robust after many replications.

Figure 8 shows that features from blocks 2 and 3 in case of the intervals and blocks 3 and 5 in case of the chords are rarely chosen. The numbers in this figure denote the feature number $\{1, \ldots, \text{dimension(feature)}\}$ of each feature group. Concerning the selected not-windowed MFCC features it can be seen in Table 4 that only the first three of 16 features are selected in each of the four instrument classes.

Table 5 shows the evaluation results for intervals with reduced blocks for all variable groups and only not-windowed MFCCs (no. 1, 2, 3). Again, in all cases except wind SVM showed the best results. For wind, RaFo was the best method. A comparison of Tables 1 and 5 shows that block reduction to just block 1 is sufficient for all instrument classes except for the string class because there the results get worse. In addition, block 4 appears to be important for strings (approx. 9 % error rate, see right hand side of Table 5).

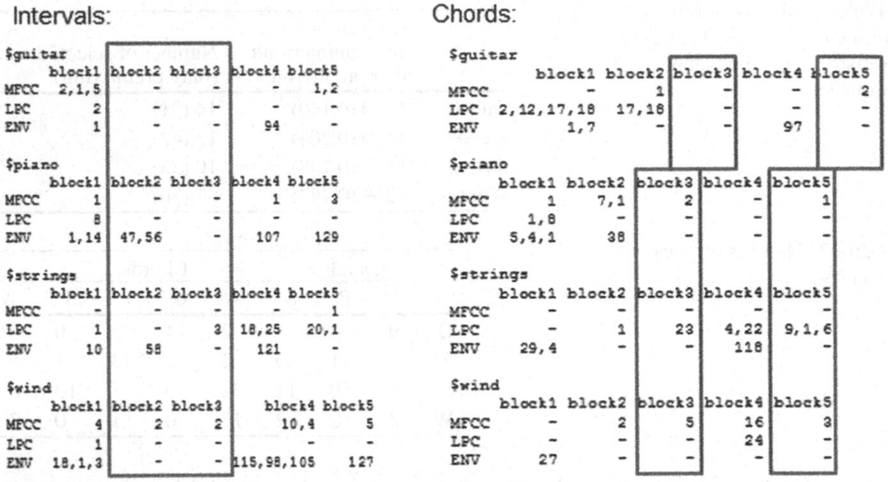

Intervals:

```
$guitar
     block1 block2 block3 block4 block5
MFCC  2,1,5      -      -      -    1,2
LPC       2      -      -      -      -
ENV       1      -      -     94      -

$piano
     block1 block2 block3 block4 block5
MFCC      1      -      -      1      3
LPC       8      -      -      -      -
ENV    1,14  47,56      -    107    129

$strings
     block1 block2 block3 block4 block5
MFCC      -      -      -      -      1
LPC       1      -      3  18,25   20,1
ENV      10     58      -    121      -

$wind
     block1 block2 block3       block4 block5
MFCC      4      2      2         10,4      5
LPC       1      -      -            -      -
ENV  18,1,3      -      - 115,98,105    127
```

Chords:

```
$guitar
                block1 block2 block3 block4 block5
MFCC                 -      1      -      -      2
LPC     2,12,17,18  17,18      -      -      -
ENV                1,7      -      -     97      -

$piano
     block1 block2 block3 block4 block5
MFCC      1    7,1      2      -      1
LPC     1,8      -      -      -      -
ENV   5,4,1     38      -      -      -

$strings
     block1 block2 block3 block4 block5
MFCC      -      -      -      -      -
LPC       -      1     23   4,22  9,1,6
ENV    29,4      -      -    118      -

$wind
     block1 block2 block3 block4 block5
MFCC      -      2      5     16      3
LPC       -      -      -     24      -
ENV      27      -      -      -      -
```

Fig. 8 Common features blockwise

Table 4 Selected not-windowed MFCC feature-numbers (1–16)

	Intervals	Chords
G	1	–
P	1	2,4
S	2	–
W	1,2,3	2

Table 5 Intervals—evaluation results and number of selected variables

	Block No. 1 (75 var.)		Blocks No. 1 and 4 (146 var.)	
	Misclass. error (best tuning result after prefiltering)	Number of selected var. (only prefiltering)	Misclass. error (best tuning result after prefiltering)	Number of selected var. (only prefiltering)
Guitar	0.094 (0.092)	10 (19)	0.097 (0.093)	10 (26)
Piano	0.118 (0.113)	11 (22)	0.119 (0.105)	10 (34)
Strings	0.201 (0.182)	7 (17)	0.091 (0.083)	14 (36)
Wind	0.109 (0.109)	7 (16)	0.107 (0.109)	11 (20)

4 Conclusion

Musical instrument recognition in intervals and chords leads to good misclassification errors of around 10 % for the intervals when using only block 1 for guitar, piano, wind and the two blocks 1 and 4 for strings. The number of selected variables is between 7 and 14.

It would be useful to carry out more than one replication of the procedure to get more robust results concerning statements of selected features for interval and chord classification and more stable misclassification error rates. Also, investigations based on statistical tests and further investigation in the classification of chords with a reduced block number will be done.

References

Akaike, H. (1974). A new look at the statistical model identification. *IEEE Transactions on Automatic Control, 19*(6), 716–723. doi:10.1109/TAC.1974.1100705.

Brown, J. C. (1999). Computer identification of musical instruments using pattern recognition with cepstral coefficients as features. *Journal of the Acoustical Society of America, 105*(3), 1933–1941.

Eichhoff, M., & Weihs, C. (2010). Musical instrument recognition by high-level features. In *Proceedings of the 34th Annual Conference of the German Classification Society (GfKl)*, July 21–23 (pp. 373–381). Berlin: Springer.

Fletcher, N. H. (2008). *The physics of musical instruments*. New York: Springer.

Goto, M., Hashigushi, H., Nishimura, T., & Oka, R. (2003). RWC music database: Music genre database and musical instrument sound database. In *ISMIR 2003 Proceedings* (pp. 229–230). Baltimore: Johns Hopkins University Press.

Hall, D. E. (2001). *Musical acoustics* (3rd ed.). Belmont: Brooks Cole.

Krey, S., & Ligges, U. (2009). SVM based instrument and timbre classification. In H. Locarek-Junge, & C. Weihs (Eds.), *Classification as a tool for research*. Berlin: Springer.

McGill University. (2010). *Master samples collection on DVD*. http://www.music.mcgill.ca/resources/mums/html.

Rabiner, L., & Juang, B. H. (1993). *Fundamentals of speech recognition*. Englewood Cliffs: Prentice Hall PTR.

University of IOWA. (2011). *Electronic music studios. Musical instrument samples*. http://theremin.music.uiowa.edu.

Wold, E., Blum, T., Keislar, D., & Wheaton, J. (1999). Classification, search and retrieval of audio. *Handbook of multimedia computing* (pp. 207–226). Boca Raton: CRC Press.

Zheng, F., Zhang, G., & Song, Z. (2001). Comparison of different implementations of MFCC. *Journal of Computer Science & Technology 16*(6), 582–589.

ANOVA and Alternatives for Causal Inferences

Sonja Hahn

Abstract Analysis of variance (ANOVA) is a procedure frequently used for analyzing experimental and quasi-experimental data in psychology. Nonetheless, there is confusion which subtype to prefer for unbalanced data. Much of this confusion can be prevented when an adequate hypothesis is formulated at first. In the present paper this is done by using a theory of causal effects. This is the starting point for the following simulation study done on unbalanced two-way designs. Simulated data sets differed in the presence of an (average) effect, the degree of interaction, total sample size, stochasticity of subsample sizes and if there was confounding between the two factors (i.e. experimental vs. quasi-experimental design). Different subtypes of ANOVA as well as other competing procedures from the research on causal effects were compared with regard to type-I-error rate and power. Results suggest that different types of ANOVA should be used with care, especially in quasi-experimental designs and when there is interaction. Procedures developed within the research on causal effects are feasible alternatives that may serve better to answer meaningful hypotheses.

1 Background

Often there are various independent variables in a single analysis. When the independent variables are categorical—indicating different groups of persons for example—usually a t-test or ANOVA is used to analyze the data. For example, in a study Pollock et al. (2002) investigated how students performed on a post test capturing understanding of technical topic. The groups differed in two aspects: Firstly,

S. Hahn (✉)
Friedrich-Schiller-Universität Jena, Institut für Psychologie, Am Steiger 3, Haus 1, 07743 Jena, Germany
e-mail: hahn.sonja@uni-jena.de

M. Spiliopoulou et al. (eds.), *Data Analysis, Machine Learning and Knowledge Discovery*, Studies in Classification, Data Analysis, and Knowledge Organization, DOI 10.1007/978-3-319-01595-8_37,

only some of the students had received prior teaching on technical topics. Secondly, the groups received a different instruction: Either a simple or a more complex text was given to the students. The results of the study indicated an interaction of medium effect size: The students with no prior teaching performed better on the post test when reading the simple text instead of the complex text. For the other students the opposite was true. Whereas Pollock et al. (2002) were mainly interested in the interaction, often the overall performance of the manipulated factor is of interest (i.e. instructional text in the example).

One possible approach to test such an overall effect is to use an ANOVA and focus on the main effect of the manipulated factor (see e.g. Eid et al. 2010 for background and terminology of ANOVA). When the other independent variable can not be manipulated, the number of observations in the subsamples may differ. This is called an unbalanced design. Most software packages provide different algorithms that lead to different results in unbalanced designs (see e.g. Langsrud (2003); Searle (1987)). In this paper I will use the distinction of Type I-IV sums of squares that was used originally in the software package SAS GLM. The different algorithms can be regarded testing different hypotheses due to different weighting of the cell means (see Searle (1987); Werner (1997)). Often resarchers using ANOVA are actually interested in causal questions. Eid et al. (2010) note that the theory of causal effects provides an adequate basis in this case (see also Wüthrich-martone (2001)).

The theory of causal effects (Steyer et al. in press) distinguishes between covariates C_X that are prior to a treatment variable X that is again prior to an outcome variable Y. In our example the covariate prior teaching is prior to the different instructional texts (x vs. x', treatment variable) and these are prior to the post test (outcome variable). Thus, although prior teaching is not a continuous variable here, it will be regarded as a covariate. Only a single covariate is used here to simplify matters. Usually various covariates will have to be regarded. The overall or average treatment effect (ATE) is defined by

$$\text{ATE} \equiv E[E^{X=x}(Y|C_x) - E^{X=x'}(Y|C_x)]$$

Referring to the example this means: First calculate the (theoretical) differences on the outcome variable between both instructional methods for each value of the covariate. Then calculate the weighted sum of these differences using the (theoretical) distribution of the covariate as weights. So the ATE takes the distribution of the covariate into account. For choosing the right procedure to test the hypothesis H_0: ATE $= 0$, the design of the study plays an important role: In a randomized experiment the distribution of the covariates does not differ systematically between the treatment groups. In a quasi-experiment usually this is not true: Here the covariates often affect both, the treatment variable and the outcome variable. This is called *confounding*. Adjustment procedures should be used in this case.

In unbalanced two-way designs, the question occurs how to test the hypothesis H_0: ATE $= 0$. Procedures like ANOVA, that are recommended in textbooks, will be compared to procedures from the causal effects research in a simulation study. Some of the hypotheses tested by the different procedures coincide when there is no confounding or no interaction. Furthermore some of the procedures use the general linear model (GLM) that assumes a fixed design matrix. Nonetheless this assumption is rarely met in many real world investigations as subsample sizes are not fixed by the design of the study. Therefore, I manipulated these aspects in the simulation study.

2 Method

2.1 Simulated Data Sets

The data from Pollock et al. (2002) was used as a starting point for the simulation study. First the covariate Z was simulated as an indicator variable (0: no prior teaching, 1: prior teaching) with $P(Z = 1) = 0.4$. The treatment variable X was also an indicator variable (0: simple text, 1: complex text). In the randomized experiment condition $P(X = 1|Z = 0) = P(X = 1|Z = 1) = 0.5$, and for the quasi-experimental condition $P(X = 1|Z = 0) = \frac{1}{3}$ and $P(X = 1|Z = 1) = \frac{3}{4}$ indicating a favorable treatment assignment in this case. For the fixed sample sizes the proportion of sample sizes corresponded exactly to these probabilities. For stochastic sample sizes binomial distributions were used to reflect deviations due to the sampling process. The simulation procedure further ensured that there were at least two observations in each cell.

The outcome variable Y was calculated as a linear combination of Z, X and the interaction XZ with additional normally distributed error terms. The linear combination was constructed similar to the data from Pollock et al. (2002) with minor changes: The size of the interaction was modified so that no interaction, an interaction like in the study (middle interaction) or the twice that interaction (strong interaction). The coefficient of X was modified so that two different values for the ATE resulted. For the simulation under the Null hypothesis the ATE was zero. For the simulation under the Alternative hypothesis the effect size like defined in Steyer and Partchev (2008, p. 40) for the ATE was 0.5. In different data sets we found effect sizes up to 1.0, strongly depending on the content of the analysis.

The total sample size N ranged between 30 and 900. The nominal α-level was set at .05 as a typical level used in psychology. As the procedures described are parametric, also other nominal α-levels are possible. The simulation study was conducted using R 2.13. For each combination 1,000 data sets were simulated. Each data set was analyzed using the following procedures.

2.2 Procedures Under Investigation

The t-test for unequal variances (Welch-test) with the treatment variable as the independent variable as well as various ANOVA approaches were investigated. The following variants of ANOVA used the covariate prior teaching and the treatment variable instructional method as factors:

ANOVA 1: ANOVA of Type I sums of squares including the treatment variable as the first and the covariate as the second factor

ANOVA 2: ANOVA of Type II sums of squares (This is equivalent to ANOVA of Type I including the covariate as the first factor.)

ANOVA 3: ANOVA of Type III sums of squares and ANOVA of Type IV sums of squares (Both procedures lead to identical results when there are no empty cells.)

The procedures are contained in the standard installation of R and in the *car* library (Fox and Weisberg 2011), respectively. A cross check assured that results equaled those of SPSS.

There are many procedures from research on causal effects that aim to correct for bias introduced by confounding variable. They stem from different research traditions. Most of the procedures presented here focus on the relation between covariate and outcome variable, whereas the last one using Matching focuses on the relation between covariate and treatment variable. For an wider overview see for example Schafer and Kang (2008).

The first procedure is an ANCOVA procedure proposed by Aiken and West (1996). This variant of the standard ANCOVA allows for an additional interaction term and conducts a mean centering on the covariate. It tests the hypothesis H_0: ATE $= 0$. Although qualitative covariates may cause problems in procedures like ANCOVA, here we can easily include them as the present covariate is dichotomous. The procedure relies on the general linear model and the assumptions therein (i.e. fixed subsample sizes, normally distributed error terms with equal variances). When these assumptions are fulfilled, this procedure can be regarded a baseline procedure for testing this hypothesis. See Schafer and Kang (2008) for more background information and also for R-Syntax.

The second procedure that extends the ANCOVA approach is EffectLite developed by Steyer and Partchev (2008). It relies on maximum likelihood estimation and uses Mplus in the background. A benefit of this method is that—beside of interaction—it allows for stochastic group sizes, heterogeneous error terms and latent variables for example.

The third procedure is a Regression Estimate approach (Regr.Est.) that was developed by Schafer and Kang (2008). The background of this procedure stems from research on missing values. It uses a special computation of standard errors. In the simulation study by Schafer and Kang (2008) it showed a very robust behavior. Additional background and R-code are provided along with the article of Schafer and Kang (2008).

Finally as a matching procedure the Match function with exact matching and ATE estimation of the Matching package of Sekhon (2011) was used. The procedure uses weights to calculate estimators and standard errors for testing the ATE. Like in the package a standard normal distribution of the test statistic was used for assessing significance. Schafer and Kang (2008) do not recommend using matching methods based on weights.

3 Results

3.1 Type-I-Error Rates

Table 1 shows the empirical type-I-error rates for the different procedures under different simulated conditions.

The t-test shows only good adherence to the nominal α-level in the experimental conditions. The ANOVA approaches behave differently: ANOVA 1 yields similar problems like the t-test, thus working only well in experimental conditions. Furthermore stochastic group sizes lead to slightly higher type-I-error rates even in the randomized experiment. ANOVA 2 displays the best behavior of all ANOVA approaches, only failing when there are both interaction and confounding or both interaction and stochastic group sizes. ANOVA 3 shows problems whenever an interaction is present.

ANCOVA with mean centering works well unless when there are strong interactions combined with stochasticity. EffectLite shows difficulties with small sample sizes in all conditions. The Regression Estimates procedure never exceeds the nominal α-level but seems to behave quite conservative. Finally, the Matching procedure sometimes excesses the nominal α-level for small sample sizes while being quite conservative in the presence of interaction.

3.2 Power

An important aspect for applied researchers is to know about the power of the different procedures (see Fig. 1).

The t-test shows the lowest power of all investigated procedures. ANOVA 1 and ANOVA 2 belong to the procedures with the highest power. ANOVA 3 was not included in the graph because of the too high type-I-error rate under the Null hypothesis.

The ANCOVA approach and EffectLite belong to the procedures with the highest power. The Regression Estimates procedure shows a rather low power, and the Matching procedure a medium power.

Table 1 Empirical type-I-error rates for the nominal α-level of 5 % under different side conditions for different procedures. Type-I-error rates exceeding the 95 % confidence interval [.037,.064] are printed in bold. In each condition 1,000 data sets were simulated

Interaction	No				Middle				Strong			
N	30	90	300	900	30	90	300	900	30	90	300	900
Randomized experiment, fixed group sizes												
t-test	.039	.040	.044	.040	.022	.030	.024	.040	.015	.014	.019	.013
ANOVA 1	.042	.050	.054	.046	.047	.047	.045	**.066**	.048	.053	.049	.046
ANOVA 2	.042	.050	.054	.046	.047	.047	.045	**.066**	.048	.053	.049	.046
ANOVA 3	.045	.045	.052	.043	.048	**.067**	**.121**	**.266**	**.076**	**.136**	**.310**	**.716**
ANCOVA	.042	.050	.054	.046	.047	.047	.045	**.066**	.048	.053	.049	.046
EffectLite	**.077**	.061	.056	.047	**.075**	.058	.048	**.069**	**.082**	.060	.051	.046
Regr.Est.	.031	.021	.023	.018	.023	.022	.014	.028	.024	.018	.018	.010
Matching	**.069**	.056	.056	.046	.054	.046	.037	.051	.043	.025	.028	.019
Quasi-experiment, fixed group sizes												
t-test	**.069**	**.124**	**.358**	**.788**	**.105**	**.309**	**.832**	**.997**	**.130**	**.504**	**.979**	**1.00**
ANOVA 1	**.076**	**.134**	**.375**	**.803**	**.155**	**.376**	**.876**	**.998**	**.272**	**.711**	**.996**	**1.00**
ANOVA 2	.050	.042	.043	.052	.052	.059	.058	.063	.053	**.066**	**.080**	**.147**
ANOVA 3	.041	.052	.039	.047	.055	**.080**	**.087**	**.217**	**.067**	**.120**	**.239**	**.589**
ANCOVA	.047	.045	.040	.055	.050	.055	.048	.048	.051	.054	.046	.041
EffectLite	**.080**	.054	.045	.055	**.087**	.063	.051	.048	**.078**	.058	.048	.043
Regr.Est.	.050	.022	.015	.018	.055	.023	.018	.021	.036	.021	.004	.012
Matching	**.082**	.059	.040	.055	**.071**	.052	.040	.042	.048	.029	.012	.017
Randomized experiment, stochastic group sizes												
t-test	.041	.054	.054	.040	.047	.056	.052	.058	.032	.064	.055	.040
ANOVA 1	.048	.064	.064	.047	**.074**	**.076**	**.078**	**.076**	**.095**	**.134**	**.129**	**.100**
ANOVA 2	.044	.057	.055	.042	.061	.052	.061	.064	.052	**.084**	**.077**	**.071**
ANOVA 3	.047	.044	.048	.043	.062	.062	**.103**	**.259**	**.068**	**.109**	**.280**	**.715**
ANCOVA	.043	.057	.054	.042	.058	.054	.060	.064	.048	**.084**	**.075**	**.073**
EffectLite	**.069**	.062	.056	.043	**.088**	.055	.058	.058	.058	**.071**	.053	.041
Regr.Est.	.040	.027	.026	.019	.039	.020	.023	.018	.030	.029	.017	.017
Matching	.063	**.066**	.056	.043	**.076**	.051	.051	.050	.042	.047	.030	.026
Quasi-experiment, stochastic group sizes												
t-test	.044	**.104**	**.322**	**.796**	.050	**.223**	**.732**	**.998**	.062	**.333**	**.917**	**1.00**
ANOVA 1	.051	**.120**	**.332**	**.809**	**.074**	**.277**	**.791**	**.999**	**.164**	**.498**	**.960**	**1.00**
ANOVA 2	.047	.046	.060	.040	.046	.045	**.076**	**.072**	**.068**	**.083**	**.103**	**.180**
ANOVA 3	.038	.044	.062	.044	.050	.054	**.117**	**.228**	.062	**.098**	**.276**	**.626**
ANCOVA	.046	.049	.062	.039	.052	.046	**.068**	.054	**.078**	**.075**	**.081**	**.076**
EffectLite	**.081**	.052	.063	.039	**.095**	.050	**.065**	.048	**.087**	**.065**	**.066**	.056
Regr.Est.	.045	.021	.029	.020	.045	.017	.029	.021	.053	.028	.024	.017
Matching	**.070**	.055	.064	.040	**.075**	.042	.058	.040	**.068**	.043	.037	.030

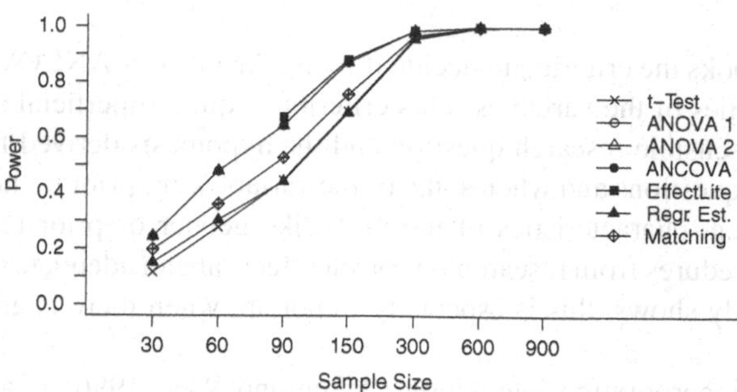

Fig. 1 Comparison of power for different procedures. Only procedures with sufficient adherence to the nominal α-level were included. Results are based on 1,000 simulated data sets

4 Discussion

Both, the very progressive behavior of the t-test in the quasi-experimental condition and the low power in the experimental condition can be explained as this procedure does not incorporate the covariate. The results show that ANOVA procedures in many cases are not appropriate to test the hypothesis about the ATE. This can be explained by comparing the hypotheses underlying the ANOVA procedures with the definition of the ATE. Depending on the respective ANOVA procedure, the hypotheses coincide only in cases with no interaction or in the experimental condition.

Alternative procedures from the research on causal effects work better than ANOVA in many cases containing an interaction. Differences between these procedures result from the different approaches. The ANCOVA approach from Aiken and West (1996) works fine unless when the group sizes are stochastic. Stochastic group sizes violate the assumption of fixed subsample sizes. The EffectLite approach works well with bigger sample sizes. The reason therefore is that the maximum likelihood estimation requires big sample sizes. The conservative behavior of the Regression Estimates approach might be due to its robustness (compare Schafer and Kang (2008)). Finally the Matching approach from Sekhon (2011) using weights has been criticized on various occasions (e.g. Schafer and Kang (2008)). The progressive behavior for small sample sizes might be avoided when a t-distribution instead of a standard normal distribution is used for significance testing.

The simulation study is limited in various ways as only a small set of conditions was investigated. In particular, this refers to the single, dichotomous covariate and the error term distribution. Further simulation studies should also focus on the stochasticity of subsample sizes, as this is quite common in applied studies.

5 Conclusion

In many textbooks the criterion to decide if to use ANOVA or ANCOVA procedures are characteristics of the variables. This criterion is quite superficial and should be replaced by a sensible research question and the hypothesis derived from it. In the case of causal questions and when some of the variables are prior to the intervention investigated (e.g. characteristics of the units like gender or prior teaching in the example), procedures from research on causal effects are an adequate choice. As the simulation study shows, this is especially important when there is are interactions in the data.

ANCOVA incorporating interaction (Aiken and West 1996) is a good choice when the group sizes are fixed by the design of the study. Other procedures like Regression Estimates (Schafer and Kang 2008) or EffectLite (Steyer and Partchev 2008) have the capability of treating frequently occurring difficulties like stochastic group sizes or variance heterogeneity. As the simulation study showed, EffectLite should be used only in sufficiently large samples (i.e. $N \geq 90$).

References

Aiken, L. S., & West, S. G. (1996). *Multiple regression: Testing and interpreting interactions.* Thousand Oaks: Sage.

Eid, M., Gollwitzer, M., & Schmitt, M. (2010). *Statistik und Forschungsmethoden: Lehrbuch.* Basel: Beltz Verlag.

Fox, J., & Weisberg, S. (2011). *An R companion to applied regression.* Los Angeles: Sage.

Langsrud, O. (2003). ANOVA for unbalanced data: Use Type II instead of Type III sums of squares. *Statistics and Computing, 13*, 163–167.

Pollock, E., Chandler, P., & Sweller, J. (2002). Assimilating complex information. *Learning and Instruction, 12*, 61–86.

Schafer, J. L., & Kang, J. D. Y. (2008). Average causal effects from observational studies: A practical guide and simulated example. *Psychological Methods, 13*, 279–313.

Sekhon, J. S. (2011). Multivariate and propensity score matching software with automated balance optimization: The Matching package for R. *Journal of Statistical Software, 42*(7), 1–52.

Searle, S. R. (1987). *Linear models for unbalanced data.* New York: Wiley.

Steyer, R., Mayer, A., Fiege, C. (in press). Causal inference on total, direct, and indirect effects. In A. C. Michalos (Ed.), *Encyclopedia of quality of life and well-being research.* Springer.

Steyer, R., & Partchev, I. (2008). EffectLite for Mplus: A program for the uni- and multivariate analysis of unconditional, conditional and average mean differences between groups. www.statlite.com. Accessed 5 May 2008.

Werner, J. (1997). *Lineare Statistik: das allgemeine lineare Modell.* Weinheim: Beltz.

Wüthrich-martone, O. (2001). *Causal modeling in psychology with qualitative independent variables.* Aachen: Shaker.

Testing Models for Medieval Settlement Location

Irmela Herzog

Abstract This contribution investigates two models for the spread of Medieval settlements in the landscape known as Bergisches Land in Germany. According to the first model, the spread was closely connected with the ancient trade routes on the ridges. The alternative model assumes that the settlements primarily developed in the fertile valleys. The models are tested in a study area for which the years are known when the small hamlets and villages were first mentioned in historical sources. It does not seem appropriate to apply straight-line distances in this context because the trade routes of that time include curves. Instead an adjusted distance metric is derived from the ancient trade routes. This metric is applied to generate a digital raster map so that each raster cell value corresponds to the adjusted distance to the nearest trade route (or fertile valley respectively). Finally, for each model a Kolmogorov–Smirnov test is applied to compare the adjusted distances of the Medieval settlements with the reference distribution derived from the appropriate raster map.

1 Introduction

Two models have been proposed for the spread of Medieval settlements in the landscape known as Bergisches Land in Germany. Some experts think that the spread was closely connected with the ancient trade routes which were already in use before the population increase in Medieval times (e.g. Nicke 1995). An alternative hypothesis assumes that the settlements primarily developed in the valleys with good soil (Kolodziej 2005). This contribution investigates the two

I. Herzog (✉)
The Rhineland Commission for Archaeological Monuments and Sites,
The Rhineland Regional Council, Endenicher Str. 133, 53115 Bonn, Germany
e-mail: i.herzog@lvr.de

M. Spiliopoulou et al. (eds.), *Data Analysis, Machine Learning and Knowledge Discovery*, Studies in Classification, Data Analysis, and Knowledge Organization,
DOI 10.1007/978-3-319-01595-8_38,

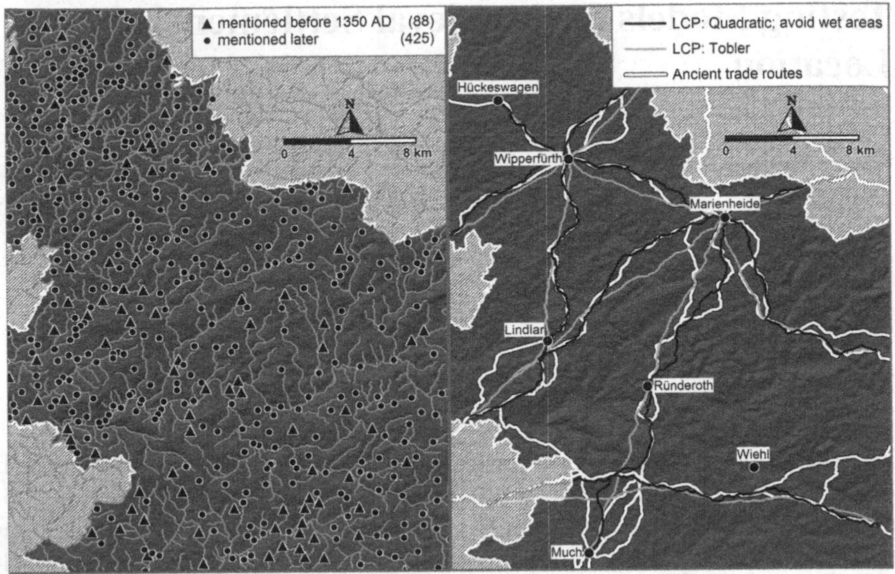

Fig. 1 *Left*: Medieval settlements in the study area, the background shows the terrain and the streams. *Right*: Ancient trade routes and LCPs

hypotheses focusing on an area covering 675 km^2 of the Bergisches Land (Fig. 1). For this study area, a publication is available (Pampus 1998) listing the years when the small hamlets and villages were first mentioned in historical sources, with a total of 513 locations mentioned in Medieval times (i.e. between 950 and 1500 AD), 88 of these were recorded before 1350 AD (early settlements are indicated by triangles in Fig. 1, left). This list of settlements is probably very close to a complete sample, because it includes a high proportion of very small places and only a small fraction of the place names in the list of Pampus could not be located on historical or modern maps. Merely a small amount of the settlements on the list were abandoned. Therefore the assumption is plausible that the sample is not seriously biased. However, a settlement might have been mentioned a long time after its first house had been built. So the history of settlement in this area is perhaps not reflected adequately in the years when the place names were first mentioned.

On the oldest available reliable maps from the study area, drawn in the middle of the nineteenth century, nearly all settlements are still very small, and for this reason it was quite easy to identify the centre of each settlement, which was marked by a dot in a geographical information system (GIS). To test the hypotheses, the distribution of these dots with respect to the targets (i.e. fertile valleys and known ancient trade routes) is compared with the background distribution. In practice, a raster map is created and for each raster cell the distances to the nearest targets are calculated.

The distance distribution of those raster cells which include the settlement dots is compared to the distribution of all raster cells in the study area.

2 Least-Cost Distance

However, the straight-line distance does not seem appropriate for testing the models, because in a hilly terrain with many creeks and small rivers, people did not walk "as the crow flies". The ancient trade routes preserve the movement patterns of the time when they developed. For the study area, these trade routes are described by Nicke (2001) who published overview maps and lists the villages along the routes. So it was possible to outline the routes roughly in a GIS. The route sections connecting the villages were digitized from maps created in the middle of the nineteenth century (Fig. 1, right), assuming that the ancient route layout is still preserved at that time. On the basis of a successful reconstruction of these trade routes by least-cost paths (LCPs), an adjusted distance measure can be derived. Different cost models result in different LCPs. A popular cost model (Tobler 1993; Herzog 2010) depends only on the slope of the terrain and estimates the walking time required:

$$\mathbf{cost}(s) = 10/\mathbf{e}^{-3.5|s+0.05|} \tag{1}$$

where s is the slope (calculated by vertical change divided by horizontal change). The slope is derived from the ASTER digital elevation model (ASTER GDEM is a product of METI and NASA). The resolution of the elevation data is about 30 m.

The goodness of fit of LCPs to the ancient trade routes was assessed by visual impression. Formal methods like that presented by Goodchild and Hunter (1997) were not applied because the improvements after adjusting the model parameters as described below were quite evident. A slope-dependent cost function for wheeled vehicles produced better results than the Tobler cost function. The movement patterns of pedestrians differ from that of carts or wagons with respect to the grade which they can climb efficiently without switchbacks. The term critical slope is used for the transition grade where it is not longer efficient to mount the grade directly. Whereas the critical slope for walkers is at about 25 % (Minetti 1995), a lower critical slope in the range between 8 and 16 % is appropriate for carts or wagons (e.g. Grewe 2004 referring to Roman roads). A simple quadratic function can be constructed with a given critical slope c (Llobera and Sluckin 2007; Herzog 2010):

$$\mathbf{cost}(s) = 1 + (s/c)^2 \tag{2}$$

where c and s are percent slope values. The LCPs resulting from a quadratic cost function with a critical slope of 13 % performed best. However the LCP results came closer to the known trade routes when the slope costs were combined with penalties for crossing wet areas. Initially modern data on streams were used, but some small creeks are not included in the stream data set, and in some cases, the river or creek changed since Medieval times due to meandering or modern construction work. So instead of the modern stream layer, the digital soil map (provided by the Geologischer Dienst, Krefeld, Germany) formed the basis for identifying wet areas. Most small creeks missing in the stream layer but depicted in the mid nineteenth century map are within wet areas indicated on the soil map. Moreover, wet soils are still present in areas where meanders of rivers like the Wupper were located in former times. The LCPs resulting from the penalty factor of 5 for the wet soils coincided better with the trade routes than the paths generated with factors 4 or 6. The best LCP results were achieved when some ford locations identified on historical maps were assigned a lower factor of 2 within the wet areas.

These LCPs are fairly close to the Medieval trade routes (Fig. 1, right). The fit of the LCPs to the trade routes is not perfect which may be caused by several factors, e.g. the inaccuracies of the digital elevation model, landscape change due to modern activities like quarries, mining and the creation of water reservoirs and the fact that additional aspects (beyond slope and avoiding wet areas) might have played a role as well.

So a new metric was defined which takes the slope, the wet areas, and the fords into account. The metric is symmetric because the slope-dependent quadratic cost function is symmetric. The Tobler cost function is not symmetric, but assuming that the same route was chosen for moving to a given target and back, the symmetric function $\text{newcost}(s) = \text{cost}(s) + \text{cost}(-s)$ should be used instead of the asymmetric function $\text{cost}(s)$ (Herzog 2010).

3 Accessibility Maps

On the basis of the distance metric described above, raster accessibility maps are created. The accessibility map with respect to the ancient routes relies on the main routes described by Nicke (2001), ignoring alternative (mostly later) route sections which are also shown in Fig. 1 (right). The digital soil map was used to identify fertile valleys: the soil map attributes include the upper limit of soil quality which indicates agrarian productivity with a maximum value of 100 in Germany, but not exceeding 75 in the Bergisches Land. A thematic map showing this attribute allowed to identify streams surrounded by fertile valleys. So two accessibility maps with two sets of line-shaped targets are created: the first set consists of Medieval trade routes, the second set is formed by streams surrounded by fertile valleys. For each raster cell of the accessibility map, the accessibility value corresponds to the least-cost distance to the nearest trade route or valley with good soils respectively (Fig. 2).

All raster cell values within the study area form the reference probability distribution, and the sample of the cells including the settlement dots will be compared with this reference distribution.

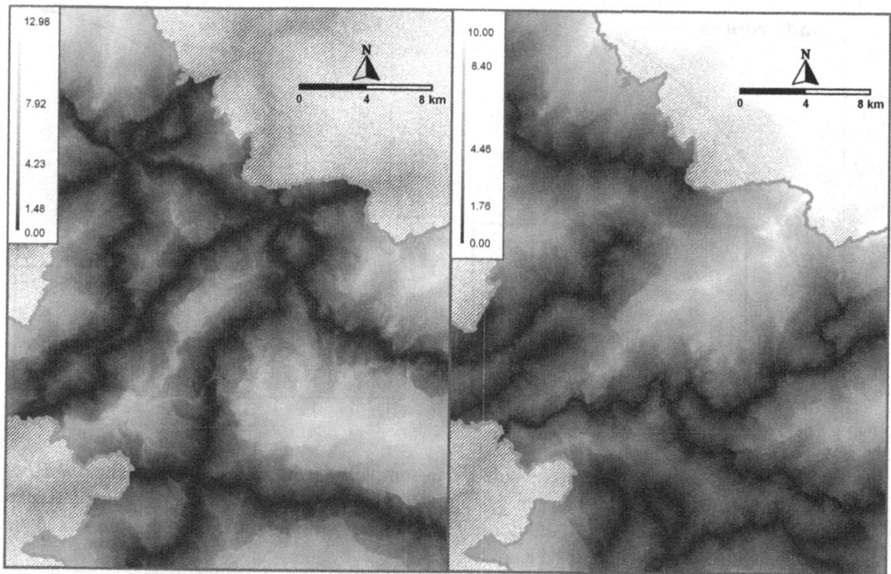

Fig. 2 Accessibility maps, with respect to ancient trade routes (*left*), and fertile valleys (*right*). One distance unit in the map corresponds to the costs of covering 1 km on dry level ground. The legend indicates the quartiles of the distance distribution

4 Testing

One-sample Kolmogorov–Smirnov tests (e.g. Conover 1971) are applied to test the hypothesis that the Medieval settlements are closer (in terms of the least-cost distance) to the linear targets than the reference distribution derived from the appropriate accessibility map. Both reference distributions are skewed to the right so that the popular t-test, which assumes normal distribution, is not applicable. The Kolmogorov–Smirnov test is a non-parametric goodness-of-fit test, which compares the empirical distribution function F_n of the independent, identical distributed (iid) sample consisting of n observations with the cumulative distribution function $F(x)$. The test considers only the maximum vertical distance between these two functions:

$$D = \max_x \; F_n(x) - F(x) \tag{3}$$

The null hypothesis of this one-sided test is rejected if D exceeds a threshold depending on n and the confidence level α. Tables listing the exact threshold values up to $n = 40$ are published (e.g. Conover 1971), for $n > 40$ conservative approximations are used. This allows presenting the test in a diagram, which shows the two distribution functions and their maximum distance (Fig. 3).

Moreover, with a given threshold an upper confidence band for the empirical distribution function can be constructed. The null hypothesis of the goodness-of-fit

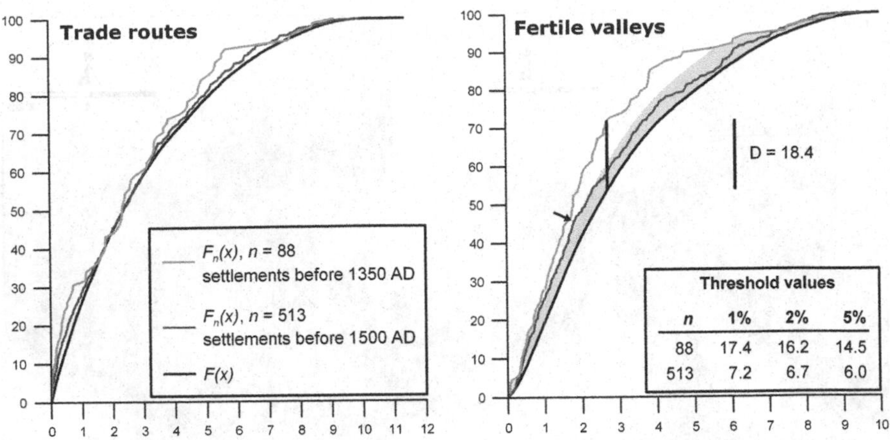

Fig. 3 Kolmogorov–Smirnov tests (one-sided, one-sample). Fertile valleys (*right*): The *black bar* indicates the maximum distance between F_{88} and F; the grey confidence band is drawn for $\alpha = 1\,\%$ and $n = 513$; F_{513} exceeds the confidence band at the position indicated by the *arrow*. On the x-axis the same distance units are used as in Fig. 2

test is rejected if at some point x, $F_n(x)$ exceeds the upper confidence band. So this test is easily intuitive for archaeologists without sound statistical background. Figure 3 (left) shows that the Medieval settlements within the study area are somewhat closer to the trade routes than the raster cells forming the reference distribution, however, the null hypothesis of the goodness-of-fit test is not rejected at the 1 % level. This applies both to the set of settlements mentioned before 1350 AD and all Medieval settlements. The least-cost distance of the Medieval settlements to the streams surrounded by fertile valleys is significantly lower than that of the reference distribution (Fig. 3, right). For $\alpha = 1\,\%$ the null hypothesis is rejected both for the settlements mentioned before 1350 AD and all Medieval settlements. This test supports the model that Medieval settlements primarily developed in the fertile valleys rather than close to ancient trade routes.

5 Discussion

The Kolmogorov–Smirnov test relies on the assumption that the sample is iid. There are some indications that Medieval settlement locations are not quite iid: (a) the distance between any two neighbouring settlements is above a certain limit (no settlement is on top of another one); (b) Siegmund (2009) discusses early Medieval settlements in the fairly flat area of the Niederrhein, Germany, and comes to the conclusion that they are regularly spaced with a distance between 2.25 and 3.25 km. However, for $n > 40$ the Kolmogorov–Smirnov test is conservative tending

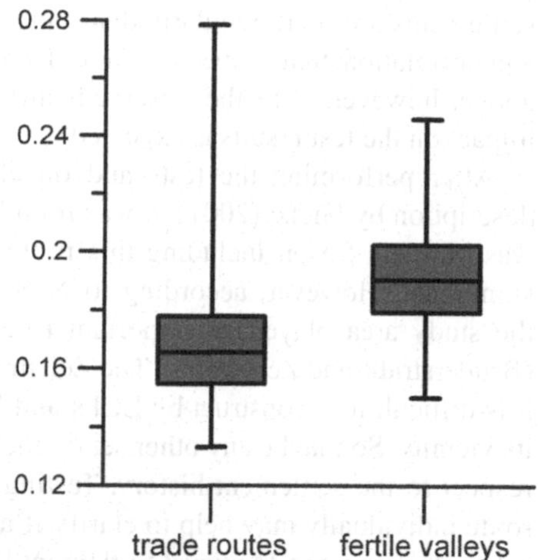

Fig. 4 Box plots of the autocorrelation coefficients (Moran's I) of 190 sets of random points in the study area

to favour the null hypothesis, and a very low confidence level was chosen. Therefore the conclusion that fertile valleys were preferred probably is still valid.

To check the effect of autocorrelation, Fotheringham et al. (2000) suggest comparing the autocorrelation coefficient of the data set investigated with those of experimental distributions. Their approach assumes that the attribute values are known only at the given spatial locations. But at any point within the study area the attribute value can be derived from the accessibility map, so a different approach is applied: 190 data sets each consisting of 513 random points within the study area were simulated and the autocorrelation coefficients (Moran's I) for the distance to trade route (or fertile valleys respectively) were calculated for each data set. The popular Moran's I measures the spatial dependency of nearby locations (Fotheringham et al. 2000):

$$I = \left(\frac{n}{\sum_i \sum_j w_{ij}}\right)\left(\frac{\sum_i \sum_j w_{ij}(x_i - \overline{x})(x_j - \overline{x})}{\sum_i (x_i - \overline{x})^2}\right) \tag{4}$$

where n is the number of samples, x_i is the value at location number i, and the weights w_{ij} are determined by the inverse distance between two locations i and j. As calculating least-cost distances is very time-consuming, the Euclidian distance was used instead. If neighbouring values are more similar than with a random distribution, a positive autocorrelation coefficient is expected. Negative autocorrelation indicates higher dissimilarities of nearby locations than with a random distribution.

According to the box plots of the Moran's I experimental distributions (Fig. 4), for Medieval settlements both Moran's I for trade routes (0.15079) and fertile valleys (0.16871) can be found in the lowest quartile of the distributions. In fact, for fertile valleys, the coefficient is within the bottom 10 % range, and for trade routes within the bottom 20 % range. This is probably the result of the fact that the settlements are more regularly distributed than a random sample. A more positive autocorrelation than expected should create a problem with the test performed above, however, with the observed fairly low values of Moran's I, no negative impact on the test results is expected.

After performing the tests and on checking the digitized routes against the description by Nicke (2001), it was found that a route (Homburgische Eisenstraße) was omitted. So on including this route, the ancient route model might become significant. However, according to Nicke (1995) only two ancient routes within the study area played an important role for the spread of Medieval settlements (Brüderstraße and Zeitstraße). The Zeitstraße seems to be very problematic, because it is difficult to reconstruct by LCPs and because there are only few settlements in its vicinity. So maybe any other set of ancient trade routes was more important with respect to the settlement history. Testing all subsets of trade routes or each trade route individually may help to clarify if any of these old roads played a significant role with respect to Medieval settlement location in the study area.

References

Conover, W. J. (1971). *Practical Nonparametric Statitistics* (pp. 293–301). New York: Wiley.

Fotheringham, A. S., Brunsdon, C., & Charlton, M. (2000). *Quantitative Geography. Perspectives on Spatial Data Analysis* (pp. 201–206). London: Sage.

Goodchild, M. F., & Hunter, G. J. (1997). A simple positional accuracy measure for linear features. *International Journal of Geographical Information Science, 11*(3), 299–306.

Grewe, K. (2004). Alle Wege führen nach Rom - Römerstraßen im Rheinland und anderswo. In H. Koschik (Ed.), 'Alle Wege führen nach Rom', Internationales Römerstraßenkolloquium Bonn. *Materialien zur Bodendenkmalpflege im Rheinland* (Vol. 16, pp. 9–42).

Herzog, I. (2010). Theory and practice of cost functions. In F. J. Melero, P. Cano, & J. Revelles (Eds.), *Fusion of Cultures. Abstracts of the XXXVIII Conference on Computer Applications and Quantitative Methods in Archaeology, CAA 2010* (pp. 431–434), Granada.

Kolodziej, H. (2005). *Herzog Wilhelm I. von Berg. 1380–1408*. Bergische Forschungen 29. VDS – Verlagsdruckerei Schmidt GmbH, Neustadt an der Aisch, 12.

Llobera, M., & Sluckin, T. J. (2007). Zigzagging: Theoretical insights on climbing strategies. *Journal of Theoretical Biology, 249*, 206–217.

Minetti, A. E. (1995). Optimum gradient of mountain paths. *Journal of Applied Physiology, 79*, 1698–1703.

Nicke, H. (1995). *Das Oberbergische Land. Ein Landschaftsportrait*. Wiehl: Martina Galunder Verlag.

Nicke, H. (2001). *Vergessene Wege*. Nümbrecht: Martina Galunder Verlag.

Pampus, K. (1998). Urkundliche Erstnennungen oberbergischer Orte. *Beiträge zur Oberbergischen Geschichte. Sonderband*, Gummersbach.

Siegmund, F. (2009). Archäologische Beiträge zur Geschichte des frühen Mittelalters im Raum zwischen Niers und Kendel. Weeze und der Raum an Niers und Kendel im Mittelalter. *Weezer Archiv, Schriftenreihe der Gemeinde Weeze 3*, 136–144.

Tobler, W. (1993). *Non-isotropic geographic modeling*. Technical Report 93–1. http://www.geodyssey.com/papers/tobler93.html (seen 06-September-2012).

Supporting Selection of Statistical Techniques

Kay F. Hildebrand

Abstract In this paper we describe the necessity for a semi-structured approach towards the selection of techniques in quantitative research. Deciding for a set of suitable techniques to work with a given dataset is a non-trivial and time-consuming task. Thus, structured support for choosing adequate data analysis techniques is required. We present a structural framework for organizing techniques and a description template to uniformly characterize techniques. We show that the former will provide an overview on all available techniques on different levels of abstraction, while the latter offers a way to assess a single method as well as compare it to others.

1 Introduction

Researchers and students engaging in quantitative analysis of their data are always faced with a set of decisions as to which techniques to use for the data set at hand. The decisions made in quantitative analysis are motivated by financial or temporal efficiency, trends in previous choices of other researchers or plain convenience. They are certainly—at least to some extent—motivated by technical and functional aspects. Up to now, there are no standards to support decisions of that kind.

It is important to mention here that our approach does not aim to make a decision for a researcher or student, but to support the decision making process. Despite the fact that some consider elegant data analysis an *art*, the aforementioned constraints often stand in the way of treating it accordingly. Especially in early stages of their education, students may not have acquired the expertise to handle data analysis problems adequately yet.

K.F. Hildebrand (✉)
European Research Center for Information System (ERCIS), Münster, Germany
e-mail: hildebrand@ercis.uni-muenster.de

M. Spiliopoulou et al. (eds.), *Data Analysis, Machine Learning and Knowledge Discovery*, Studies in Classification, Data Analysis, and Knowledge Organization, DOI 10.1007/978-3-319-01595-8_39,

Furthermore, it can be difficult to identify suitable techniques, because there is little effort from experts developing new quantitative methods to make them available to non-expert communities in an understandable way. Being understandable in this context does not mean to strip the techniques from their core contents and present them in a superficial fashion. On the contrary, we are going to propose an approach that will preserve all necessary complexity but only reveal it in a stepwise manner. Thus, more complexity is only introduced when required for decision in favor of or against a data analysis technique.

The remainder of the paper will be structured as follows: We will first analyze, what previous work has been done with regard to a framework for data analysis and a description structure for data analysis techniques (Sect. 2). We will then present our approaches to these two items in Sects. 3 and 4. A discussion as well as an outlook on upcoming research will conclude the paper.

2 Related Work

With regard to providing a structure that allows researchers or students to identify the most suitable technique for the problem at hand, very little research has been done. Grob and Bensberg provide a placement of data mining itself in contrast to knowledge discovery. They also propose a very rough outline of a process model for data mining (Grob and Bensberg 2009). Their process model lacks structure and is only applicable to data mining instances of data analysis. Another approach has been presented by Feelders et al. (2000). They describe a high-level process model which offers the possibility to go back to the previous step, if required. The mentioned steps are Problem Definition, Acquisition of Background Knowledge, Selection of Data, Pre-processing of Data, Analysis and Interpretation as well as Reporting and Use. Each step is described in further detail. However, they do not address the issue of technique selection.

A more detailed approach is presented by Jackson (2002). While she, too, focuses on data mining only, the paper has a much closer look on the concepts around data mining. There is a list of actors in the data mining process, including a description of each. Furthermore, recurring tasks and techniques are identified. These are data summarization, segmentation, classification, prediction and dependency analysis. She provides a matrix that maps nine groups of techniques to the aforementioned tasks. She does not provide different levels of abstraction nor does she extend the list of dimensions. The process model she describes is a sequence of six steps—Business Understanding, Data Understanding, Data Preparation, Modeling, Evaluation, Deployment—each of which is refined into three to five sub steps. The process is motivated by industry environments and their demand in terms of efficiency and speed.

It should be noted at this point that none of the available literature focuses on research and studying environments and all papers restrict themselves to data mining. We could not find anything that specifically deals with the selection of techniques for data mining or statistical analysis. The communities that constantly develop and improve available techniques, such as statisticians, mathematicians or machine learners, fail to make their results available to others in a way that allows researchers from unrelated fields to benefit from the advances. A non-expert of data analysis, however, often is overwhelmed by the complexity that is prominent in technical literature. Since there is hardly any work on building a bridge between these two groups in the form of a framework that provides manageable complexity for technique selection, we could not find any publication that goes one step further and tries to provide a description structure for a single technique.

3 A Framework for Placing Data Analysis Techniques

Choosing a technique to apply to a given data set in order to achieve a certain result is a complex task. We propose to support this process by graphically structuring the tools available. "*Placing* a technique" as used in the heading of Sect. 3, means defining which areas in a layered series of graphics one has to select in order to arrive at a certain technique.

To get a better understanding of the approach, it should be noted right at the beginning, that the final artifact is going to be a web application/portal that can be used by students and researchers alike to carry out their tool selection process. The process requires them to select an area on the first layer that their case is concerned with. They are then presented with a more detailed graphical representation of that area and again have to choose, which section applies to them. Eventually, a set of suitable techniques—according to their choices in the different layers—is presented to them.

At the same time, the set of incorporated techniques can be updated by researchers who have developed a new tool. At the same time, the description of the framework will make it clear that this application neither can nor is meant to make a decision for or against a tool, but is merely a way to structure the selection process and thus make it manageable.

For this graphical structure to be helpful, it needs to cover the possible use cases in data analysis. The framework should also unveil complexity in a stepwise manner so as to not overstrain researchers' and student's willingness to engage in the selection process. When using the framework, the first steps should be intuitive for users from different communities. Later on, interaction can be less generic and complexity may increase. Given this, we will introduce the first of a number of structuring layers of the framework. This layer should be simple for the reasons mentioned before. It should hold a high level of abstraction and allow for a quick overview. The segment interesting to the researcher should be identifiable rapidly.

Figure 1 shows the result of our qualitative research on how to represent the areas of data analysis graphically. It is what a user of the application will be first confronted with, when using the framework. We assume that every user is able to categorize their data analysis problem with respect to its general objective into either

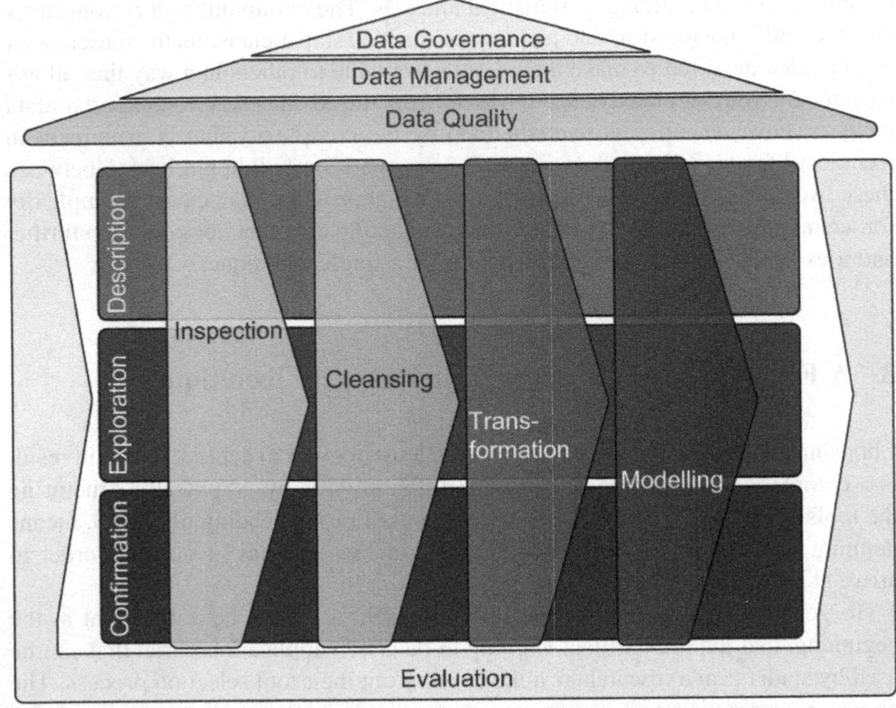

Fig. 1 Prototypical statistical framework

Confirmation, Exploration or Description (the three horizontal bars in the center of Fig. 1). Cases that do not fit into one of these categories should easily be assigned to a category of the "roof" of the framework (i.e. Data Governance, Data Management or Data Quality) or the Evaluation frame. In case, the problem at hand is located at the center, the user has to decide which step of the process he or she is interested in.

At this point, we have identified Inspection, Cleansing, Transformation and Modeling as the steps that should be featured in a simple overview as this. Eventually, choosing a segment from this first layer will trigger opening a new layer which further details the selected segment. We still have to determine the number of abstraction layers that will enable an educated, yet not overwhelming selection of an appropriate technique. An interactive approach as we propose it here requires, of course, some form of animation in the framework as well as predefined actions. One option to implement this framework is an animated website. This way would also allow remote access as well as parallel interaction. Suitable techniques might be JavaScript, Flash or HTML5. This is yet to be evaluated. Complementing the framework will be a community of users, both providing and obtaining content.

Table 1 Exemplary heading sequences of data analysis techniques

Technique	First level heading	Second level heading	Third level heading	Source
Standard deviation	Preparation and description of data	Measures of dispersion	Variance, [...] and variance coefficient	Hartung et al. (2009)
Additive regression	Transformations: engineering the input and output	Combining multiple models	–	Witten et al. (2011)

3.1 Methodology

In order to make a conscious decision on what dimensions to include in the framework and which categories to separate them into, we used a semi-structured qualitative research approach. Fundamental to our analysis were the tables of contents of 14 respective textbooks from the areas of stochastic, data analysis and data mining (Dekking et al. 2010; Fahrmeir et al. 2007; Feelders et al. 2000; Feller 1968, 1971; Georgii 2007; Han et al. 2011; Hand et al. 2001; Handl 2002; Hartung et al. 2009; Hastie et al. 2008; Härdle and Simar 2011; Witten et al. 2011). The tables of contents have been examined for headings that indicate the description of one single technique in the following section. For these headings, all higher-level headings have been extracted together with the name of the technique itself. The result was a 451-item list of techniques and their placement in the area of data analysis according to the authors of the book.

In a second step, the terminology of different authors had to be harmonized in order to compare the categorizations of one method by different authors. Methods and headings in languages other than English were translated. In other cases we had to make use of our knowledge of the domain and content of the books, respectively (e.g. Rank Correlation and Spearman's Correlation were mapped to the same technique). In a third step, we examined sequences of headings in order to find similarities. At this point, we were only interested in the highest level of abstraction because our goal was to extract a canonical structure for generally placing methods in an overview framework. Common first level headings were condensed into the segments that can be found in Fig. 1. Two examples for a heading sequence can be found in Table 1.

4 A Description Structure for Data Analysis Techniques

While the framework described above can be used to identify a set of techniques that may be suitable to tackle a certain data analysis problem, there still remains the problem of which technique to eventually choose. As a remedy to that we propose to link to each of the techniques that have previously been entered into the framework, a profile that shortly describes the essence of that technique. The important thing being, that the profile does not need to convey an in-depth specification. It should merely give a hint to the user, which technique to investigate further. In order to do that, we have developed an outline that consists of six points which condense the core of a technique.

Point number one—Objective—puts into one sentence what applying this technique to data will eventually achieve. Upon reading it, a researcher should be able to generally decide whether it is worthwhile for him or her to consider this method. It is by no way meant to advertise the technique. An impartial short description is required.

The second item—Assumptions—describes the requirements to the data or the underlying model the technique makes (e.g. error terms are normally distributed or attributes are at least ordinal in scale). By putting them at the beginning of the description structure, one can quickly decide whether the technique is actually applicable to the data at hand.

Point three—Specifications—hints at the amount of work that has to be done before the method can be applied. Typical examples for specifications are underlying distributions, thresholds, parameters, sample sizes, exit conditions, etc. But there can also be aspects not as obvious: in description techniques, for instance, colors, shapes, sizes and the like can be specified.

Item four—Algorithm—comprises a list of steps, not necessarily into deepest detail, that communicates a rough outline of the actions that have to be carried out when applying the technique. It can be pseudo-code in the case of very structured techniques, but it may also be a free text description of the approach.

The second to last point—Result—explains what one receives from a technique and in which way it is interpretable. This field helps to separate the purely algorithmic outcome of a method from the interpretation a researcher attributes to it. Results can be classifiers, clusterings, probabilities, parameters or diagrams, plots and graphs.

The last item—Evaluation—describes which features of a result can be reviewed and where potential pitfalls are. It may be required to check the fit of a regression curve, examine a diagram for the message it transports, inspect clusters regarding internal homogeneity or interpret correlation values.

In Table 2, two exemplary techniques have been submitted to the proposed structure. On the one hand, there is a pie chart, which has been chosen deliberately to emphasize the broad applicability of the structure, even to techniques that are not inherently algorithmic. On the other hand, Support Vector Machines are a complicated method that needs to be reduced in complexity in order to fit into the structure and in order to be understandable without previous knowledge. The square brackets in Table 2 denote omissions that have been made due to lack of space. Detailed specifications, algorithms or symbols have been omitted on purpose. This way, complexity is reduced and the selection process is supported. It should be noted at this point, that the simplification of methods is merely done for during the process of choosing a technique. Eventually, users will have to embark on the complexity of a method in order make an informed decision and prevent a poor choice of techniques.

Table 2 Applied description structure for data analysis techniques

Technique	Colored pie chart	Support vector machines
Objective	Visualize the composition of a population/sample	Binary linear classification of input data
Assumptions	The manifestation of variable is known for each item in the population/sample or attributed to a dummy category (e.g. unknown). Variables are either nominal, categorical or binned metric	Objects need to be classified in a binary fashion
Specifications	Reserve as many colors for visual representation as there are manifestations	If the data is not linearly separable to begin with: A kernel function to project objects into higher-dimensional space. A number of dimensions for the higher-dimensional space. A set of training data. [...]
Algorithm	Determine the overall number of objects. Determine the number of objects per manifestation. Compute relative frequencies for each manifestation. Assign each manifestation a sector of a circle according to its relative frequency as well as a color. Compile a full circle from the sectors specified above. Compile a key that assigns colors to manifestations	Determine a maximum-margin hyperplane to separate the data points. In order to do that, determine the two parallel hyperplanes that have the greatest distance from each other but still separate the data. [...] Input new data and determine on which side of the hyperplane an object lies in order to decide its class
Result	A colored pie chart representing the given population/sample	A maximum margin hyperplane depending on training data and kernel
Evaluation	Check pie charts adequacy in terms of visualizing the population/sample: Are sectors distinguishable with respect to size and color? If not, reassign colors or re-evaluate choice of technique	Check the size of margins, do new objects often lie inside them? Is there a chance of overfitting on the training data?

5 Conclusion and Outlook

In this paper, we have presented a first step towards more educated selection of techniques in data analysis by students and researchers. We started by introducing a first layer of a framework that will eventually support the complete selection process for data analysis techniques. Finally, a description structure for data analysis methods was described which will become the lowest level of the aforementioned framework. It allows for a homogeneous description of various techniques.

As for the decision framework, we are fully aware of the fact that the approaches presented here can only be a very first attempt to improve the decision process with respect to quantitative methods. Upcoming work will include a refinement of the framework: further layers to structure complexity need to be added. The empirical analysis of tables of content will be evaluated to provide input on how to structure levels of finer granularity of the framework. Other ways of introducing structure (e.g. ontologies) will be examined. We are planning on implementing a web-based prototype once the framework has been outlined in its entirety. This prototype should be complemented by a community concept so as to allow researchers to add their artifacts and others to retrieve suitable techniques. Introducing a community also aims at avoiding the pitfall of merely copying textbook knowledge in a more accessible format. Wiki-like discussion about the content of description structures can force contributors to re-evaluate their opinions on techniques.

The description structure presented here only contains technical information on the tool it describes at the moment. It may be worthwhile to extend its six sections and introduce a section that holds meta information on related tools, original resources (papers, articles, etc.) and links to available implementations or even an interface to an implementation (e.g. in form of a web service). This section can improve user experience of the web application. From a technical point of view, it will be crucial to establish connections between approaches that aim at the same kind of problem. Thus, experts can specify pros and cons of techniques.

References

Dekking, F. M., Kraaikamp, C., Lopuhaa, H. P., & Meester, L. E. (2010). *Modern introduction to probability and statistics*. London: Springer.

Fahrmeir, L., Künstler, R., Pigeot, I., & Tutz, G. (2007). Statistik: der Weg zur Datenanalyse; mit 25 Tabellen *New York*.

Feelders, A., Daniels, H., & Holsheimer, M. (2000). Methodological and practical aspects of data mining. *Information & Management, 37*, 271–281.

Feller, W. (1971). *An introduction to probability theory and its applications* (2nd ed.) (Vol. 2, p. 669). New York: Wiley.

Feller, W. (1968). *An introduction to probability theory and its applications* (3rd ed.) (Vol. 1, p. 509). New York: Wiley.

Georgii, H.-O. (2007). *Stochastik*. (3rd ed.) (p. 378). Gruyter.

Grob, H. L., & Bensberg, F. (2009). Das data-mining-konzept. *Computergestütztes Controlling*. Münster: Institut für Wirtschaftsinformatik.

Härdle, W., & Simar, L. (2011). *Applied multivariate statistical analysis*. Berlin: Springer.

Han, J., Kamber, M., & Pei, J. (2011). *Data mining: concepts and techniques*. (3rd ed.) (p. 744). Los Altos: Morgan Kaufmann.

Hand, D. J., Mannila, H., & Smyth, P. (2001). Principles of data mining.

Handl, A. (2002). *Multivariate analysemethoden* (p. 464). Berlin: Springer.

Hartung, J., Elpelt, B., & Klösener, K.-H. (2009). Statistik. Lehr- und Handbuch der angewandten Statistik. *Biometrical Journal, 15*, 1145.

Hastie, T., Tibshirani, R., & Friedman, J. (2008). The elements of statistical learning: Data mining, inference and prediction. *International Statistical Review* (2nd ed.) (Vol. 77, p. 745). Berlin: Springer.

Jackson, J. (2002). Data mining: A conceptual overview. *Communications of the Association for Information Systems, 8*, 267–296.

Witten, I. H., Frank, E., & Hall, M. A. (2011). Data mining: Practical machine learning tools and techniques. *Annals of Physics* (Vol. 54, p. 664). Los Altos: Morgan Kaufmann.

Alignment Methods for Folk Tune Classification

Ruben Hillewaere, Bernard Manderick, and Darrell Conklin

Abstract This paper studies the performance of alignment methods for folk music classification. An edit distance approach is applied to three datasets with different associated classification tasks (tune family, geographic region, and dance type), and compared with a baseline n-gram classifier. Experimental results show that the edit distance performs well for the specific task of tune family classification, yielding similar results to an n-gram model with a pitch interval representation. However, for more general classification tasks, where tunes within the same class are heterogeneous, the n-gram model is recommended.

1 Introduction

With the growth of the Music Information Retrieval field and the expansion of data mining methods, folk music analysis has regained attention through the past decades. Folk music archives represent a cultural heritage, therefore they need to be categorized and structured to be more easily consulted and searched. The retrieval of similar tunes from a folk tune database has been the subject of several MIREX contests, and alignment methods have proven to be the most successful at this task (Urbano et al. 2011). Various melodic similarity measures have been investigated

R. Hillewaere (✉) · B. Manderick
Artificial Intelligence Lab, Department of Computing, Vrije Universiteit Brussel, Brussels, Belgium
e-mail: rhillewa@vub.ac.be; bmanderi@vub.ac.be

D. Conklin
Department of Computer Science and AI, Universidad del País Vasco UPV/EHU, San Sebastián, Spain

Ikerbasque, Basque Foundation for Science, Bilbao, Spain
e-mail: conklin@ikerbasque.org

M. Spiliopoulou et al. (eds.), *Data Analysis, Machine Learning and Knowledge Discovery*, Studies in Classification, Data Analysis, and Knowledge Organization,
DOI 10.1007/978-3-319-01595-8_40,

for the exploration of a folk song database, and they have been combined in the attempt to find an optimal measure (Müllensiefen and Frieler 2004).

Music classification has become a broad subfield of the computational music research area, with many challenges and possible approaches (Weihs et al. 2007). In a recent study, alignment methods have been applied to the specific task of *tune family classification* (van Kranenburg et al. 2013), a tune family being an ensemble of folk songs which are variations of the same ancestral tune. In that work, alignment methods with various features were compared with several global feature models, in which a melody is represented as a vector of global feature values. It was shown that alignment methods achieve remarkable accuracies for tune family classification in comparison with the global feature models, regardless which features were used to represent the data. An open question, however, is how alignment methods perform on other types of folk tune classification tasks where tunes within a class do not present detectable melodic similarity.

The n-gram model is another machine learning technique that can be applied to both music classification (Hillewaere et al. 2009) and music retrieval (Uitdenbogerd and Zobel 1999). In this study, we investigate the performance of a simple alignment method, the edit distance, versus an n-gram classifier for the following tasks:

(a) tune family classification, in order to verify that the edit distance achieves similar results to the alignment methods reported by van Kranenburg et al. (2013);
(b) two fundamentally different folk music classification tasks. The first task is *geographic region classification*, which we have thoroughly studied in our previous work (Hillewaere et al. 2009). The second task is folk tune *genre classification*, where the genres are the dance types of the tunes (Hillewaere et al. 2012).

Given the excellent results with alignment methods in the study by van Kranenburg et al. (2013), they might also perform well on the different classification tasks proposed in (b). However, we do not expect this to happen and hypothesize the high performance is due to high similarity within tune families and that n-gram models over the same representations will perform equally well.

Since folk music is orally transmitted, traditionally by people singing during their social activities or work, over time multiple variations arise in the tunes. This phenomenon has led to the notion of tune family, i.e. an ensemble of tunes that all derive from the same ancestral tune. This is a hypothetical concept, since we generally cannot trace the historical evolution of a folk song. Given this definition of a tune family, it is obvious that songs of the same tune family are very similar, although numerous musical variations between them are possible, we only cite a few (van Kranenburg et al. 2007): melodic contour, rhythmic changes, insertion and deletion of parts, range, and number of phrases. We illustrate this with a score example in Fig. 1, which shows the first phrases of three tunes belonging to the tune family called "Heer". Clearly, the first two phrases are highly similar, and the third phrase repeats the same melodic motif. This is typical for the tune family concept,

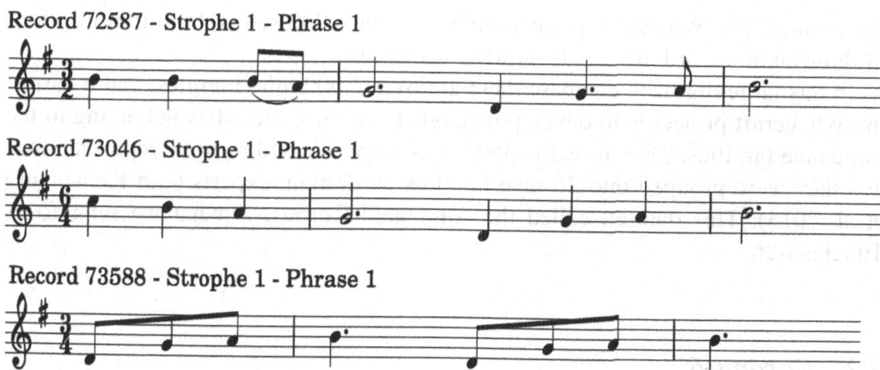

Fig. 1 Three tunes from the same tune family are very similar

and it is evident that tunes from the same geographic region or with the same dance type generally differ a lot more, which makes these classification tasks harder.

To verify our hypothesis, an edit distance method is applied to three folk music datasets with three different classification tasks, which will be described in the next section. For each of the folk tune collections, the pieces are encoded in melodic and rhythmic representations: as strings of pitch intervals, and as strings of duration ratios. These basic representations have been chosen to compare the predictive power of models based on melodic information versus rhythmic information. For each data collection and for each representation, pairwise edit distances are computed and the classification is done with a one nearest neighbour algorithm, which is similar to the approach used by van Kranenburg et al. (2013). A tenfold cross validation scheme is used to assess the performances in terms of classification accuracies.

2 Data Sets

In our experiments we use three folk tune datasets in MIDI format with different associated classification tasks, which we detail in this section.

2.1 *TuneFam-26*

This dataset of 360 songs is the tune family dataset used in the study of van Kranenburg et al. (2013). The source of this dataset is a larger collection called "Onder de groene linde", which is hosted at the Meertens Institute in Amsterdam. It contains over 7,000 audio recordings of folk songs that were tape-recorded all over the country. The Witchcraft project digitized over 6,000 songs, both transcriptions of those audio recordings and from written sources.

In this large digitized database there are over 2,000 tune families, and a part of the Witchcraft project is to develop methods to retrieve melodies belonging to the same tune families. Therefore, the 360 songs were selected as to be representative, and they were grouped into 26 tune families by domain experts (van Kranenburg et al. 2013). This dataset, called the Annotated Corpus, is what we refer to as TuneFam-26.[1]

2.2 Europa-6

This is a collection of folk music from six geographic regions of Europe (England, France, South Eastern Europe, Ireland, Scotland and Scandinavia), for which the classification task is to assign unseen folk songs to their region of origin. Li et al. (2006) studied this problem with factored language models, and they selected 3,724 pieces from a collection of 14,000 folk songs transcribed in the ABC format.

Their collection was pruned to 3,367 pieces by filtering out duplicate files, and by removing files where the region of origin was ambiguous. In order to end up with core melodies that fit for our research purpose, a preprocessing in two steps was carried out: the first step ensures that all pieces are purely monophonic by retaining only the highest note of double stops which occurred in some of the tunes, and in the second step we removed all performance information such as grace notes, trills, staccato, etc. Repeated sections and tempo indications were also ignored. Finally, by means of abc2midi we generated a clean quantized MIDI corpus, and removed all dynamic (velocity) indications generated by the style interpretation mechanism of abc2midi. In our previous work, we have shown that n-gram models outperform global feature models on this corpus (Hillewaere et al. 2009).

With a total of 3,367 pieces, Europa-6 is a larger dataset than TuneFam-26, and another contrast is that this dataset not only contains sung music, but also folk dances for example.

2.3 Dance-9

The corpus Dance-9 is a large subset of Europa-6: 2,198 folk tunes subdivided into nine dance type categories, the largest ones being jigs, reels and polskas. The associated classification task is to predict the dance type of an unseen tune. Several

[1] We would like to thank Peter van Kranenburg for sharing the Annotated Corpus and for the kind correspondence.

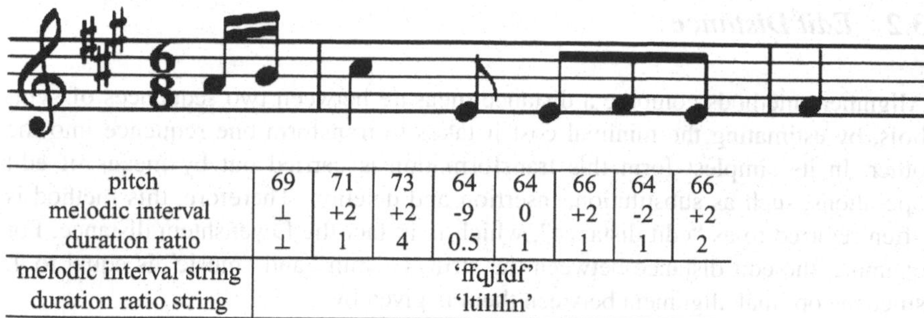

pitch	69	71	73	64	64	66	64	66
melodic interval	⊥	+2	+2	-9	0	+2	-2	+2
duration ratio	⊥	1	4	0.5	1	1	1	2
melodic interval string	'ffqjfdf'							
duration ratio string	'ltillm'							

Fig. 2 Excerpt of the Scottish jig "With a hundred pipers", illustrating the event feature sequences and the string representation

string methods have been compared with global feature models and event models using this data set (Hillewaere et al. 2012).

To construct Dance-9 we only selected the ABC-files with an unambiguous dance type annotation. Furthermore, we discarded all dance types that occurred insufficiently by putting a minimal threshold of 50 tunes, because it would lead to a highly unbalanced dataset. To the remaining 2,198 pieces, the same preprocessing steps as for Europa-6 have been applied, as is for the conversion to MIDI.

3 Methods

3.1 Music Representation

A piece of music can be viewed as an ordered sequence of events, and every event is represented by an event feature of one's choice. In our case of monophonic folk music, the music events are note objects, with pitch and duration as basic event features. In this paper, we use two derived event features to represent the pieces: the first is the melodic interval in semitones between the current and the previous note, and the second is the duration ratio, i.e. the duration of the current note divided by the duration of the previous note. The obtained event features can be mapped onto characters, choosing a distinct character for each distinct feature value. Thereby the event feature sequence is mapped onto an ASCII symbol string, in which case we talk about a string representation. This is illustrated in Fig. 2 on a short excerpt of a jig.

3.2 Edit Distance

Alignment methods compute a distance measure between two sequences of symbols, by estimating the minimal cost it takes to transform one sequence into the other. In its simplest form this transformation is carried out by means of edit operations, such as substitution, insertion and deletion. Therefore, this method is often referred to as "edit distance", which is in fact the Levenshtein distance. For example, the edit distance between the strings "ismir" and "music" is equal to 4, since the optimal alignment between them is given by

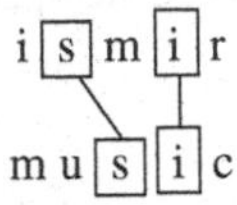

i s m i r

m u s i c

which means four edit operations are needed: two substitutions ("i" to "m" and "r" to "c"), one insertion (the "u") and one deletion (the "m"). For the purpose of our current research, we have used WEKA's implementation of the edit distance (http://www.cs.waikato.ac.nz/ml/weka/).

The edit distance defines a pairwise distance metric, therefore the classification can be performed with an instance-based nearest neighbour classifier. Given a test instance, the prediction of its class label is solely based on the training instance which is closest with respect to the edit distance (instead of the usual Euclidean distance).

3.3 *n*-Gram Models

An n-gram model is a generative model for sequences which computes the probability of an entire sequence as the product of the probability of individual events within the sequence. Each event is conditioned on $n-1$ previous events and these conditional probabilities are estimated from a training corpus, with smoothing applied in order to avoid zero probabilities (Manning and Schutze 1999).

The n-gram model can be used for classification, by constructing a separate model for every class, and classifying a new sequence according to the model which generates the sequence with highest probability. To apply the model to music, every piece of the data set is transformed into an event feature sequence according to a feature of choice (e.g., duration ratio or melodic interval, see Sect. 3.1), and for each class the n-grams occurring in the class are compiled.

It is important to mention that the music representation is basically the same as for the edit distance approach, but the essential difference between these methods is that an n-gram model aims to model the transitions for a given class, whereas the edit distance computes a global pairwise similarity measure between pieces.

Table 1 The tenfold cross validation classification accuracies for our experiments on the three datasets

	Melodic int		Pitch	Duration ratio		Duration
	Alignment	Pentagram	Global	Alignment	Pentagram	Global
TuneFam-26	94.4 (3.9) *92.0*	90.8 (3.7)	73.6 (8.9) *74.0*	80.3 (5.9) *74.0*	70.6 (5.9)	55.0 (7.5) *55.0*
Europa-6	49.5 (2.0)	64.1 (3.0)		47.5 (3.1)	55.1 (2.2)	
Dance-9	50.0 (2.6)	66.1 (2.2)		63.2 (1.4)	74.4 (2.0)	

Numbers in parentheses are the standard deviation over the ten folds. For comparison, the numbers italicized are the results by van Kranenburg et al. (2013)

4 Results and Discussion

In this section, the experimental results of the edit distance on the three datasets are reported for both the interval and the duration ratio representations. They are compared with a pentagram model (an n-gram model with $n = 5$), over the same representations. To assess the performance of both methods, we have set up a tenfold cross validation scheme to compute classification accuracies. The folds were taken in a stratified way, which is especially important for the results with TuneFam-26, to avoid that an entire tune family would be contained in the test fold, in which case a correct prediction would be impossible. We also ensured that the exact same folds were used for all experiments to do an impartial comparison.

The classification accuracies are reported in Table 1. The column on the left shows the melodic interval results, and the right column contains the duration ratio performances. The edit distance results are reported in the alignment columns, and for comparison we also include the results reported by van Kranenburg et al. (2013) (italicized numbers).

First of all, we observe higher accuracies on TuneFam-26 than on the other corpora. The edit distance approach classifies the tune family dataset with a high accuracy of 94.4 % on pitch intervals, which is very similar to the 92 % reported by van Kranenburg et al. (2013). This is remarkable since the edit distance is a simpler method than that used by van Kranenburg et al. (2013), which uses gap opening and extension weights in the computation. The edit distance slightly outperforms the pentagram model that still achieves an accuracy of 90.8 %, in other words there are only 13 more misclassified pieces.

With the duration ratios, the edit distance performs again very well on the tune family dataset with an accuracy of 80.3 %, outperforming both the pentagram model and the alignment method on duration ratio reported by van Kranenburg et al. (2013), though the high standard deviation of the accuracy estimate on both approaches should be noted (Table 1).

For the classification of geographic region or genre, the pentagram models clearly yield higher accuracies than the edit distance, with approximately 15 % difference for both datasets with the melodic interval representation. We remind the reader that 1 % on Europa-6 or Dance-9 corresponds to a larger amount of pieces due to the difference in the sizes of the data sets, so this result shows that alignment methods are suitable for the specific task of tune family classification, but obtain much lower accuracies on more general types of classification tasks.

To summarize, all these results indicate that the tune family classification task is relatively easy. This finding specifically contradicts the statement of van Kranenburg et al. (2013) that the tune family classification task is more difficult than the region classification on Europa-6. They suggest that the heterogeneity of tunes between regions makes the task easier, but it appears in our results this is not the case. On the contrary, there is more heterogeneity within one region than there is in one tune family, which makes the region classification significantly harder.

We have also constructed two global feature models on TuneFam-26, based on global features derived from pitch on the one hand and duration on the other hand, similarly as in our previous work (Hillewaere et al. 2012). The accuracies obtained with an SVM classifier (with parameters tuned by a grid search) are reported in the respective columns of Table 1, and compared with the global feature results found by van Kranenburg et al. (2013). These accuracies confirm their statement that global feature approaches are of limited use for tune family classification.

5 Conclusions

In this paper we have investigated how a simple alignment method, called the edit distance, performs on three different folk music classification tasks: (a) classification of tune families, (b) classification of geographic region of origin, and (c) classification of dance types. Three folk tune datasets are used to assess the performance of the edit distance method in comparison with a pentagram model. Experimental results have shown the following:

- the edit distance approach performs well on the tune family dataset, yielding similar results to those reported by van Kranenburg et al. (2013);
- for edit distance, the tune family classification task is easier than classification of geographic region or dance type;
- for geographic region or dance type classification, an n-gram model is more appropriate.

We believe that these findings are due to the intrinsic concept of a tune family, since highly similar tunes are present within any tune family. Music retrieval methods using local sequential information, such as alignment methods and n-gram models, are capable of capturing this similarity and therefore lead to high performances. When pieces within classes are highly similar, alignment methods will achieve good classification results. On the other hand, when classes are more heterogenous the n-gram model is more appropriate.

References

Hillewaere, R., Manderick, B., & Conklin, D. (2009). Global feature versus event models for folk song classification. In *Proceedings of the 10th International Society for Music Information Retrieval Conference* (pp. 729–733). Kobe, Japan.

Hillewaere, R., Manderick, B., & Conklin, D. (2012). String methods for folk tune genre classification. In *Proceedings of the 13th International Society for Music Information Retrieval Conference* (pp. 217–222). Porto, Portugal.

Van Kranenburg, P., Garbers, J., Volk, A., Wiering, F., Grijp, L., & Veltkamp, R. (2007). Towards integration of MIR and folk song research. In *Proceedings of the 8th International Conference on Music Information Retrieval* (pp. 505–508). Vienna, Austria.

Van Kranenburg, P., Volk, A., & Wiering, F. (2013). A comparison between global and local features for computational classification of folk song melodies. *Journal of New Music Research, 42*(1), 1–18.

Li, X., Ji, G., & Bilmes, J. (2006). A factored language model of quantized pitch and duration. In *International Computer Music Conference* (pp. 556–563). New Orleans, USA.

Manning, C., & Schutze, H. (1999). *Foundations of statistical natural language processing*. Cambridge: MIT Press.

Müllensiefen, D., & Frieler, K. (2004). Optimizing measures of melodic similarity for the exploration of a large folk song database. In *Proceedings of the 5th International Conference on Music Information Retrieval*. Barcelona, Spain.

Uitdenbogerd, A. L., & Zobel, J. (1999). Matching techniques for large music databases. In *Proceedings of the ACM Multimedia Conference* (pp. 57–66). Orlando, Florida.

Urbano, J., Lloréns, J., Morato, J., & Sánchez-Cuadrado, S. (2011). Melodic similarity through shape similarity. In S. Ystad, M. Aramaki, R. Kronland-Martinet, K. Jensen (Eds.), *Exploring music contents* (pp. 338–355). Berlin: Springer.

Weihs, C., Ligges, U., Mörchen, F., & Müllensiefen, D. (2007). Classification in music research. *Advances in Data Analysis and Classification, 1*(3), 255–291.

Comparing Regression Approaches in Modelling Compensatory and Noncompensatory Judgment Formation

Thomas Hörstermann and Sabine Krolak-Schwerdt

Abstract Applied research on judgment formation, e.g. in education, is interested in identifying the underlying judgment rules from empirical judgment data. Psychological theories and empirical results on human judgment formation support the assumption of compensatory strategies, e.g. (weighted) linear models, as well as noncompensatory (heuristic) strategies as underlying judgment rules. Previous research repeatedly demonstrated that linear regression models well fitted empirical judgment data, leading to the conclusion that the underlying cognitive judgment rules were also linear and compensatory. This simulation study investigated whether a good fit of a linear regression model is a valid indicator of a compensatory cognitive judgment formation process. Simulated judgment data sets with underlying compensatory and noncompensatory judgment rules were generated to reflect typical judgment data from applied educational research. Results indicated that linear regression models well fitted even judgment data with underlying noncompensatory judgment rules, thus impairing the validity of the fit of the linear model as an indicator of compensatory cognitive judgment processes.

1 Theories of (Non-)compensatory Human Judgment Formation

Human judgment formation can be considered as the integration process of multiple pieces of information, either perceived in the environment or retrieved from memory, into a usually unidimensional judgment. Reliance on human judgment is common throughout a variety of professional domains, e.g. education, therapy,

T. Hörstermann (✉) · S. Krolak-Schwerdt
University of Luxembourg, Route de Diekirch, 7220 Walferdange, Luxembourg
e-mail: thomas.hoerstermann.001@student.uni.lu; sabine.krolak@uni.lu

M. Spiliopoulou et al. (eds.), *Data Analysis, Machine Learning and Knowledge Discovery*, Studies in Classification, Data Analysis, and Knowledge Organization, DOI 10.1007/978-3-319-01595-8_41,

or medicine. Besides the aim of investigating the impact of available information on judgment, central interest of research on human judgment formation rests on the judgment rules underlying the integration process. Psychological theories of human judgment formation stated different assumptions on this integration process. A prominent and persistent theoretical approach considers human judgments to be a weighted linear integration of the available information (Anderson 1974; Anderson and Butzin 1974; Brunswik 1952). According to Gigerenzer and Goldstein (1996), this judgment rule can be classified as a compensatory judgment rule, because the influence of a piece of information can be outweighed by the effect of other pieces of information. In contrast, a series of influential theories of noncompensatory judgment rules has been proposed (Gigerenzer 2008), e.g. the Take-The-Best-Heuristic (TTB) (Gigerenzer and Goldstein 1996), a stepwise search rule for discriminating pieces of information. According to the TTB, available pieces of information are initially sorted by their subjective validity regarding the judgment to be done. In the judgment process, the pieces of information are serially elaborated if the piece of information is discriminative for the judgment. As the first discriminative piece of information is found, the judgment formation process is aborted and the judgment is exclusively based on the first discriminating piece of information. TTB well illustrates noncompensatory judgment rules as all pieces of information less valid than the first discriminating piece are neglected in judgment formation.

2 Empirical Modelling of Judgment Formation Processes

Various studies illustrate attempts to model judgment rules of empirically observed judgment data. In general, these studies selected plausible possible judgment rules from theory and compared their empirical fit in explaining the actual judgments observed. Dhami and Harries (2001) compared a compensatory linear regression model and a noncompensatory cut-off value model on lipid-lowering drug prescription judgments of general psychologists, finding a nearly equivalent fit of both models to the observed judgment data. Hunter et al. (2003) investigated weather-related judgments of professional pilots and reported that among various compensatory (linear integration) and noncompensatory (cut-off value model, minimum model) judgment rules, an unweighted additive model best predicted pilots' weather-related judgments. Across various studies, it turned out "…that decision makers (insofar as they are behaving appropriately) are paramorphically well represented by linear models" (Dawes and Corrigan 1974, p. 105). In the framework of compensatory theories of judgment, the generally adequate fit of linear models can be interpreted as a confirmation that the underlying cognitive judgment formation process is also linear and compensatory. In contrast, Gigerenzer and Goldstein (1996) reported that, for the case of binary pieces of information, predicted judgments by a weighted linear model are nearly indistinguishable from judgments predicted by the noncompensatory TTB. Furthermore, Bröder (2000, 2002) demonstrated that the TTB rule itself can be expressed in terms of a weighted linear model, if the β-weight of a predictor exceeds the sum of the β-weights of all less valid pieces of information (see (1)). The I's in (1) refer to the values of the binary pieces of information, whereas the β-weights reflect the impact of the corresponding piece of information on the judgment J.

$$J = \beta_1 I_1 + \beta_2 I_2 + \ldots + \beta_n I_n, \quad \beta_i > \sum_{j=i+1}^{n} \beta_j, \quad I_1 \ldots I_n \in \{0, 1\} \tag{1}$$

Given this point of view, the validity of the conclusion that the generally adequate fit of linear models indicates an underlying compensatory judgment formation process is impaired. Instead, the adequate fit rather indicates a high flexibility of linear models in fitting empirical judgment data resulting from various compensatory or noncompensatory cognitive judgment rules.

3 Research Question

The present study aimed at assessing the capability of linear models' fit in discriminating between compensatory and noncompensatory judgment rules within an applied setting of judgment modelling, namely the assessment of student achievement in educational research. In contrast to Gigerenzer and Goldstein's (1996) binary data analysis, criteria and information in educational judgments are usually multi-leveled (e.g. grades, test scores). Furthermore, findings of educational research indicate that many achievement criteria resulting from human judgments, including grades, can be considered to be transformations of within-class rank orders of the students (Ingenkamp 1971; Rheinberg 2001). Therefore, the study applied linear regression models to simulated data sets of rank-order judgments resulting from the compensatory or noncompensatory integration of multi-level information, and investigated whether the fit of the regression model might be considered as a valid indicator of an underlying compensatory judgment rule.

4 Data Simulation

In this study, 100 data sets were simulated to reflect typical data sets of applied educational research. Each data set contained data of 50 fictitious teachers and judgments for 25 students per teacher were simulated, corresponding to common class sizes. As the basis for the judgments, five pieces of information were simulated for each student. To reflect student information in educational research, the grade distribution of 1,802 Luxembourgish 6th grade primary school students (see (2)) was taken as the reference distribution for the construction of the simulated student information. The construction of the simulated student information was implemented by an ordinal regression model (see (2)), in which the odd for a grade less or equal than x for a student k was determined by a threshold parameter t_x and a standardized normal distributed student ability parameter u_k. The threshold parameters were set to values that result in grade probabilities according to the reference distribution for a student ability parameter of $u_k = 0$, whereas the weighting γ of u_k was chosen to result in inter-grade correlations of $r \sim 0.60$, as observed in the actual Luxembourgish grade data.

$$p(x) = \begin{cases} 0.39 & \text{for } x = 1 \\ 0.32 & \text{for } x = 2 \\ 0.20 & \text{for } x = 3 \\ 0.07 & \text{for } x = 4 \\ 0.02 & \text{for } x = 5 \\ 0 & \text{else} \end{cases} \qquad ln\left(\frac{P(X \leq x)}{1 - P(X \leq x)}\right) = \gamma u_k + t_x \tag{2}$$

For each of the simulated students, teacher judgments with different underlying judgment rules were constructed. As a compensatory judgment rule, a weighted linear judgment rule was implemented. The weighting coefficients of the linear judgment rule were chosen to imply a decreasing impact from the first to the last piece of student information (see (3)). For noncompensatory judgment rules, (a) an adapted TTB judgment rule and (b) a minimum value judgment rule were implemented. The TTB judgment rule was implemented according to Bröder (2000) by a linear model with weighting coefficients that did not allow the influence of a piece of information to be outweighed by less valid pieces of information (see (4)). For this, the restriction $\beta_i > \sum_{j=i+1}^{n} \beta_j$ was adapted to $\beta_i > \sum_{j=i+1}^{n} \beta_j m_j$, with m_j representing the range of values of information j ($m_j = 5 - 1 = 4$), to take into account that the simulated student information is multi-leveled. In the TTB judgment rule, the validity of information decreased from the first to the last piece of information, thus resembling the relative impact of each piece of information in the weighted linear judgment rule. The minimum value judgment rule, representing teacher judgments solely based on the worst piece of student information, was actually implemented as a maximum function of the pieces of information (see (5)), due to the scale of Luxembourgish grades in which lower grades indicate higher achievement.

$$J = 0.30I_1 + 0.25I_2 + 0.20I_3 + 0.15I_4 + 0.10I_5 \tag{3}$$

$$J = 5^4 I_1 + 5^3 I_2 + 5^2 I_3 + 5^1 I_4 + 5^0 I_5 \tag{4}$$

$$J = max(I_1, I_2, I_3, I_4, I_5) \tag{5}$$

For each of the 50 teachers in the simulated data sets, the judgments for his 25 students were then rank transformed, resulting in a rank order of the students according to the respective judgment rule. Ranks were assigned ascendingly, that means lower ranks indicated higher achievement of a student. In case of tied ranks, the average rank was assigned to the students.

5 Results

Linear regression models with the five pieces of information as predictors were fitted to the individual rank order judgments of each teacher resulting from the weighted linear judgment rule, the TTB judgment rule, and the minimum value judgment rule, using the `lm` function of the `R 2.10.1` software package. The ratio of explained variance R^2 was computed for each regression model as an indicator to which extent an individual teacher's rank order, given a compensatory or noncompensatory underlying judgment rule, could be explained by the linear model. Figure 1 displays the teachers' mean R^2s for the 100 simulated data sets for the three underlying judgment rules.

For judgments with an underlying weighted linear judgment rule, the regression models showed a nearly perfect fit throughout all data sets with a mean explained variance of 93.6 % ($SD = 0.58$). The linear regression models on average explained 92.1 % ($SD = 0.61$) of the variance of the TTB judgments and 84.6 % ($SD = 0.82$) of the variance of the minimum value judgments. Thus, linear regression models fitted well the rank order judgments resulting from all three underlying judgment rules, and showed a nearly indistinguishable fit between the weighted linear and the TTB judgment rule. Because merely considering the fit of the regression model tended not to be a valid indicator for an underlying compensatory linear judgment rule, the inspection of the models' regression coefficients was done to investigate whether the pattern of coefficients could reveal the underlying judgment rules. The mean regression coefficients for the three underlying judgment rules are displayed in Fig. 1. As to be expected, the regression coefficients for the underlying weighted linear judgment rule well reflected the actual weighting of the pieces of information, whereas the underlying minimum value judgment rule resulted in nearly equal regression coefficients for all pieces of information. For the TTB judgment rule, strongly exaggerated regression coefficients ($M = 3.82$ (TTB) vs. $M = 1.84$ (linear)) resulted for the most valid first piece of information. However, the coefficient pattern for the TTB judgment rule did not fulfill the restrictions according to Bröder (2000) (see (1) and (4)), required for the expression of a TTB judgment rule in terms of a weighted linear model. In none of the simulated data sets, the coefficients for the first two pieces of information fulfilled the restrictions, whereas the restriction for the coefficient of the third piece was fulfilled in 1 of 100 data sets and for the fourth piece in 18 of 100 data sets. Thus, an underlying TTB judgment rule could not be validly discriminated from a weighted linear

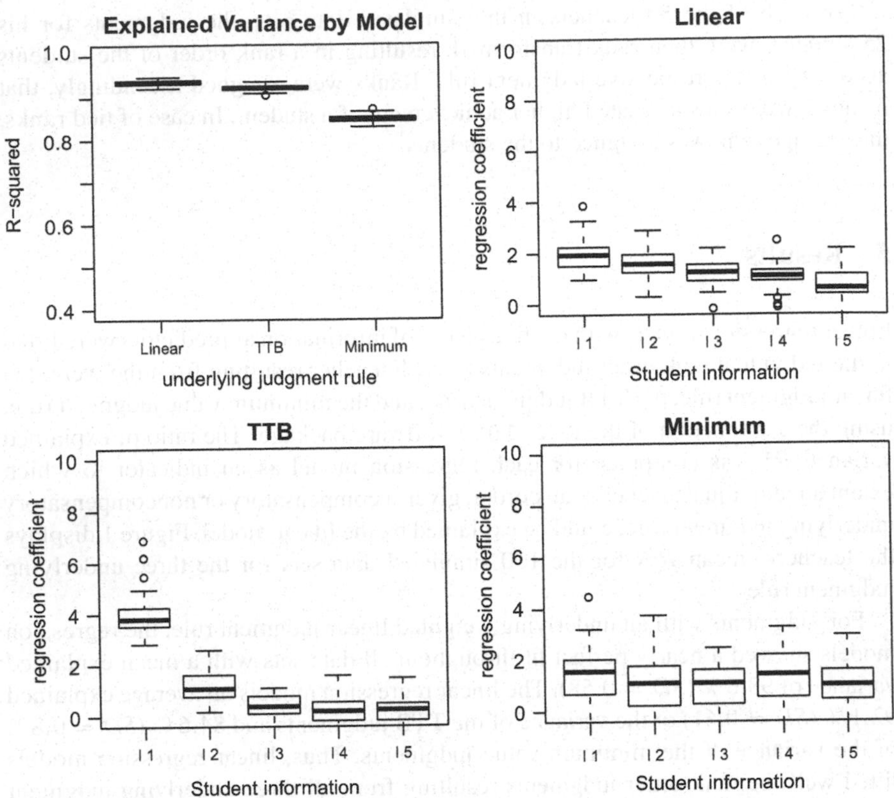

Fig. 1 Mean R^2s and mean regression coefficients by underlying judgment rule for $r = 0.60$

judgment rule, neither by the fit of the regression model nor by the pattern of regression coefficients.

To investigate whether the results were rather due to the partial multicollinearity of the pieces of information than due to an insensitivity of the linear regression model to noncompensatory judgment rules, the simulation was repeated and 100 additional data sets with uncorrelated pieces of information were simulated. The mean ratios of explained variance and the regression coefficients are displayed in Fig. 2. The regression models again showed a nearly perfect fit for the linear ($M = 95.1\,\%$, $SD = 0.51$) and TTB judgment rule ($M = 93.7\,\%$, $SD = 0.57$). The fit of the regression models for the minimum value judgment rule decreased to a mean explained variance of 63.9 % ($SD = 1.64$). Similar to the findings for the intercorrelated pieces of information, the pattern of regression coefficients for the TTB judgment rule (see Fig. 2) did not indicate an underlying noncompensatory judgment rule. As observed for correlated pieces of information, the TTB restrictions for the coefficients of the first two pieces of information were

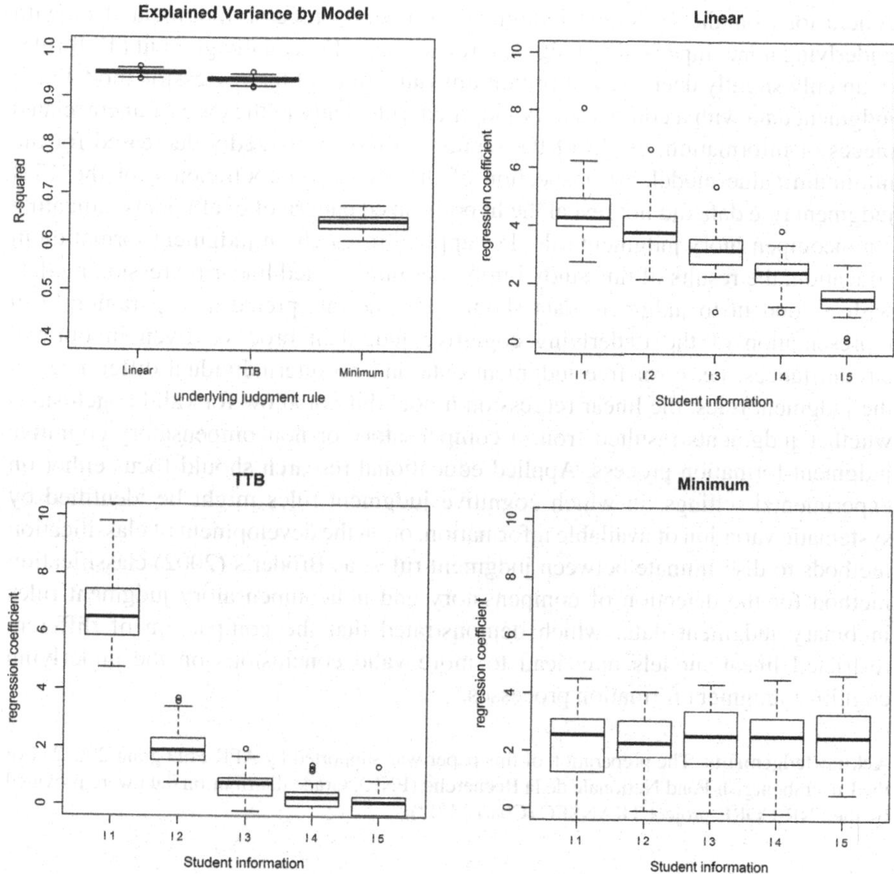

Fig. 2 Mean R^2s and mean regression coefficients by underlying judgment rule for $r = 0.00$

fulfilled in none of the data sets. The restriction was fulfilled in 12 data sets for the third piece of information and in 47 data sets for the fourth piece of information.

6 Discussion

The study investigated whether the fit of a linear regression model to empirical judgment data is a valid indicator for an underlying linear and compensatory cognitive process of judgment formation. Linear regression models were applied to simulated data sets resembling typical judgment data from applied educational research. The simulated judgments were either based on a linear compensatory judgment rule or on two noncompensatory judgment rules, namely a TTB judgment rule or a minimum value judgment rule. In sum, the results of the study indicated that the fit of the linear regression model is not a sufficient indicator for valid conclusions on the cognitive judgment formation process. Judgment data with underlying noncompensatory judgment rules resulted in equally good fit (TTB rule) or an only slightly decreased fit (minimum value rule) of the regression model than judgment data with a compensatory judgment rule. Only in the case of uncorrelated pieces of information, the fit of the regression model markedly decreased for the minimum value model. An inspection of the regression coefficients for the TTB judgment rule data did not reveal the hypothesized pattern of coefficients indicating a noncompensatory judgment rule. For applied research on judgment formation in education, the results of this study imply that unrestricted linear regression models with a good fit to judgment data should only be interpreted as a paramorphical representation of the underlying cognitive judgment process. Even in optimal circumstances, i.e. error-free judgment data and no interindividual differences in the judgment rules, the linear regression model did not allow for valid conclusions whether judgments resulted from a compensatory or noncompensatory cognitive judgment formation process. Applied educational research should focus either on experimental settings, in which cognitive judgment rules might be identified by systematic variation of available information, or on the development of classification methods to discriminate between judgment rules, as Bröder's (2002) classification method for the detection of compensatory and noncompensatory judgment rules in binary judgment data, which demonstrated that the comparison of different restricted linear models may lead to more valid conclusions on the underlying cognitive judgment formation processes.

Acknowledgements The preparation of this paper was supported by AFR PhD grant 2962244 of the Luxembourgish Fond Nationale de la Recherche (FNR). Grade distribution data were provided by the FNR-CORE-project TRANSEC (C08/LM/02).

References

Anderson, N. H. (1974). Cognitive algebra. Integration theory applied to social attribution. In L. Berkowitz (Ed.), *Advances in experimental social psychology* (Vol. 7, pp. 1–100). New York: Academic.

Anderson, N. H., & Butzin, C. A. (1974). Performance = Motivation × Ability. An integration-theoretical analysis. *Journal of Personality and Social Psychology, 30*, 598–604.

Bröder, A. (2000). Assessing the empirical validity of the "take-the-best" heuristic as a model of human probabilistic inference. *Journal of Experimental Psychology: Learning, Memory and Cognition, 26*, 1332–1346.

Bröder, A. (2002). Take the best, Dawes' rule, and compensatory decision strategies. A regression-based classification method. *Quality & Quantity, 36, 219–238.*

Brunswik, E. (1952). *Conceptual framework of psychology*. Chicago: University of Chicago Press.

Dawes, R. M., & Corrigan, B. (1974). Linear models in decision making. *Psychological Bulletin, 81*, 95–106.

Dhami, M. K., & Harries, C. (2001). Fast and frugal versus regression models of human judgment. *Thinking and Reasoning, 7*, 5–27.

Gigerenzer, G. (2008). Why heuristics work. *Perspectives on Psychological Science, 3*, 20–29.

Gigerenzer, G., & Goldstein, D. G. (1996). Reasoning the fast and frugal way: Models of bounded rationality. *Psychological Review, 103*, 650–669.

Hunter, D. R., Martinussen, M., & Wiggins, M. (2003). Understanding how pilots make weather-related decisions. *International Journal of Aviation Psychology, 13*, 73–87.

Ingenkamp, K. (1971). *Die Fragwürdigkeit der Zensurengebung* [The dubiosness of grading]. Weinheim: Beltz.

Rheinberg, F. (2001). Bezugsnormen und schulische Leistungsbeurteilung [Reference norms and assessment of academic achievement]. In F. E. Weinert (Ed.), *Leistungsmessung in Schulen* (pp. 59–72). Weinheim: Beltz.

Sensitivity Analyses for the Mixed Coefficients Multinomial Logit Model

Daniel Kasper, Ali Ünlü, and Bernhard Gschrey

Abstract For scaling items and persons in large scale assessment studies such as Programme for International Student Assessment (PISA; OECD, PISA 2009 Technical Report. OECD Publishing, Paris, 2012) or Progress in International Reading Literacy Study (PIRLS; Martin et al., PIRLS 2006 Technical Report. TIMSS & PIRLS International Study Center, Chestnut Hill, 2007) variants of the Rasch model (Fischer and Molenaar (Eds.), Rasch models: Foundations, recent developments, and applications. Springer, New York, 1995) are used. However, goodness-of-fit statistics for the overall fit of the models under varying conditions as well as specific statistics for the various testable consequences of the models (Steyer and Eid, Messen und Testen [Measuring and Testing]. Springer, Berlin, 2001) are rarely, if at all, presented in the published reports.

In this paper, we apply the mixed coefficients multinomial logit model (Adams et al., The multidimensional random coefficients multinomial logit model. Applied Psychological Measurement, 21, 1–23, 1997) to PISA data under varying conditions for dealing with missing data. On the basis of various overall and specific fit statistics, we compare how sensitive this model is, across changing conditions. The results of our study will help in quantifying how meaningful the findings from large scale assessment studies can be. In particular, we report that the proportion of missing values and the mechanism behind missingness are relevant factors for estimation accuracy, and that imputing missing values in large scale assessment settings may not lead to more precise results.

D. Kasper · A. Ünlü (✉) · B. Gschrey
Chair for Methods in Empirical Educational Research, TUM School of Education, and Centre for International Student Assessment (ZIB), Technische Universität München, Arcisstrasse 21, 80333 Munich, Germany
e-mail: daniel.kasper@tum.de; ali.uenlue@tum.de; bernhard.gschrey@tum.de

M. Spiliopoulou et al. (eds.), *Data Analysis, Machine Learning and Knowledge Discovery*, Studies in Classification, Data Analysis, and Knowledge Organization, DOI 10.1007/978-3-319-01595-8_42,

1 Introduction

To analyze data obtained from large scale assessment studies such as Programme for International Student Assessment (PISA; OECD 2012) or Progress in International Reading Literacy Study (PIRLS; Martin et al. 2007) different versions of the Rasch model (Fischer and Molenaar 1995) are applied. For instance, in PISA the mixed coefficients multinomial logit model (Adams et al. 1997) has been established, to scale items and persons. One may be interested in how well the model fits the data. But goodness-of-fit statistics for the overall fit of this model under varying conditions are rarely presented in the published reports, if they are presented at all (OECD 2002, 2005, 2009, 2012).

One special characteristic of the PISA assessment data is the presence of missing values. Missing values can occur due to missing by design as well as item non-response (OECD 2012). The handling of missing values seems to be crucial, because an improper treatment of missing values may result in invalid statistical inferences (Huisman and Molenaar 2001). In this paper, we apply the mixed coefficients multinomial logit model to PISA data for varying forms of appearance of missing values. Based on various overall and specific fit statistics, we compare how sensitive this model is, across changing conditions.

In the mixed coefficients multinomial logit model, the items are described by a fixed set of unknown parameters, $\boldsymbol{\xi}$, and the student outcome levels (the latent variable), $\boldsymbol{\theta}$, are random effects. Three parts of this models can be distinguished: the conditional item response model, the population model, and the unconditional, or marginal, item response model (for technical details, see Adams et al. 1997).

In addition to the afore mentioned components, a posterior distribution for the latent variable for each individual n is specified by

$$h_\theta(\boldsymbol{\theta}_n; \boldsymbol{w}_n, \boldsymbol{\xi}, \boldsymbol{\Gamma}, \boldsymbol{\Sigma}|\boldsymbol{x}_n) = \frac{f_x(\boldsymbol{x}_n; \boldsymbol{\xi}|\boldsymbol{\theta}_n) f_\theta(\boldsymbol{\theta}_n)}{\int_\theta f_x(\boldsymbol{x}_n; \boldsymbol{\xi}|\boldsymbol{\theta}) f_\theta(\boldsymbol{\theta})},$$

where $\boldsymbol{x}_n$ is the response vector, and $\boldsymbol{\Gamma}$, $\boldsymbol{w}_n$, and $\boldsymbol{\Sigma}$ are parametrizing the postulated multivariate normal distribution for $\boldsymbol{\theta}$ (OECD 2012, p. 131). Estimates for $\boldsymbol{\theta}$ are random draws from this posterior distribution for each student, and these are referred to as plausible values (see Mislevy 1991; Mislevy et al. 1992).

The mixed coefficients multinomial logit model is used in PISA for three purposes: national calibration, international scaling, and student score generation (estimation of students' plausible values). Multidimensional versions of this model have been fitted to PISA data; for instance, a three-dimensional version has had reading, science, and mathematics as its (correlated) dimensions. For estimating the parameters of this model, the software *ConQuest*® can be used (Wu et al. 2007).

Missing values in PISA can occur due to missing by design (different students are administered different test items) as well as by item non-response. Usually three mechanisms producing item non-response are distinguished: Missing Completely At Random (MCAR), Missing At Random (MAR), and Not Missing At Random

(NMAR) (Little and Rubin 2002; Schafer 1997). When MCAR, missing item scores form a simple random sample from all scores in the data, that is, there is no relation to the value of the item score that is missing, or to any other variable. If missingness is related to one or more observed variables in the data, the process is called MAR. NMAR means that missingness is related to the value that is missing or to unobserved variables.

To control for item non-response, different procedures are studied in the statistical literature (Huisman and Molenaar 2001). One popular technique is *imputation*. Using this technique, missing responses are estimated, and the estimates are substituted for the missing entries. However, a number of imputation techniques are available (e.g., see van Ginkel et al. 2007a), so the question is what methods are to be preferred.

Huisman and Molenaar (2001) compared six imputation methods for dealing with missing values. They used four real complete data sets with different sample sizes, and missing values were created in the samples using three different mechanisms resulting in MCAR, MAR, and NMAR. The proportion of created missing values was $P = 0.05$, $P = 0.10$, and $P = 0.20$. In general, model based imputation techniques perform better than randomization approaches. But this effect can only be observed for a missing value proportion of at least $P = 0.10$ and when missingness is due to MAR or NMAR. An effect due to sample size could not be observed.

Van Ginkel et al. (2010) used two-way imputation with error and compared it with listwise deletion. The method of two-way imputation is based on a two-way ANOVA model. It produces relatively unbiased results regarding such measures as Cronbach's alpha, the mean of squares in ANOVA, item means, mean test score, or the loadings from principal components analysis. A description of the two-way ANOVA model can be found in van Ginkel et al. (2007c). Missingness was introduced into a real complete data set using the mechanisms MCAR, MAR, and NMAR. The data set consisted of ten unidimensional items. The method of two-way imputation with error outperformed listwise deletion with respect to different criteria (e.g., Cronbach's alpha and mean test score). The results were almost as good as those obtained from the complete data set.

The strength of the method of two-way imputation with error (TW+e) was also shown in several other studies (van der Ark and Sijtsma 2005; van Ginkel et al. 2007b, 2007c). This method may also be useful for large scale assessment studies such as PISA. Another imputation method considered is multiple imputation by chained equations (MICE), which is a multiple imputation technique that operates in the context of regression models. In general, missing values are replaced by plausible substitutes based on the distribution of the data. The MICE procedure contains a series of regression models, where each variable with missing data is modeled conditional on the other variables in the data. Iterations then yield multiple imputations (for a detailed explanation of the method, see Azur et al. 2011).

Because of large number of variables (more than 200) and respondents (around half a million) sophisticated methods of imputation such as the multiple imputation by chained equations (van Buuren et al. 2006; van Buuren 2007) possibly may not be applicable. Unfortunately, information about how different methods for dealing with missing values perform in the context of PISA are lacking so far. In this regard, the present paper will study whether the application of these imputation methods may lead to improved estimates. The afore mentioned studies can only serve as a reference, for how sensitive the mixed coefficients multinomial logit model may "react" to missing values or different imputation methods. The reason for this is, that none of the studies have investigated the sensitivity of multidimensional versions of Rasch type models for missing value analyses. Moreover, crucial criteria such as the goodness-of-fit of these models or the accuracy of the item parameter estimates have not been investigated in those studies.

2 Study Design

To study the estimation accuracy of the mixed coefficients multinomial logit model under varying conditions for missing values, we analyzed data from the PISA 2009 study (OECD 2012). We used a complete data set of 338 German students on the mathematics and science test items of Booklet Nr. 9. Missing values for this data set were created using the mechanisms MCAR, MAR, and NMAR. For MCAR, each data point had the same probability of being coded as missing value. Under the condition of MAR, missingness was associated with gender: for men, the probability of a missing value was nine times higher than for women. To reach NMAR, in addition to the correlation of missingness with gender, the probability of a missing value was eight times higher for incorrect answers (that is, for zero entries) than for correct answers.

Three proportions of missing values were considered: $P = 0.01$, $P = 0.03$, and $P = 0.05$. These proportions capture the usual amount of missingness in the PISA test booklets. As imputation methods, we used two-way imputation with error (TW+e) and multiple imputation by chained equations (MICE). Each of the imputation methods was applied one time to every data set, so for any imputation method, missing condition, and proportion of missing values, there is one imputed data set.

All of the $2 \times 3 \times 3 \times 1$ imputed data sets, the nine missing data sets (MD), and the complete data set were analyzed with the mixed coefficients multinomial logit model, whereat the mathematical items were allocated to one dimension and the science items to another dimension. As criteria for the sensitivity of this model, the item fit statistic MNSQ and the item parameter estimates were used. MNSQ quantifies how well the model fits the data. This fit statistic is applicable especially for large numbers of observations. A perfect value of MNSQ is 1.0, whereas values less than 1.0 indicate an overfit, values greater than 1.0 an underfit. In general, mean squares in a near vicinity of 1.0 indicate little distortion. On the other hand, the item parameters may be interpreted as the difficulties or discrimination intensities of the items, and theoretically, they can range in the reals or subsets thereof. As parameters of the mixed coefficients multinomial logit model, they can be estimated by maximum likelihood procedures.

For both statistics, we calculated the differences between the estimates obtained from the complete data sets and the estimates for the missing values and imputed data sets. The absolute values of these differences were averaged and the standard deviations were calculated. In addition, ANOVA models were applied.

3 Results

The means of the absolute differences for MNSQ between the estimates from the complete data sets and the estimates for the missingness and imputed data sets are summarized in Table 1. As can be seen, the mean differences in MNSQ between the complete data sets and the imputed as well as the missing data sets are small. As expected, the difference is larger when the proportion of missing values increases. The mechanisms behind missingness obviously influence the estimation accuracy in the case of NMAR. The effect of imputation methods on estimation accuracy is small. In general, using the missing data set (MD) for the analysis results in the least biased estimates.

As the results of the ANOVA show, the small effects of imputation methods on estimation accuracy in terms of MNSQ are statistically significant (Table 2). Also the effects of the proportion of missing values and NMAR on the estimation accuracy in terms of MNSQ are statistically significant. In addition, all two-way interaction terms were included in the model, but were not significant.

The means of the absolute differences for the estimated item parameters between the complete data sets and the missingness and imputed data sets are summarized in Table 3. Generally, these results are similar to the previous findings. We observe small differences between the estimates obtained from the complete data sets and the imputed as well as the missing data sets. The difference is larger when the proportion of missing values increases, and an effect of the mechanisms underlying missingness can be observed for NMAR.

As the results of the ANOVA show, the effects of the proportion of missing values ($P = 0.05$), NMAR, TW+e, and MICE on the estimation accuracy in terms of the item parameter estimates are statistically significant (Table 4). In addition, all two-way interaction terms were included in the model, but were not significant.

4 Discussion

For scaling items and persons in PISA, the mixed coefficients multinomial logit model is used. However, statistics for the fit of this model under varying conditions for dealing with missing values are rarely, if at all, presented in the published reports. We have applied the mixed coefficients multinomial logit model to PISA

Table 1 Means of the absolute differences for MNSQ

	MCAR			MAR			NMAR		
	P			*P*			*P*		
Imputation Method	0.01	0.03	0.05	0.01	0.03	0.05	0.01	0.03	0.05
TW+e	0.008	0.005	0.012	0.006	0.011	0.013	0.008	0.013	0.015
MICE	0.007	0.008	0.011	0.006	0.013	0.014	0.006	0.014	0.018
MD	0.006	0.007	0.009	0.006	0.009	0.012	0.004	0.011	0.011

TW+e two-way imputation with error, *MICE* multiple imputation by chained equations, *MD* missing data set (no imputation)

Table 2 ANOVA for mean difference for MNSQ

Effect	*F*	*df*1	*df*2	*p*
TW+e[a]	4.51	1	795	0.03
MICE[a]	5.60	1	795	0.02
P3[b]	0.83	1	795	0.36
P5[b]	75.51	1	795	0.00
MAR[c]	1.00	1	795	0.32
NMAR[c]	16.80	1	795	0.00
TW+e*MAR	0.59	1	795	0.44
MICE*MAR	0.00	1	795	0.97
TW+e*NMAR	1.18	1	795	0.28
MICE*NMAR	1.41	1	795	0.24
TW+e*P3	1.17	1	795	0.28
TW+e*P5	0.09	1	795	0.76
MICE*P3	0.56	1	795	0.46
MICE*P5	2.93	1	795	0.09

[a] Reference category was no imputation
[b] Reference category was $P = 0.01$
[c] Reference category was MCAR

Table 3 Means of the absolute differences for estimated item parameters

	MCAR			MAR			NMAR		
	P			*P*			*P*		
Imputation Method	0.01	0.03	0.05	0.01	0.03	0.05	0.01	0.03	0.05
TW+e	0.013	0.018	0.035	0.012	0.028	0.035	0.017	0.032	0.045
MICE	0.011	0.024	0.031	0.011	0.020	0.045	0.015	0.034	0.043
MD	0.008	0.014	0.018	0.009	0.023	0.029	0.012	0.019	0.025

TW+e two-way imputation with error, *MICE* multiple imputation by chained equations, *MD* missing data set (no imputation)

data under varying conditions for missing values. Based on various fit statistics, we have compared how sensitive this model is, across changing conditions.

With respect to the fit criterion MNSQ, we have shown that the proportion of missing values obviously influences estimation accuracy; less accurate estimates are observed for higher proportions of missing values. The mechanisms behind

Table 4 ANOVA for mean difference for estimated item parameters

Effect	F	$df1$	$df2$	p
TW+e[a]	19.39	1	741	0.00
MICE[a]	5.87	1	741	0.02
P3[b]	0.14	1	741	0.72
P5[b]	126.36	1	741	0.00
MAR[c]	0.07	1	741	0.78
NMAR[c]	16.59	1	741	0.00
TW+e*MAR	2.12	1	741	0.15
MICE*MAR	0.55	1	741	0.46
TW+e*NMAR	0.83	1	741	0.36
MICE*NMAR	0.06	1	741	0.80
TW+e*P3	0.27	1	741	0.60
TW+e*P5	3.69	1	741	0.06
MICE*P3	0.05	1	741	0.83
MICE*P5	3.12	1	741	0.08

[a] Reference category was no imputation
[b] Reference category was $P = 0.01$
[c] Reference category was MCAR

missingness also appear to be relevant for estimation accuracy. As the study of this paper corroborates, imputing missing values does not lead to more precise results in general. In future research, it would be interesting to investigate the effects of imputation techniques in matters of higher proportions of missing values, as well as of appropriate modifications of the mixed coefficients multinomial logit model for the lower proportions in PISA.

Generally, the pattern of results for the estimated item parameters resembles the results for MNSQ. Again, the proportion of missingness, the imputation methods, and the mechanisms creating the missing values have an influence on the estimation accuracy. It seems that the imputation methods considered here do not lead to more accurate results regarding the fit criterion and the item parameters, at least under the conditions studied in this paper.

Which of the imputation techniques should be preferred in educational large scale assessment studies such as PISA? The findings of this paper cannot favor one over the other, of the two analyzed imputation techniques MICE and TW+e. Similar results were obtained for both methods. In some cases, the TW+e method led to better results, and in other cases, MICE performed better.

Nonetheless, the considered missingness proportions were relatively small, and the investigation of the influence of missing values on such other criteria as the important students' plausible values in PISA would have exceeded the scope of this paper. These topics must be pursued in future research.

References

Adams, R., Wilson, M., & Wang, W. (1997). The multidimensional random coefficients multinomial logit model. *Applied Psychological Measurement, 21*, 1–23.

Azur, M. J., Stuart, E. A., Frangakis, C., & Leaf, P. J. (2011). Multiple imputation by chained equations: What is it and how does it work? *International Journal of Methods in Psychiatric Research, 20*, 40–49.

Fischer, G. H., & Molenaar, I. W. (Eds.), (1995). *Rasch models: Foundations, recent developments, and applications*. New York: Springer.

Huisman, M., & Molenaar, I. W. (2001). Imputation of missing scale data with item response models. In A. Boomsma, M. van Duijn, & T. Snijders (Eds.), *Essays on item response theory* (pp. 221–244). New York: Springer.

Little, R., & Rubin, D. (2002). *Statistical analysis with missing data*. New York: Wiley.

Martin, M. O., Mullis, I. V. S., & Kennedy, A. M. (2007). *PIRLS 2006 Technical Report*. Chestnut Hill: TIMSS & PIRLS International Study Center.

Mislevy, R. (1991). Randomization-based inference about latent variables from complex samples. *Psychometrika, 56*, 177–196.

Mislevy, R., Beaton, A., Kaplan, B., & Sheehan, K. (1992). Estimating population characteristics from sparse matrix samples of item responses. *Journal of Educational Measurement, 29*, 133–161.

OECD (2002). *PISA 2000 Technical Report*. Paris: OECD Publishing.

OECD (2005). *PISA 2003 Technical Report*. Paris: OECD Publishing.

OECD (2009). *PISA 2006 Technical Report*. Paris: OECD Publishing.

OECD (2012). *PISA 2009 Technical Report*. Paris: OECD Publishing.

Schafer, J. (1997). *Analysis of incomplete multivariate data*. London: Chapman & Hall.

Steyer, R., & Eid, M. (2001). *Messen und Testen [Measuring and Testing]*. Berlin: Springer.

Van Buuren, S. (2007). Multiple imputation of discrete and continuous data by fully conditional specification. *Statistical Methods in Medical Research, 16*, 219–242.

Van Buuren, S., Brand, J., Groothuis-Oudshoorn, C., & Rubin, D. (2006). Fully conditional specification in multivariate imputation. *Journal of Statistical Computation and Simulation, 76*, 1049–1064.

Van der Ark, L. A., & Sijtsma, K. (2005). The effect of missing data imputation on Mokken scale analysis. In L. A. van der Ark, M. A. Croon, & K. Sijtsma (Eds.), *New developments in categorical data analysis for the social and behavioral sciences* (pp. 147–166). Mahwah: Erlbaum.

Van Ginkel, J., Sijtsma, K., Van der Ark, L. A., & Vermunt, J. (2010). Incidence of missing item scores in personality measurement, and simple item-score imputation. *Methodology, 6*, 17–30.

Van Ginkel, J., Van der Ark, L. A., & Sijtsma, K. (2007a). Multiple imputation for item scores in test and questionnaire data, and influence on psychometric results. *Multivariate Behavioral Research, 42*, 387–414.

Van Ginkel, J., Van der Ark, L. A., & Sijtsma, K. (2007b). Multiple Imputation for item scores when test data are factorially complex. *British Journal of Mathematical and Statistical Psychology, 60*, 315–337.

Van Ginkel, J., Van der Ark, L. A., Sijtsma, K., & Vermunt, J. (2007c). Two-way imputation: A Bayesian method for estimating missing scores in tests and questionnaires, and an accurate approximation. *Computational Statistics & Data Analysis, 51*, 4013–4027.

Wu, M., Adams, R., Wilson, M., & Haldane, S. (2007). *ACER ConQuest: Generalised item response modelling software*. Camberwell: ACER Press.

Confidence Measures in Automatic Music Classification

Hanna Lukashevich

Abstract Automatic music classification receives a steady attention in the research community. Music can be classified, for instance, according to music genre, style, mood, or played instruments. Automatically retrieved class labels can be used for searching and browsing within large digital music collections. However, due to the variability and complexity of music data and due to the imprecise class definitions, the classification of the real-world music remains error-prone. The reliability of automatic class decisions is essential for many applications. The goal of this work is to enhance the automatic class labels with confidence measures that provide an estimation of the probability of correct classification. We explore state-of-the-art classification techniques in application to automatic music genre classification and investigate to what extend posterior class probabilities can be used as confidence measures. The experimental results demonstrate some inadequacy of these confidence measures, which is very important for practical applications.

1 Introduction

Automatic music classification is one of the most often addressed fields in Music Information Retrieval (MIR). The classifiers are aimed to learn high-level music concepts such as music genre, music style, mood, or played instruments. These high-level concepts are really handy for practical application and so are of a particular interest of the research community. State-of-the-art automatic music classification systems already reach quite high accuracy values. Methods for music classification involve various machine learning techniques such as Gaussian Mixture Models

H. Lukashevich (✉)
Fraunhofer IDMT, Ehrenbergstr. 31, 98693 Ilmenau, Germany
e-mail: lkh@idmt.fraunhofer.de

M. Spiliopoulou et al. (eds.), *Data Analysis, Machine Learning and Knowledge Discovery*, Studies in Classification, Data Analysis, and Knowledge Organization, DOI 10.1007/978-3-319-01595-8_43,

(GMM) and Support Vector Machines (SVM). Once trained, the classifiers can predict class labels for the unseen data.

However, the classification of the real-world music remains error-prone. First, the feature representations of musical signals are imperfect and incomplete. Second, the fusion of music styles causes natural blurring of the class borders. Third, it is impossible to collect representative training data to cover the entire variability of music over the world.

In this paper we address the topic of so called confidence measures that accompany a classification decision and give insight into the reliability of this decision. The remainder of the paper is structured as follows. Section 2 introduces a concept of confidence measure in a classification system. Section 3 describes our experimental setup including experimental dataset and classification systems. The results are presented and discussed in Sect. 4. The last section concludes the paper and discusses some points for the further research.

2 Confidence Measures in Classification Systems

Confidence measure (CM) is a score (preferably between 0 and 1) that accompanies a classification decision. This score indicates the reliability of a decision made by the classifier. A higher CM corresponds to the more reliable decision.

2.1 Definition and Requirements

For many practical applications it is not sufficient to be able to rank classification decisions according to their CMs. The absolute CM values become important once it is necessary: (1) to compare and/or to combine classification decisions originated by various classifiers; (2) to introduce a reject option, which is based on CM values; or (3) to interpret the classification results.

In this paper we define a CM as a score between 0 and 1, which accompanies a classification decision and fulfills the requirements proposed by Duin and Tax (1998):

1. On the average a fraction c of all objects with confidence c should be classified correctly.
2. Reliably classified objects should have larger confidences than objects close to the decision boundary.

Such CMs are easy to interpret. For instance, if we rank classification decisions of some system according to their confidence and get 100 decisions with CMs around 0.7, we can expect about 70 of them to be correct. Such CMs are helpful in many practical applications: they can serve as a hint for the end user and can be used to filter out insecure decisions or to fuse classification decisions emitted by several classification systems.

To measure the quality of CMs, Duin and Tax (1998) introduced confidence error and normalized confidence error scores. Confidence error q is the mean square error between the confidence estimate $c(\mathrm{x})$ and the correctness indicator $t(\mathrm{x})$: $q = E\{(t(\mathrm{x}) - c(\mathrm{x}))^2\}$. Normalized confidence error ρ is independent of the classification error ε

$$\rho = \frac{q}{\varepsilon} - \frac{1}{2}\,,\; 0 \leq \rho \leq \frac{1}{2}\,. \tag{1}$$

The reader should refer to Duin and Tax (1998) for more insight into the range of ρ in (1). For the interpretation of results it is sufficient to keep in mind that (1) $\rho = 0$ indicates an ideal case where for all correct decisions c is equal 1 and for all incorrect decisions c is equal 0; (2) $\rho = \frac{1}{2}$ corresponds to a random assignment of c values.

2.2 *Estimation of Confidence Measures*

Estimation of CMs differs for generative and discriminative classifiers. *Generative classifiers* are based on class conditional probability density functions $p(\mathrm{x}|\omega_i)$ for each class ω_i. One of the commonly used generative classifiers is GMM. Here, a confidence measure $c(\mathrm{x})$ for sample x can be defined as the posterior class probability $p(\omega_i|\mathrm{x})$:

$$c(\mathrm{x}) = p(\omega_i|\mathrm{x}) = \frac{p(\mathrm{x}|\omega_i)p(\omega_i)}{p(\mathrm{x})} = \frac{p(\mathrm{x}|\omega_i)p(\omega_i)}{\sum_i p(\mathrm{x}|\omega_i)p(\omega_i)}\,. \tag{2}$$

Discriminative classifiers directly model the decision boundary between classes. Let $f(\mathrm{x})$ be a function indicating a distance from x to a decision boundary. The confidence $c(\mathrm{x})$ can be defined via a mapping of $f(x)$:

$$c(\mathrm{x}) = \frac{1}{1 + \exp(Af(\mathrm{x}) + B)}\,, \tag{3}$$

where A and B are some constants that need to be estimated experimentally. The sigmoid function in (3) is successfully applied by Platt (1999) for one of the most used binary discriminant classifiers, namely SVM. In case of a multi-class classification the probabilistic output for SVM can be estimated by pairwise coupling as suggested by Wu et al. (2004).

3 Experimental Setup

In this paper we evaluate two state-of-the-art classification systems in application to the automatic music genre classification task. The experimental setup for this evaluation is described in the following subsections. It includes experimental dataset, feature extraction, classification, and evaluation measures.

3.1 Dataset

The experimental dataset includes 1,458 full-length music tracks from a public ISMIR 2004 Audio Description Dataset.[1] This dataset hast equally sized fixed training and test parts. Each track in the dataset is labeled with one of six genres: `Classical` (638 tracks), `Electronic` (230), `Jazz & Blues` (52), `Metal & Punk` (90), `Rock & Pop` (202), and `World` (246). The task is to learn classification rules from the music tracks in the training part in order to predict genre labels in the test part.

3.2 Feature Extraction

We utilize a broad palette of low-level acoustic features and several mid-level representations. The set of extracted acoustic features is nearly identical to the one used in Lukashevich (2012). The features can be subdivided in three categories by covering timbral, rhythmic and tonal aspects of sound. *Timbre features* comprise low-level features, such as Mel-Frequency Cepstral Coefficients, Octave Spectral Contrast, Audio Spectrum Centroid, Spectral Crest Factor, Spectral Flatness Measure, and their mid-level representations. These mid-level features are computed on 5.12 s (with 2.56 s hop size) excerpts and observe the evolution of the low-level features. The mid-level representations help to capture the timbre texture by descriptive statistics, see Dittmar et al. (2007) for details. *Rhythmic features* are derived from the energy slope in excerpts of the frequency-bands of the Audio Spectrum Envelope. Further mid-level rhythmic features are gained from the Auto-Correlation Function (Dittmar et al. 2007). *Tonality features* are based on the Enhanced Pitch Class Profiles feature proposed in Lee (2006).

3.3 Frame-Wise Classification

Extracted feature vectors and corresponding genre labels are used for the frame-wise classification. The temporal resolution of frame-wise classification it is equal to 2.56 s. Here we apply two classification systems:

System 1 Performs dimensionality reduction via Linear Discriminant Analysis (LDA) and subsequent classification with GMM.

System 2 Does variable selection by Inertia Ratio Maximization using Feature Space Projection (IRMFSP) and classifies with SVM.

[1] http://ismir2004.ismir.net/genre_contest/index.htm

The following paragraph provides brief information on the applied techniques. Variable selection method *IRMFSP* is proposed by Peeters and Rodet (2003). During each iteration of the algorithm, we look for a feature dimension (variable) maximizing the ratio of between-class inertia to the total-class inertia. *LDA* is one of the most often used supervised dimensionality reduction techniques (Fukunaga 1990). Original feature vectors are linearly mapped into new feature space guaranteeing a maximal linear separability by maximization of the ratio of between-class variance to the within-class variance. *SVM* is a discriminative classifier, attempting to generate an optimal decision plane between feature vectors of the training classes (Vapnik 1998). We use a SVM with a Radial Basis Function kernel in this paper. Posterior class probabilities are estimated as proposed by Platt (1999). Furthermore we additionally apply the pairwise coupling algorithm suggested by Wu et al. (2004) to support the multi-class case. *GMM* is a commonly used generative classifier. Single data samples of each class are interpreted as being generated from various sources and each source is modeled by a single multivariate Gaussian. The parameters of a GMM are estimated using the Expectation-Maximization algorithm proposed by Dempster et al. (1977).

3.4 Song-Wise Classification

Although the frame-wise classification is helpful for some applications, often the song-wise class decisions are required. Song-wise decisions are estimated by majority voting (MV) within all frames of a song. We apply both hard and soft MV. In the *hard MV* the most frequent class decision over all frames within a song is taken as a song-wise decision. The corresponding CM is set to its normalized frequency. In the *soft MV* the class with a highest mean confidence over all frames within a song is taken as a song-wise decision. This highest mean confidence becomes a confidence of a song-wise decision.

3.5 Evaluation Measures

To measure the performance of the classification system trained for M classes, we define a *confusion matrix* $\mathrm{C} \in \mathbb{R}^{M \times M}$. Matrix element c_{ij} is the number of samples with true class label ω_i classified as class ω_j. In addition we define a *normalized confusion matrix* $\mathrm{C}_{\mathrm{norm}}$ where each row of C is normalized to a sum of 1. The *overall accuracy* is defined as a fraction of a number of correctly classified samples to a number of all samples: $A = \mathrm{trace}(\mathrm{C})/\sum_{ij} c_{ij}$. The *error rate* is a fraction of incorrectly classified samples: $\varepsilon = 1 - A$.

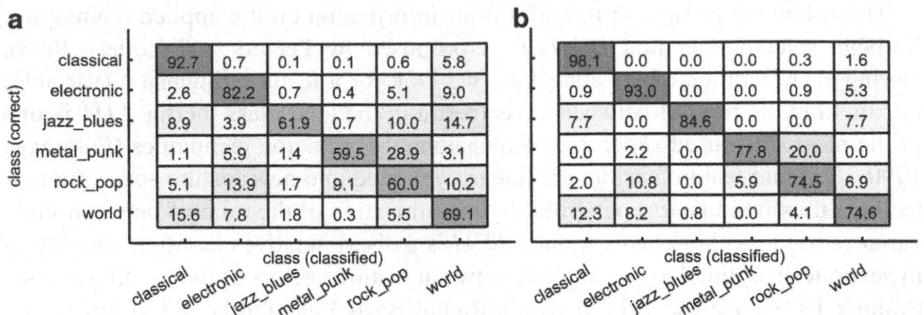

Fig. 1 Confusion matrices show frame-wise and song-wise classification results for system 2 (IRMFSP + SVM), all values are given in percent. (**a**) Frame-wise, IRMFSP + SVM. (**b**) Song-wise, IRMFSP + SVM

4 Results

Figure 1 presents detailed results of frame-wise and song-wise classification for classification System 2 in a form of confusion matrices, here the overall accuracy reaches 0.78 and 0.88 for frame-wise and song-wise classification, respectively. System 1 achieves comparable overall accuracy of 0.73 and 0.87, correspondingly. Both classification systems show solid results that are comparable to the state of the art for this public dataset. Song-wise classification is more reliable in comparison to the frame-wise classification. The confusion matrices show that some genres are confused more often that the others. So `Metal & Punk` is often confused with `Rock & Pop`. These confusions could partly be explained by the natural fusion of these music genres.

In this work we are interested in the question to which extend the posterior class probabilities can be used as confidence measures. Table 1 shows the normalized confidence error ρ as defined in Sect. 2.1. The values of ρ for frame-wise classification equal to 0.10 and 0.12 signify the meaningfulness of proposed CMs. Remind the range of ρ in (1). The song-wise classification via MV aggregation improves the overall accuracy, the error rate ε decreases. However, the aggregation via MV increases the normalized confidence error ρ. It is interesting to observe that the hard and soft MV lead to close error rate values but have a strong influence on the confidence measures. In case of hard MV the CMs are more reasonable. Especially in case of the LDA + GMM classification system the CMs become close to random ones for soft MV.

According to the definition of a CM, on the average a fraction c of all objects with confidence c should be classified correctly. Do our CMs meet this requirement? Figure 2 depicts the relation between the CMs and accuracy values for frame-wise classification. Here we collect classified frames within equidistant confidence bins and take a look to their accuracy. As expected from the low values of normalized

Table 1 Normalized confidence error, as defined in Sect. 2.1

Time resolution (aggregation)	Frame-wise (none)		Song-wise (hard MV)		Song-wise (soft MV)	
Measure	ε	ρ	ε	ρ	ε	ρ
Method						
LDA + GMM	0.27	0.10	0.13	0.25	0.13	0.43
IRMFSP + SVM	0.22	0.12	0.12	0.20	0.12	0.29

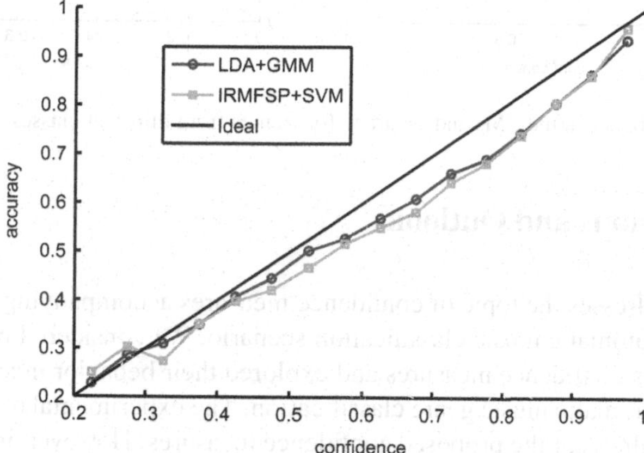

Fig. 2 Relation between the CMs and accuracy values for frame-wise classification

confidence error ρ, both classification systems output reasonable CMs, close to the desired ones. In general, the CMs values are higher than corresponding accuracy values. This tendency is only different for CMs lower than 0.3. Note, that CM values of a winning class in this range in the experiment with six classes correspond to nearly equiprobable class decisions. In our experiment only a few classification decisions fall into the range of low confidence values so that we do not focus on this part of the accuracy vs. confidence curve.

However, this nearly ideal relation between CM values and accuracy does not hold true if we examine it for frames from each class individually. As it can be seen in Fig. 3, the frames from class `Classical` with CMs of about 0.8 reach accuracy of about 0.9. In contrast, the frames from class `Rock & Pop` with CMs of about 0.8 exhibit accuracy of only 0.6. This behavior of CMs violates the requirements introduced in Sect. 2.1 and therefore hampers the interpretation of classification results. We can observe that the relation between CMs and accuracy depends on the accuracy reached for a particular class. So for classes with accuracy higher than an overall accuracy (see `Classical`) the CMs values are lower than the reached accuracy. In contrast, for classes with lower class accuracies (see `Metal & Punk` and `Rock & Pop`) the CMs are higher than the accuracy values. Our observations correspond to the findings of Fumera et al. (2000) where the authors discuss the effects of the imperfect estimation of posterior class probabilities.

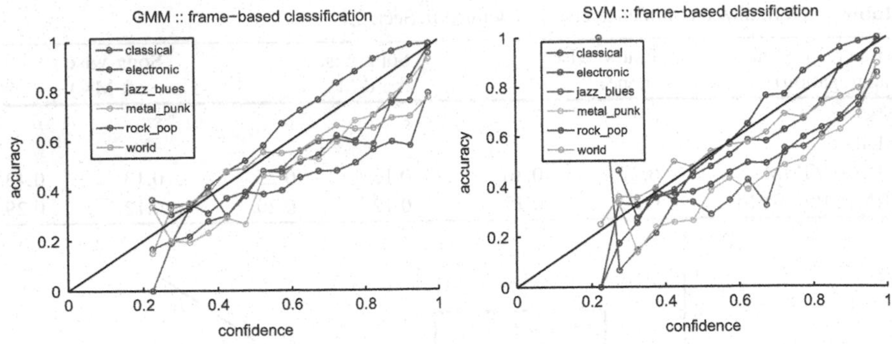

Fig. 3 Relation between the CMs and accuracy for frames from different classes

5 Conclusions and Outlook

This paper addresses the topic of confidence measures accompanying classification decisions in automatic music classification scenario. We considered posterior class probabilities as confidence measures and explored their behavior in state-of-the-art systems for automatic music genre classification. The experimental results signified the meaningfulness of the proposed confidence measures. However, in reflection of classification results for distinct classes the confidence measures appeared to be not optimal and need to be corrected.

Our future research will be directed towards further investigation on the dependencies for class accuracies and confidence measures. Moreover, we are interested in the possibilities to correct the confidences measures via a mapping function so that they fulfill the theoretical requirements. It would be significant to understand the behavior of the confidence measures during hard and soft MV and to answer the question if the correction of the frame-level confidence measures can improve the accuracy of the song-level classification.

Acknowledgements This research work is a part of the SyncGlobal project.[2] It is a 2-year collaborative research project between Piranha Music & IT from Berlin and Bach Technology GmbH, 4FriendsOnly AG, and Fraunhofer IDMT in Ilmenau, Germany. The project is co-financed by the German Ministry of Education and Research in the frame of an SME innovation program (FKZ 01/S11007).

[2] http://www.syncglobal.de/

References

Dempster, A. P., Laird, N. M., & Rdin, D. B. (1977). Maximum likelihood from incomplete data via the EM algorithm. *Journal of the Royal Statistical Society, Series B, 39*, 1–38.

Dittmar, C., Bastuck, C., & Gruhne, M. (2007). Novel mid-level audio features for music similarity. In E. Schubert, K. Buckley, R. Eliott, B. Koboroff, J. Chen, & C. Stevens (Eds.), *Proceedings of the International Conference on Music Communication Science (ICOMCS)* (pp. 38–41). Sydney: University of Western Sydney.

Duin, R. P. W., & Tax, D. M. J. (1998). Classifier conditional posterior probabilities. *Advances in pattern recognition* (pp. 611–619). Berlin: Springer.

Fukunaga, K. (1990). *Introduction to statistical pattern recognition* (2nd ed.). *Computer science and scientific computing series*. Boston: Academic Press.

Fumera, G., Roli, F., & Giacinto, G. (2000). Multiple reject thresholds for improving classification reliability. In F. Ferri, J. Iñesta, A. Amin, & P. Pudil (Eds.), *Advances in pattern recognition* (Vol. 1876). *Lecture notes in computer science* (pp. 863–871). Berlin: Springer.

Lee, K. (2006). Automatic chord recognition from audio using enhanced pitch class profile. In *Proceedings of the International Computer Music Conference (ICMC)*. New Orleans: Tulane University.

Lukashevich, H. (2012). Applying multiple kernel learning to automatic genre classification. In W. Gaul, A. Geyer-Schulz, L. Schmidt-Thieme, & J. Kunze (Eds.), *Challenges at the interface of data analysis, computer science, and optimization* (pp. 393–400). Berlin: Springer.

Peeters, G., & Rodet, X. (2003). Hierarchical Gaussian tree with inertia ratio maximization for the classification of large musical instruments databases. In *Proceedings of the 6th International Conference on Digital Audio Effects (DAFx)*. London: Queen Mary, University of London.

Platt, J. C. (1999). Probabilistic outputs for support vector machines and comparisons to regularized likelihood methods. *Advances in Large Margin Classifiers, 10*(3), 61–74.

Vapnik, V. (1998). *Statistical learning theory*. New York: Wiley.

Wu, T. F., Lin, C. J., & Weng, R. C. (2004). Probability estimates for multi-class classification by pairwise coupling. *Journal of Machine Learning Research, 5*(16), 975–1005.

Using Latent Class Models with Random Effects for Investigating Local Dependence

Matthias Trendtel, Ali Ünlü, Daniel Kasper, and Sina Stubben

Abstract In psychometric latent variable modeling approaches such as item response theory one of the most central assumptions is local independence (LI), i.e. stochastic independence of test items given a latent ability variable (e.g., Hambleton et al., Fundamentals of item response theory, 1991). This strong assumption, however, is often violated in practice resulting, for instance, in biased parameter estimation. To visualize the local item dependencies, we derive a measure quantifying the degree of such dependence for pairs of items. This measure can be viewed as a dissimilarity function in the sense of psychophysical scaling (Dzhafarov and Colonius, Journal of Mathematical Psychology 51:290–304, 2007), which allows us to represent the local dependencies graphically in the Euclidean 2D space. To avoid problems caused by violation of the local independence assumption, in this paper, we apply a more general concept of "local independence" to psychometric items. Latent class models with random effects (LCMRE; Qu et al., Biometrics 52:797–810, 1996) are used to formulate a generalized local independence (GLI) assumption held more frequently in reality. It includes LI as a special case. We illustrate our approach by investigating the local dependence structures in item types and instances of large scale assessment data from the Programme for International Student Assessment (PISA; OECD, PISA 2009 Technical Report, 2012).

M. Trendtel (✉) · A. Ünlü · D. Kasper · S. Stubben
Chair for Methods in Empirical Educational Research, TUM School of Education, and Centre for International Student Assessment (ZIB), Technische Universität München, Lothstrasse 17, 80335 Munich, Germany
e-mail: matthias.trendtel@tum.de; ali.uenlue@tum.de; daniel.kasper@tum.de; sina.stubben@tum.de

M. Spiliopoulou et al. (eds.), *Data Analysis, Machine Learning and Knowledge Discovery*, Studies in Classification, Data Analysis, and Knowledge Organization, DOI 10.1007/978-3-319-01595-8_44,

1 Introduction

In educational measurement, models are often concerned with connecting observed responses of an individual to test items with the individual's latent ability. Traditionally, for instance in item response theory (IRT; e.g. Mcdonald 1999), ability is represented by a (unidimensional or multidimensional) latent continuous variable. This variable is assumed to explain all influences on the examinee's responses to the test items. An alternative approach to model the examinee's ability and responses is based on discrete structures called knowledge spaces or learning spaces (Falmagne and Doignon 2011). Here, latent ability is represented by a latent discrete variable, which also determines the examinee's response pattern completely. Thus, in both approaches the latent variable represents the whole latent space, i.e. accounts for all (systematic) associations between test items. This means, both approaches postulate local independence given the latent variable.

More technically (see Hambleton et al. 1991), for n items, the binary random variable representing the (correct or incorrect) response of an examinee to item i is denoted by S_i. Furthermore, the probability of a response of an examinee with latent ability θ is denoted by $P(S_i|\theta)$, where $P(S_i = 1|\theta)$ and $P(S_i = 0|\theta)$ are the probabilities of a correct and incorrect response to item i, respectively. Local independence (LI) is satisfied iff

$$P(S_1, S_2, \ldots, S_n|\theta) = P(S_1|\theta)P(S_2|\theta)\ldots P(S_n|\theta) = \prod_{i=1}^{n} P(S_i|\theta), \quad (1)$$

where $P(S_1, S_2, \ldots, S_n|\theta)$ is the joint probability for all responses $S_1, S_2, \ldots, S_n$ in the test, given the latent ability θ.

We show that in PISA 2009 for reading items within one item unit (OECD 2012), LI may not be satisfied when assuming a model with a one-dimensional latent continuous variable explaining test performance. Since in PISA such latent variable models are used for scaling items and persons, the parameter estimation can be biased (e.g., see Chen and Wang 2007).

In this paper, we provide means to model and, thus, control for local item dependencies via visualization and a generalized concept of "local independence." First, we visualize the occurring local item dependencies by deriving a metric based on the dissimilarity between item pairs using Rosenbaum's test for LI (Rosenbaum 1984). Second, we demonstrate that a more general form of local independence derived from the latent class model with random effects (LCMRE), which was introduced by Qu et al. (1996), yields considerably fewer violations of the LI assumption; in the sense that we reject the null hypothesis "LI is satisfied" less frequently. As discussed by Ünlü (2006), the LCMRE can be applied to learning space theory (LST) and viewed as a generalization of the LST fundamental basic local independence model (BLIM). This generalized approach provides a classification of the sample in a mastery and a non-mastery class, and it allows for estimating the misclassification rates of the items to prevent fallacies. That application in LST is instructive, and we therefore give a brief introduction to LST before we elaborate on local dependence under the LCMRE.

2 Basics of Learning Space Theory

For a comprehensive introduction to LST, the reader is referred to Falmagne and Doignon (2011). For the purposes of this paper, our discussion is restricted to the very main concepts and ideas of LST. LST utilizes set theory to describe the possible states of knowledge of a learner. Knowledge is represented by the set of test items an individual is able to master under ideal conditions. More formally, given a set of items Q, the subset $K \subseteq Q$ of items a learner with a certain ability should answer correctly is called her/his *knowledge state*. Any family of subsets of Q containing the empty set $\emptyset$ and Q builds a so-called *knowledge structure* $\mathcal{K}$ (on Q).

As an example consider the set Q of five items $Q = \{a, b, c, d, e\}$. Then

$$\mathcal{K} = \{\emptyset, \{a\}, \{b\}, \{a, b\}, \{a, b, c\}, \{a, b, d\}, \{a, b, c, d\}, \{a, b, c, d, e\}\}$$

forms a knowledge structure and, for instance, a person with knowledge state $\{a, b, d\}$ is capable of answering items a, b, d correctly, but not items c and e.

Since lucky guesses and careless errors are possible, a learner's responses $R \subseteq Q$, i.e. the set of items she/he answers correctly in reality, is not generally equal to her/his knowledge state K. In LST, the probability of a response R for a learner with knowledge state K is modeled by the basic local independence model (BLIM). The BLIM assumes LI, so that probabilities factor into the probabilities of slipping (or not slipping) on items $i \in K$ and guessing (or not guessing) on item $i \notin K$.

Definition 1. A quadruple $(Q, \mathcal{K}, p, r)$ is called a BLIM iff

1. $\mathcal{K}$ is a knowledge structure on Q,
2. p is a probability distribution on $\mathcal{K}$, i.e., $p : \mathcal{K} \to]0, 1[, K \mapsto p(K)$, with $p(K) > 0$ for any $K \in \mathcal{K}$, and $\sum_{K \in \mathcal{K}} p(K) = 1$,
3. r is a response function for $(Q, \mathcal{K}, p)$, i.e., $r : 2^Q \times \mathcal{K} \to [0, 1], (R, K) \mapsto r(R, K)$, with $r(R, K) \geq 0$ for any $R \in 2^Q$ and $K \in \mathcal{K}$, and $\sum_{R \in 2^Q} r(R, K) = 1$ for any $K \in \mathcal{K}$,
4. r satisfies LI, i.e.,
$r(R, K) = \prod_{i \in K \setminus R} \beta_i \cdot \prod_{i \in K \cap R} (1 - \beta_i) \cdot \prod_{i \in R \setminus K} \eta_i \cdot \prod_{i \in Q \setminus (R \cup K)} (1 - \eta_i)$,
with two constants $\beta_i, \eta_i \in [0, 1[$ for each item $i \in Q$ called the item's careless error and lucky guess probabilities, respectively.

The BLIM assumes that the manifest multinomial probability distribution on the response patterns is governed by the latent state proportions and response error rates. Postulating that the response patterns are independent from each other and that the occurrence probability of any response pattern $R \in 2^Q$, $\rho(R)$, is constant across the subjects, $\rho(R)$ can be parameterized as $\rho(R) = \sum_{K \in \mathcal{K}} r(R, K)p(K)$.

3 Latent Class Model with Random Effects

In the following, we consider the simplest knowledge structure $\mathcal{K}$ on Q, i.e. $\mathcal{K} = \{\emptyset, Q\}$. In this case, there are only two knowledge states and the BLIM degenerates to the classical unrestricted 2-classes latent class model. We connect LST notation with the concept of binary response variables S_i and a latent ability variable θ by regarding a set Q of n items. Then, the item response variable S_i is equal to one for all item answered correctly, i.e. $i \in R$, and $S_i = 0$ for items $i \notin R$. Further, the two knowledge states result in a binary ability variable $\theta = 1$ for $K = Q$ and $\theta = 0$ for $K = \emptyset$. Hence, the assumption of LI is denoted as $P(S_1 = s_1, S_2 = s_2, \ldots, S_n = s_n | \theta = \vartheta) = \prod_{i=1}^{n} P(S_i = s_i | \theta = \vartheta)$ for any $s_i, \vartheta \in \{0, 1\}$ for all i. Note that θ is discrete in this case.

Since LI may not be justified in general, we regard a different model extended to allow for local dependence using random effects (see Qu et al. 1996). The latent class model with random effects (LCMRE) assumes that an individual's responses are not completely governed by the knowledge state K. In addition, there might be a subject- and/or item-specific random factor $T \sim \mathrm{N}(0, 1)$ influencing the examinee's item performance. Under the LCMRE, the latent (discrete) ability θ supplemented by the latent (continuous) random factor T represents the complete latent space having influence on the item performance.

Under this model, general "local independence" (GLI) is satisfied iff, given θ and T, the factorization of probabilities holds

$$P(S_1 = s_1, \ldots, S_n = s_n | \theta = \vartheta, T = t) = \prod_{i=1}^{n} P(S_i = s_i | \theta = \vartheta, T = t)$$

for any $s_i, \vartheta \in \{0, 1\}$ $(1 \leq i \leq n)$ and $t \in \mathbb{R}$.

The latent random variables θ and T are assumed to be independent and for any item $i \in Q$, the conditional probability $P(S_i = 1 | \theta = \vartheta, T = t)$ for solving the item, given $\theta = \vartheta$ and $T = t$, is parametrized through a probit regression (Qu et al. 1996): $\Phi(a_{i\vartheta} + b_{i\vartheta} t)$.

4 Testing Local Independence

In practice, the local independence assumption (1) may not hold. Therefore, several tests are known to test for pairwise local dependence of an item pair. In this paper, we consider a test introduced by Rosenbaum (1984) to derive a visualization of local item dependencies. Regard a set Q of n items. For an item pair $(i, j) \in Q \times Q$, an individual can obtain a total score k of correct responses to the $n-2$ items in $Q \setminus \{i, j\}$ with $k \in \{0, \ldots, n-2\}$. For any k, the sample distribution of (S_i, S_j) can be tabulated in a 2×2 contingency table. Let m_{uvk} be the expected counts of the cell (u, v) in the kth contingency table for $u, v \in \{0, 1\}$ and $k \in \{0, \ldots, n-2\}$. The odds ratio R_k in the kth table is defined as $R_k = \frac{m_{00k} m_{11k}}{m_{10k} m_{01k}}$. LI implies the null hypothesis

$$H_0 : R_k \geq 1, \text{ for } k = 0, \ldots, n-2.$$

In Rosenbaum's approximate test, a high p-value indicates LI between the considered items and a low p-value indicates local dependence.

To visualize local item dependencies, the p-value is used as a dissimilarity measure for any item pair. Thereby, a small p-value indicating higher local dependence represents a higher similarity and a higher p-value represents a higher dissimilarity. In this way, we derive a dissimilarity function on the set of items. Given a dissimilarity function, *Fechnerian scaling* (see, e.g., Dzhafarov and Colonius 2007) can be applied to impose a metric on the item set by calculating the so-called *Fechnerian distances* (FD). Based on this metric, multidimensional scaling can be applied to project the items onto the Euclidean two-dimensional space. Two items close to each other on the plot are more locally dependent than two items further apart. Hence, the local dependence structure is visualized.

5 Local Independence and Latent Classes in PISA

As an application, we analyzed the local independence of reading items in the German sample of the PISA 2009 paper-and-pencil test. Regarding the traditional concept of LI (1), we expect to detect local item dependencies via Rosenbaum's test for LI since the continuous latent ability variable may not fully explain associations between items. We visualize occurring dependencies as described in Sect. 4.

Local independence is satisfied if and only if there is no other influencial dimension on an individual's response besides the model's latent variables. In the LCMRE the individual's responses are influenced by both, one continuous latent variable (the random effect) and one discrete variable (the latent ability) leading to a more general concept of local independence (GLI). If we test for GLI by holding fixed the latent state too, the test should not indicate LD when the data follow the LCMRE.

Table 1 Sample sizes, class sizes and number of violations of LI as a result of the test for LI for an alpha level of 0.05 based on the whole sample of the respective item unit and based on the estimated latent classes obtained from the LCMRE

	Sample size			Significant violations		
Unit label	Whole sample	Class 1	Class 2	Whole sample	Class 1	Class 2
R055	1,431	1,118	313	6	2	1
R083	1,462	314	1,148	6	5	5
R227	763	232	531	5	0	0
R412	1,434	1,266	168	4	5	5
R414	1,442	708	734	6	0	0
R447	1,461	708	753	6	1	0
R452	1,449	688	761	6	4	2
R453	1,445	1,081	364	6	2	2

In PISA 2009, items in the same so-called *item unit* are based around a common stimulus (e.g., OECD 2012). This motivates the assumption that items in the same unit can be seen as equally informative and the underlying knowledge structure is the simplest one, i.e. we can a assume a binary ability variable θ representing the two possible knowledge states (see Sect. 3). There is a total number of 37 units comprising a total of 131 items. We choose units containing exactly four items and excluded response patterns missing responses for at least one item within the unit. Units with less than 300 complete response patterns were skipped.

We analyzed nine item units and applied the test for LI for every item pair in each item unit for all response patterns. The test for LI always gives a significant result (0.00) for an item pair consisting of the same items and the test is symmetric with respect to the order of the item pair. Therefore, we take into account the respective 6 item pairs consisting of two different items only.

After applying the test for LI (see Sect. 4) in each considered item unit at least 4 item pairs show significant (alpha level of $p < 0.05$) local dependencies (see Table 1). This indicates that one continuous latent variable is not sufficient to explain the item performance for these units.

Exemplarily, in Table 2, the results of the test and the calculated Fechnerian distances (FDs, see Sect. 4) are shown for the item unit "R414". Figure 1a shows the plot of the multidimensional scaled FDs between the items of the item unit "R414" calculated based on the pairwise dissimilarities arising from the test for LI for the whole sample. Figure 1a and Table 2 show that items are strongly locally dependent. Within the unit, items "R414Q02", "R414Q06" and "R414Q09" show higher local dependencies than the item "R414Q11" with the other items.

In the next step, we applied the LCMRE using the R-package randomLCA for computations and partitioned the item responses according to the class probabilities for the two knowledge states derived from the latent class analysis. Thereby, a response pattern was allocated to the knowledge state with the highest estimated probability. Afterwards we applied the test for LI for every class separately for each item unit, hence testing for GLI. The results are summarized in Table 1. For item

Table 2 FDs and p-values (in parentheses) between the items of the unit "R414"

	R414Q02	R414Q06	R414Q09	R414Q11
R414Q02	0.000e+00 (0.000e+00*)	4.441e−16 (2.220e−16*)	6.661e−16 (4.984e−05*)	5.700e−13 (3.706e−04*)
R414Q06	4.441e−16 (2.220e−16*)	0.000e+00 (0.000e+00*)	2.220e−16 (1.110e−16*)	5.695e−13 (2.848e−13*)
R414Q09	6.661e−16 (4.984e−05*)	2.220e−16 (1.110e−16*)	0.000e+00 (0.000e+00*)	5.698e−13 (2.723e−05*)
R414Q11	5.700e−13 (3.706e−04*)	5.695e−13 (2.848e−13*)	5.698e−13 (2.723e−05*)	0.000e+00 (0.000e+00*)

An asterisk marks a significant result, i.e. $p < 0.05$

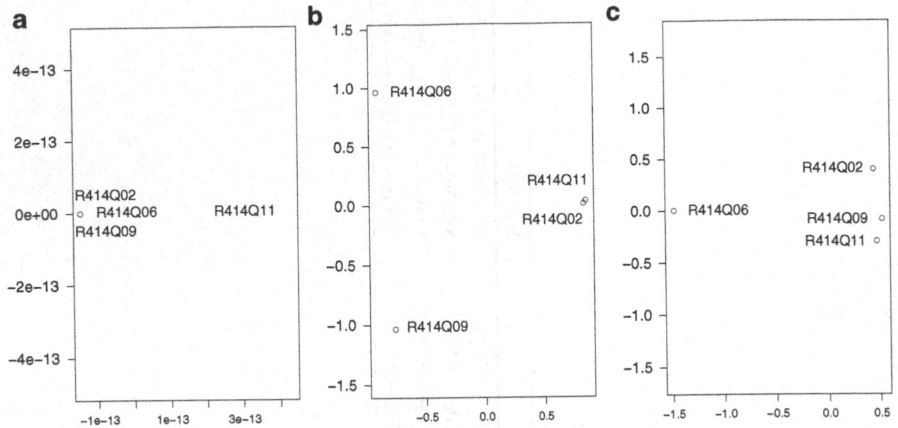

Fig. 1 LD structure of the items of the unit "R414". Part (**a**) shows the structure within the whole sample, (**b**) within knowledge state $K = Q$ and (**c**) within knowledge state $K = \emptyset$

units "R083", "R412" and "R452", the number of violations were still too high to accept the LCMRE with two latent classes as a suitable model. For all other item units, however, the reduction of violations was striking. There were no more than 2 violations of GLI in each class and item unit. The units "R227" and "R414" do not show any violation of GLI at all. As this stands against 5 and 6 violations, respectively, under a model with one continuous latent variable, the LCMRE is strongly preferable.

Table 3 shows the results of the test for GLI and the calculated Fechnerian distances between the items of the item unit "R414" for both knowledge states, $K = Q$ and $K = \emptyset$. Further, Fig. 1b, c visualize the local dependencies.

6 Conclusion

We have proposed a method to visualize violations of the traditional concept of local independence (LI) in a psychometric context using Fechnerian scaling. A generalized form of "local independence" has been derived from the latent class model with random effects. That resulted in fewer violations of LI, and thus, is assumed to prevent biased estimation (important in PISA). Possible limitations of our approach lie within the assumption of (standard) normally distributed random factor T. So far, we have no theory that could imply the normal distribution. Moreover, at least theoretically, the link function could be any distribution function (other than Φ). For the models juxtaposed in this paper, the impact of fewer violations on the estimation of the model parameters must be quantified in simulations. Additionally, analyses have to be done investigating the performance of tests of LI in other modeling approaches, e.g. the Q-statistics (Yen 1984). Considering the results of the analyses

Table 3 FDs and p-values (in parentheses) between the items of the unit "R414" of the first (upper part) and the second class (lower part)

$K = Q$	R414Q02	R414Q06	R414Q09	R414Q11
R414Q02	0.000e+00 (0.000e+00*)	2.000e+00 (1.000e+00)	1.962e+00 (9.808e−01)	7.351e−01 (3.676e−01)
R414Q06	2.000e+00 (1.000e+00)	0.000e+00 (0.000e+00*)	1.988e+00 (9.942e−01)	2.000e+00 (1.000e+00)
R414Q09	1.962e+00 (9.808e−01)	1.988e+00 (9.942e−01)	0.000e+00 (0.000e+00*)	1.980e+00 (9.901e−01)
R414Q11	7.351e−01 (3.676e−01)	2.000e+00 (1.000e+00)	1.980e+00 (9.901e−01)	0.000e+00 (0.000e+00*)
$K = \emptyset$	R414Q02	R414Q06	R414Q09	R414Q11
R414Q02	0.000e+00 (0.000e+00*)	1.999e+00 (9.997e−01)	4.916e−01 (2.458e−01)	7.059e−01 (9.811e−01)
R414Q06	1.999e+00 (9.997e−01)	0.000e+00 (0.000e+00*)	2.000e+00 (1.000e+00)	2.000e+00 (9.999e−01)
R414Q09	4.916e−01 (2.458e−01)	2.000e+00 (1.000e+00)	0.000e+00 (0.000e+00*)	2.143e−01 (1.071e−01)
R414Q11	7.059e−01 (9.811e−01)	2.000e+00 (9.999e−01)	2.143e−01 (1.071e−01)	0.000e+00 (0.000e+00*)

An asterisk marks a significant result, i.e. $p < 0.05$

of the PISA 2009 data, a 2-class latent class model with random effects seems to be sufficient to model dependencies for most item units. In terms of learning space theory, examinees can be divided into a mastery and a non-mastery group. In a next step, individual items may be analyzed to identify specific characteristics of the items, as a possible explanation for this kind of data.

References

Chen, C. T., & Wang, W. C. (2007). Effects of ignoring item interaction on item parameter estimation and detection of interacting items. *Applied Psychological Measurement, 31*, 388–411.

Dzhafarov, E. N., & Colonius, H. (2007). Dissimilarity cumulation theory and subjective metrics. *Journal of Mathematical Psychology, 51*, 290–304.

Falmagne, J.-C., & Doignon, J.-P. (2011). *Learning spaces*. Berlin: Springer

Hambleton, R. K., Swaminathan, H., & Rogers, H. J. (1991). *Fundamentals of item response theory*. Newbury Park: Sage Publications.

McDonald, R. P. (1999). *Test theory: A unified treatment*. Mahwah: Erlbaum.

OECD. (2012). *PISA 2009 Technical Report*. Paris: OECD Publishing.

Qu, Y., Tan, M., & Kutner, M. H. (1996). Random effects models in latent class analysis for evaluating accuracy of diagnostic tests. *Biometrics, 52*, 797–810.

Rosenbaum, P. R. (1984). Testing the conditional independence and monotonicity assumptions of item response theory. *Psychometrika, 49*, 425–435.

Ünlü, A. (2006). Estimation of careless error and lucky guess probabilities for dichotomous test items: A psychometric application of a biometric latent class model with random effects. *Journal of Mathematical Psychology, 50*, 309–328.

Yen, W. M. (1984). Effects of local item dependence on the fit and equating performance of the three-parameter logistic model. *Applied Psychological Measurement, 8*, 125–145.

The OECD's Programme for International Student Assessment (PISA) Study: A Review of Its Basic Psychometric Concepts

Ali Ünlü, Daniel Kasper, Matthias Trendtel, and Michael Schurig

Abstract The Programme for International Student Assessment (PISA; e.g., OECD, Sample tasks from the PISA 2000 assessment, 2002a; OECD, Learning for tomorrow's world: first results from PISA 2003, 2004; OECD, PISA 2006: Science competencies for tomorrow's world, 2007; OECD, PISA 2009 Technical Report, 2012) is an international large scale assessment study that aims to assess the skills and knowledge of 15-year-old students, and based on the results, to compare education systems across the participating (about 70) countries (with a minimum number of approx. 4,500 tested students per country). Initiator of this Programme is the Organisation for Economic Co-operation and Development (OECD; www.pisa.oecd.org). We review the main methodological techniques of the PISA study. Primarily, we focus on the psychometric procedure applied for scaling items and persons. PISA proficiency scale construction and proficiency levels derived based on discretization of the continua are discussed. For a balanced reflection of the PISA methodology, questions and suggestions on the reproduction of international item parameters, as well as on scoring, classifying and reporting, are raised. We hope that along these lines the PISA analyses can be better understood and evaluated, and if necessary, possibly be improved.

A. Ünlü (✉) · D. Kasper · M. Trendtel · M. Schurig
Chair for Methods in Empirical Educational Research, TUM School of Education, and Centre for International Student Assessment (ZIB), Technische Universität München, Arcisstrasse 21, 80333 Munich, Germany
e-mail: ali.uenlue@tum.de; daniel.kasper@tum.de; matthias.trendtel@tum.de; michael.schurig@tum.de

M. Spiliopoulou et al. (eds.), *Data Analysis, Machine Learning and Knowledge Discovery*, Studies in Classification, Data Analysis, and Knowledge Organization, DOI 10.1007/978-3-319-01595-8_45,

1 Introduction

PISA is an international large scale educational assessment study conducted by member countries of the OECD (2001, 2002a, 2004, 2007, 2010) and investigates how well 15-year-old students approaching the end of compulsory schooling are prepared to meet the challenges of today's knowledge societies (OECD 2005, 2012). The study does not focus on the students' achievement regarding a specific school curriculum but rather aims at measuring the students' ability to use their knowledge and skills to meet real-life challenges (OECD 2009a).

PISA started in 2000 and takes place every 3 years. Proficiencies in the domains reading, mathematics, and science are assessed. In each assessment cycle, one of these domains is chosen as the major domain under fine-grained investigation; reading in 2000, followed by mathematics in 2003, science in 2006, etc. The definitions of the domains can be found in the respective *assessment frameworks* (e.g., OECD 2009a). In addition to these domains, further competencies may also be assessed by a participating OECD member; for example digital reading in 2009 (OECD 2012). Besides the actual test in PISA, student and school questionnaires are used to provide additional background information (e.g., about the socio-economic status of a student). In PISA 2009 for instance, in addition to these questionnaires, in 14 countries parents were asked to fill in an optional questionnaire. The background information are used as so-called conditioning variables for the scaling of the PISA cognitive (i.e., test) data.

The number of countries (and economies) participating in PISA continues to increase (e.g., 32 and 65 countries for PISA 2000 and 2009, respectively). In each participating country, a sample of at least 150 schools (or all schools) were drawn. In each participating school, 35 students were drawn (in schools with less than 35 eligible students, all students were selected).

The PISA study involves a number of technical challenges; for example, the development of test design and measurement instruments, of survey and questionnaire scales. Accurate sampling designs, including both school sampling and student sampling, must be developed. The multilingual and multicultural nature of the assessment must be taken into account, and various operational control and validation procedures have to be applied. Focused on in this paper, the scaling and analysis of the data require sophisticated psychometric methods, and PISA employs a scaling model based on *item response theory* (IRT; e.g., Adams et al. 1997; Fischer and Molenaar 1995; van der Linden and Hambleton 1997). The proficiency scales and levels, which are the basic tool in reporting PISA outcomes, are derived through IRT analyses.

The PISA *Technical Report* describes those methodologies (OECD 2002b, 2005, 2009b, 2012). The description is provided at a level that allows for review and, *potentially*, replication of the implemented procedures. In this paper, we recapitulate the scaling procedure that is used in PISA (Sect. 2). We discuss the construction of proficiency scale and proficiency levels and explain how the results are reported and interpreted in PISA (Sect. 3). We comment on whether information provided in the

Technical Report is sufficient to replicate the sampling and scaling procedures and central results for PISA, on classification procedures and alternatives thereof, and on other, for instance more automated, ways for reporting in the PISA Technical Report (Sect. 4). Overall, limitations of PISA and some reflections and suggestions for improvement are described and scattered throughout the paper.

2 Scaling Procedure

To scale the PISA cognitive data, the *mixed coefficients multinomial logit model* (MCMLM; Adams et al. 1997) is applied (OECD 2012, Chap. 9). This model is a generalized form of the Rasch model (Rasch 1980) in IRT. In the MCMLM, the items are characterized by a *fixed* set of unknown parameters, $\boldsymbol{\xi}$, and the student outcome levels, the latent random variable $\boldsymbol{\theta}$, are assumed to be *random effects*.

2.1 Notation

Assume I items (indexed $i = 1, \ldots, I$) with $K_i + 1$ possible response categories $(0, 1, \ldots, K_i)$ for an item i. The vector-valued random variable, for a sampled person, $\boldsymbol{X}_i' = (X_{i1}, X_{i2}, \ldots, X_{iK_i})$ of order $1 \times K_i$, with $X_{ij} = 1$ if the response of the person to item i is in category j, or $X_{ij} = 0$ otherwise, indicates the $K_i + 1$ possible responses of the person to item i. The zero category of an item is denoted with a vector consisting of zeros, making the zero category a reference category, for model identification. Collecting the $\boldsymbol{X}_i'$'s together into a vector $\boldsymbol{X}' = (\boldsymbol{X}_1', \boldsymbol{X}_2', \ldots, \boldsymbol{X}_I')$ of order $1 \times t$ $(t = K_1 + \cdots + K_I)$ gives the response vector, or response pattern, of the person on the whole test.

In addition to the response vector $\boldsymbol{X}$ (person level), assume an $1 \times p$ vector $\boldsymbol{\xi}' = (\xi_1, \xi_2, \ldots, \xi_p)$ of p parameters $(p \geq I)$ describing the I items. These are often interpreted as the items' difficulties. In the response probability model, linear combinations of these parameters are used, to describe the empirical characteristics of the response categories of each item. To define these linear combinations, a set of design vectors $\boldsymbol{a}_{ij}$ $(i = 1, \ldots, I; j = 1, \ldots, K_i)$, each of length p, can be collected to form an $p \times t$ design matrix $\boldsymbol{A}' = (\boldsymbol{a}_{11}, \ldots, \boldsymbol{a}_{1K_1}, \boldsymbol{a}_{21}, \ldots, \boldsymbol{a}_{2K_2}, \ldots, \boldsymbol{a}_{I1}, \ldots, \boldsymbol{a}_{IK_I})$, and the linear combinations are calculated by $\boldsymbol{A}\boldsymbol{\xi}$ (of order $t \times 1$). In the multidimensional version of the model it is assumed that $D \geq 2$ latent traits underlie the persons' responses. The scores of the individuals on these latent traits are collected in the $D \times 1$ vector $\boldsymbol{\theta} = (\theta_1, \ldots, \theta_D)'$, where the θ's are real-valued and often interpreted as the persons' abilities.

In the model also the notion of a response score b_{ijd} is introduced, which gives the performance level of an observed response in category j of item i with respect to dimension d $(d = 1, \ldots, D)$. For dimension d and item i, the response scores across the K_i categories of item i can be collected in an $K_i \times 1$ vector $\boldsymbol{b}_{id} = (b_{i1d}, \ldots, b_{iK_i d})'$ and across the D dimensions in the $K_i \times D$ scoring sub-matrix $\boldsymbol{B}_i = (\boldsymbol{b}_{i1}, \ldots, \boldsymbol{b}_{iD})$. For all items, the response scores can be collected in an $t \times D$ scoring matrix $\boldsymbol{B} = (\boldsymbol{B}'_1, \ldots, \boldsymbol{B}'_I)'$.

2.2 MCMLM

The probability $\Pr(X_{ij} = 1; \boldsymbol{A}, \boldsymbol{B}, \boldsymbol{\xi} | \boldsymbol{\theta})$ of a response in category j of item i, given an ability vector $\boldsymbol{\theta}$, is $\exp(\boldsymbol{b}_{ij}\boldsymbol{\theta} + \boldsymbol{a}'_{ij}\boldsymbol{\xi})/(1 + \sum_{q=1}^{K_i} \exp(\boldsymbol{b}_{iq}\boldsymbol{\theta} + \boldsymbol{a}'_{iq}\boldsymbol{\xi})$, where $\boldsymbol{b}_{iq}$ is the qth row of the corresponding matrix $\boldsymbol{B}_i$, and $\boldsymbol{a}'_{iq}$ is the $(\sum_{l=1}^{i-1} K_l + q)$th row of the matrix $\boldsymbol{A}$. The *conditional item response model* (conditional on a person's ability $\boldsymbol{\theta}$) then can be expressed by $f_x(\boldsymbol{x}; \boldsymbol{\xi} | \boldsymbol{\theta}) = \exp\left[\boldsymbol{x}' \left(\boldsymbol{B}\boldsymbol{\theta} + \boldsymbol{A}\boldsymbol{\xi}\right)\right] / \sum_{z} \exp\left[\boldsymbol{z}' \left(\boldsymbol{B}\boldsymbol{\theta} + \boldsymbol{A}\boldsymbol{\xi}\right)\right]$, where $\boldsymbol{x}$ is a realization of $\mathbf{X}$ and $\sum$ is over of all possible response vectors $\boldsymbol{z}$.

In the conditional item response model, $\boldsymbol{\theta}$ is given. The unconditional, or marginal, item response model requires the specification of a density, $f_\theta(\boldsymbol{\theta})$. In the PISA scaling procedure, students are assumed to have been sampled from a multivariate normal population with mean vector $\boldsymbol{\mu}$ and variance-covariance matrix $\boldsymbol{\Sigma}$, that is, $f_\theta(\boldsymbol{\theta}) = ((2\pi)^D |\boldsymbol{\Sigma}|)^{-1/2} \exp\left[-\left(\boldsymbol{\theta} - \boldsymbol{\mu}\right)' \boldsymbol{\Sigma}^{-1} \left(\boldsymbol{\theta} - \boldsymbol{\mu}\right)/2\right]$. Moreover, this mean vector is parametrized, $\boldsymbol{\mu} = \boldsymbol{\Gamma}'\boldsymbol{w}$, so that $\boldsymbol{\theta} = \boldsymbol{\Gamma}'\boldsymbol{w} + \boldsymbol{e}$, where $\boldsymbol{w}$ is an $u \times 1$ vector of u fixed and known background values for a student, $\boldsymbol{\Gamma}$ is an $u \times D$ matrix of regression coefficients, and the error term $\boldsymbol{e}$ is $N(\boldsymbol{0}, \boldsymbol{\Sigma})$. In PISA, $\boldsymbol{\theta} = \boldsymbol{\Gamma}'\boldsymbol{w} + \boldsymbol{e}$ is referred to as *latent regression*, and $\boldsymbol{w}$ comprises the so-called *conditioning variables* (e.g., gender, grade, or school size). This is the *population model*.

The conditional item response model and the population model are combined to obtain the *unconditional*, or *marginal*, *item response model*, which incorporates not only performance on the items but also information about the students' background: $f(\boldsymbol{x}; \boldsymbol{\xi}, \boldsymbol{\Gamma}, \boldsymbol{w}, \Sigma) = \int_\theta f_x(\boldsymbol{x}; \boldsymbol{\xi} | \boldsymbol{\theta}) f_\theta(\boldsymbol{\theta}; \boldsymbol{\Gamma}, \boldsymbol{w}, \Sigma) d\boldsymbol{\theta}$. The parameters of this MCMLM are $\boldsymbol{\Gamma}$, Σ, and $\boldsymbol{\xi}$. They can be estimated using the software *ConQuest*® (Wu et al. 1997; see also Adams et al. 1997).

Parametrizing a multivariate mean of a prior distribution for the person ability can also be applied to the broader family of *multidimensional item response models* (e.g., Reckase 2009). Alternative models capable of capturing the multidimensional aspects of the data, and at the same time, allowing for the incorporation of covariate information are *explanatory item response models* (e.g., De Boeck and Wilson 2004). The scaling procedure in PISA may be performed using those models. In further research, it would be interesting to compare the different approaches to scaling the PISA cognitive data.

2.3 Student Score Generation

For each student (response pattern) it is possible to specify a posterior distribution for the latent variable $\boldsymbol{\theta}$, which is given by $h_\theta(\boldsymbol{\theta}; \boldsymbol{w}, \boldsymbol{\xi}, \boldsymbol{\Gamma}, \Sigma | \boldsymbol{x}) = f_x(\boldsymbol{x}; \boldsymbol{\xi} | \boldsymbol{\theta}) f_\theta(\boldsymbol{\theta}; \boldsymbol{\Gamma}, \boldsymbol{w}, \Sigma) / \int_\theta f_x(\boldsymbol{x}; \boldsymbol{\xi} | \boldsymbol{\theta}) f_\theta(\boldsymbol{\theta}; \boldsymbol{\Gamma}, \boldsymbol{w}, \Sigma)$. Estimates for $\boldsymbol{\theta}$ are random draws from this posterior distribution, and they are called *plausible values* (e.g., see Mislevy 1991).

Plausible values are drawn in PISA as follows. M vector-valued random deviates $(\boldsymbol{\varphi}_{mn})_{m=1,\ldots,M}$ are sampled from the parametrized multivariate normal distribution, for each individual n. For PISA, the value $M = 2{,}000$ has been specified (OECD 2012). These vectors are used to approximate the integral in the equation for the posterior distribution, using Monte-Carlo integration: $\int_\theta f_x(\boldsymbol{x}; \boldsymbol{\xi} | \boldsymbol{\theta}) f_\theta(\boldsymbol{\theta}; \boldsymbol{\Gamma}, \boldsymbol{w}, \Sigma) d\boldsymbol{\theta} \approx \frac{1}{M} \sum_{m=1}^{M} f_x(\boldsymbol{x}_n; \boldsymbol{\xi} | \boldsymbol{\varphi}_{mn}) = \Im$. The values $p_{mn} = f_x(\boldsymbol{x}_n; \boldsymbol{\xi} | \boldsymbol{\varphi}_{mn}) f_\theta(\boldsymbol{\varphi}_{mn}; \boldsymbol{\Gamma}, \boldsymbol{w}, \Sigma)$ are calculated, and the set of pairs $(\boldsymbol{\varphi}_{mn}, p_{mn}/\Im)_{m=1,\ldots,M}$ can be used as an approximation of the posterior density; and the probability that $\boldsymbol{\varphi}_{jn}$ could be drawn from this density is given by $q_{jn} = p_{jn} / \sum_{m=1}^{M} p_{mn}$. L uniformly distributed random numbers $(\eta_i)_{i=1}^{L}$ are generated; and for each random draw, the vector, $\boldsymbol{\varphi}_{i_0 n}$, for which the condition $\sum_{s=1}^{i_0-1} q_{sn} < \eta_i \le \sum_{s=1}^{i_0} q_{sn}$ is satisfied, is selected as a *plausible vector*.

A computational question that remains unclear at this point concerns the mode of drawing plausible values. A perfect reproduction of the generated PISA plausible values is not possible. It also remains unclear which of the plausible values (for a student, generally five values are generated for each dimension), if the means of those values, or if even aggregations of individual results (computed one for each plausible value), were used for "classifying" individuals into the proficiency levels.

The MCMLM is fitted to each national data set, based on the international item parameters and national conditioning variables. However, the random sub-sample of students across the participating nations and economies used for estimating the parameters, is not identifiable (e.g., OECD 2009b, p. 197). Hence, the item parameters cannot be reproduced with certainty as well.

3 Proficiency Scale Construction and Proficiency Levels

In addition to plausible values, PISA also reports proficiency (scale) levels. The proficiency scales developed in PISA do not describe what students at a given level on the PISA "*performance* scale" *actually did* in a test situation, rather they describe what students at a given level on the PISA "*proficiency* scale" *typically know and can do*. Through the scaling procedure discussed in previous section, it is possible to locate student ability and item difficulty on "performance continua" $\boldsymbol{\theta}$ and $\boldsymbol{\xi}$, respectively. These continua are discretized in a specific way to yield the proficiency scales with their discrete levels.

Level	Score points on the PISA scale
6	Higher than 698.32
5	Higher than 625.61 and less than or equal to 698.32
4	Higher than 552.89 and less than or equal to 625.61
3	Higher than 480.18 and less than or equal to 552.89
2	Higher than 407.47 and less than or equal to 480.18
1a	Higher than 334.75 and less than or equal to 407.47
1b	262.04 to less than or equal to 334.75

Fig. 1 Print reading proficiency scale and levels (taken from OECD 2012, p. 266). PISA scales were linear transformations of the natural logit metrics that result from the PISA scaling procedure. Transformations were chosen so that mean and standard deviation of the PISA scores were 500 and 100, respectively (OECD 2012, p. 143)

The methodology to construct proficiency scales and to associate students with their levels was developed and used for PISA 2000, and it was essentially retained for PISA 2009. In the PISA 2000 cycle, defining the proficiency levels progressed in two broad phases. In the first phase, a substantive analysis of the PISA items in relation to the aspects of literacy that underpinned each test domain was carried out. This analysis produced a detailed summary of the cognitive demands of the PISA items, and together with information about the items' difficulty, descriptions of increasing proficiency. In the second phase, decisions about where to set *cut-off points* to construct the levels and how to associate students with each level were made.

For implementing these principles, a method has been developed that links three variables (for details, see OECD 2012, Chap. 15): the expected success of a student at a particular proficiency level on items that are uniformly spread across that level (proposed is a minimum of 50 % for students at the bottom of the level and higher for other students at that level); the width of a level in the scale (determined largely by substantive considerations of the cognitive demands of items at that level and observations of student performance on the items); and the probability that a student in the middle of the level would correctly answer an item of average difficulty for this level (referred to as the "RP-value" for the scale, where "RP" indicates "response probability").

As an example, for print reading in PISA 2009, seven levels of proficiency were defined; see Fig. 1.

A description of the sixth proficiency level can be found in Fig. 2.

The PISA study provides a basis for international collaboration in defining and implementing educational policies. The described proficiency scales and the distributions of proficiency levels in the different countries play a central role in the reporting of the PISA results. For example, in all international reports the percentage of students performing at each of the proficiency levels is presented (see OECD 2001, 2004, 2007, 2010). Therefore, it is essential to determine the proficiency scales and levels reliably.

Level	Lower score limit	Percentage of students able to perform tasks at this level or above	Characteristics of tasks
6	698	0.8% of students across the OECD can perform tasks at least at Level 6 on the reading scale	Tasks at this level typically require the reader to make multiple inferences, comparisons and contrasts that are both detailed and precise. They require demonstration of a full and detailed understanding of one or more texts and may involve integrating information from more than one text. Tasks may require the reader to deal with unfamiliar ideas, in the presence of prominent competing information, and to generate abstract categories for interpretations. Reflect and evaluate tasks may require the reader to hypothesise about or critically evaluate a complex text on an unfamiliar topic, taking into account multiple criteria or perspectives, and applying sophisticated understandings from beyond the text. A salient condition for access and retrieve tasks at this level is precision of analysis and fine attention to detail that is inconspicuous in the texts.

Fig. 2 Summary description of the sixth proficiency level on the print reading proficiency scale (taken from OECD 2012, p. 267)

Are there alternatives? It is important to note that specification of the proficiency levels and classification based on the proficiency scale depend on qualitative expert judgments. Statistical statements about the reliability of the PISA classifications (e.g., using numerical misclassification rates) are not possible in general, in the sense of a principled psychometric theory. Such a theory can be based on (order) restricted latent class models (see Sect. 4).

4 Conclusion

The basic psychometric concepts underlying the PISA surveys are elaborate. Complex statistical methods are applied to simultaneously scale persons and items in categorical large scale assessment data based on latent variables.

A number of questions remain unanswered when it comes to trying to replicate the PISA scaling results. For example, for student score generation international item parameters are used. These parameters are estimated based on a sub-sample of the international student sample. Although all international data sets are freely available (www.oecd.org/pisa/pisaproducts), it is not evident which students were contained in that sub-sample. It would have been easy to add a filter variable, or at least, to describe the randomization process more precisely. Regarding the reproduction of the plausible values it seems appropriate that, at least, the random number seeds are tabulated. It should also be reported clearly whether the plausible values themselves are aggregated before, for instance, the PISA scores are calculated, or whether the PISA scores are computed separately for any plausible value and aggregated. Indeed, the sequence of averaging may matter (e.g., von Davier et al. 2009).

An interesting alternative to the "two-step discretization approach" in PISA for the construction of proficiency scales and levels are psychometric model-based classification methods such as the *cognitive diagnosis models* (e.g., DiBello et al. 2007; Rupp et al. 2010; von Davier 2010). The latter are discrete latent variable models (restricted latent class models), so no discretization (e.g., based on subjective expert judgments) is necessary, and classification based on these diagnostic models is purely statistical. We expect that such an approach may improve on the error of classification.

It may be useful to automatize the reporting in PISA. One way to implement that, is by utilizing *Sweave* (Leisch 2002). *Sweave* is a tool that allows to embed *R* code for complete data analyses in LaTeX documents. The purpose is to create *dynamic* reports, which can be updated *automatically* if data or analysis change. This tool can facilitate the reporting in PISA. Interestingly, different educational large scale assessment studies may then be compared, *heuristically*, data or text mining their Technical Reports.

References

Adams, R. J., Wilson, M., & Wang, W. (1997). The multidimensional random coefficients multinomial logit model. *Applied Psychological Measurement, 21*, 1–23.

De Boeck, P., & Wilson, M. (Eds.). (2004). *Explanatory item response models: A generalized linear and nonlinear approach*. New York: Springer.

Dibello, L. V., Roussos, L. A., & Stout, W. F. (2007). Review of cognitively diagnostic assessment and a summary of psychometric models. In C. R. Rao & S. Sinharay (Eds.), *Handbook of statistics* (pp. 979–1030). Amsterdam: Elsevier.

Fischer, G. H., & Molenaar, I. W. (Eds.). (1995). *Rasch models: Foundations, recent developments, and applications*. New York: Springer.

Leisch, F. (2002). Sweave: Dynamic generation of statistical reports using literate data analysis. In W. Härdle & B. Rönz (Eds.), *Compstat 2002—Proceedings in Computational Statistics* (pp. 575–580). Heidelberg: Physica Verlag.

Mislevy, R. J. (1991). Randomization-based inference about latent variables from complex samples. *Psychometrika, 56*, 177–196.

OECD (2001). *Knowledge and skills for life. First results from the OECD Programme for International Student Assessment (PISA) 2000*. Paris: OECD Publishing.

OECD (2002a). *Sample tasks from the PISA 2000 assessment*. Paris: OECD Publishing.

OECD (2002b). *PISA 2000 Technical Report*. Paris: OECD Publishing.

OECD (2004). *Learning for tomorrow's world: First results from PISA 2003*. Paris: OECD Publishing.

OECD (2005). *PISA 2003 Technical Report*. Paris: OECD Publishing.

OECD (2007). *PISA 2006: Science competencies for tomorrow's world*. Paris: OECD Publishing.

OECD (2009a). *PISA 2009 assessment framework: Key competencies in reading, mathematics and science*. Paris: OECD Publishing.

OECD (2009b). *PISA 2006 Technical Report*. Paris: OECD Publishing.

OECD (2010). *PISA 2009 results: Overcoming social background—Equity learning opportunities and outcomes*. Paris: OECD Publishing.

OECD (2012). *PISA 2009 Technical Report*. Paris: OECD Publishing.

Rasch, G. (1980). *Probabilistic models for some intelligence and attainment tests*. Chicago: University of Chicago Press.

Reckase, M. D. (2009). *Multidimensional item response theory*. New York: Springer.

Rupp, A. A., Templin, J. L., & Henson, R. A. (2010). *Diagnostic measurement: Theory, methods, and applications*. New York: The Guilford Press.

van der Linden, W. J., & Hambleton, R. K. (Eds.). (1997). *Handbook of modern item response theory*. New York: Springer.

von Davier, M. (2010). Hierarchical mixtures of diagnostic models. *Psychological Test and Assessment Modeling, 52*, 8–28.

von Davier, M., Gonzales, E., & Mislevy, R. J. (2009). What are plausible values and why are they useful? *Issues and Methodologies in Large-Scale Assessments, 2*, 9–37.

Wu, M. L., Adams, R. J., & Wilson, M. R. (1997). *ConQuest®: Multi-aspect test software* [Computer program manual]. Camberwell: Australian Council for Educational Research.

Music Genre Prediction by Low-Level and High-Level Characteristics

Igor Vatolkin, Günther Rötter, and Claus Weihs

Abstract For music genre prediction typically low-level audio signal features from time, spectral or cepstral domains are taken into account. Another way is to use community-based statistics such as Last.FM tags. Whereas the first feature group often can not be clearly interpreted by listeners, the second one lacks in erroneous or not available data for less popular songs. We propose a two-level approach combining the specific advantages of the both groups: at first we create high-level descriptors which describe instrumental and harmonic characteristics of music content, some of them derived from low-level features by supervised classification or from analysis of extended chroma and chord features. The experiments show that each categorization task requires its own feature set.

1 Introduction

Music classification plays an important role among other research fields in music information retrieval (Weihs et al. 2007; Ahrendt 2006). The target is here to identify high-level music descriptors such as harmony, rhythm or instrumentation as well as categorize music collections into genres, substyles or personal listener preferences. For the latter task very often large sets of audio features are previously

---

I. Vatolkin (✉)
TU Dortmund, Chair of Algorithm Engineering, Dortmund, Germany
e-mail: igor.vatolkin@tu-dortmund.de

G. Rötter
TU Dortmund, Institute for Music and Music Science, Dortmund, Germany
e-mail: guenther.roetter@tu-dortmund.de

C. Weihs
TU Dortmund, Chair of Computational Statistics, Dortmund, Germany
e-mail: claus.weihs@tu-dortmund.de

M. Spiliopoulou et al. (eds.), *Data Analysis, Machine Learning and Knowledge Discovery*, Studies in Classification, Data Analysis, and Knowledge Organization, DOI 10.1007/978-3-319-01595-8_46,

extracted from songs and are then used for building of classification models. Some recent works confirm the assumption that the classification quality may be improved by combination of features from the different sources—e.g. audio, cultural and symbolic characteristics in Mckay and Fujinaga (2008). However the significant advantage of the audio features still remains that they can be extracted from any existing song whereas the score is not always available, and the metadata or community statistics are often erroneous, inconsistent or missing for less popular music.

An interesting proposal to improve the classification performance based only on audio features is to extract more high-level audio characteristics as discussed in Berenzweig et al. (2003). Here the classification approach consists of two steps: at first a set of music theory-inspired music characteristics (semantic features) is predicted from low-level audio features by pattern classification. In the second step the music categories can be identified by high-level, or both high-level and low-level features. Another possibility is to predict community tags from audio, which may describe different high-level characteristics of songs, as discussed in Bertin-Mahieux et al. (2008). Our previous study (Rötter et al. 2011) confirmed also the relevance of the high-level features for personal music categories defined for a very limited number of songs.

The integration of such high-level information provides a better understanding of music categorization models. A description like "high vocal rate, no organ, a lot of guitar and almost always minor key" provide a listener or a music scientist a more clear explanation of a song than "high values of angles in phase domain, spectral centroid below 1,000 Hz and mel frequency coefficient 5 above a certain value". Another hope is that model creation from high-level characteristics may require less amount of significant features.

In our work we classify audio songs into six different genres and eight styles based on AllMusicGuide taxonomy and compare different large and up-to-date low-level, high-level and combined feature sets. The next section provides the details of the integrated feature sets. Section 3 describes the results of the experimental study. We conclude with the summary and ideas for further developments.

2 Audio Features

A rather difficult or even impossible task is to distinguish clearly between low-level and high-level features, since the boundaries between the both groups are sometimes imprecise. There is also no consensus in literature, since many different opinions exist: some authors relate to feature statistics estimated from other features (e.g. derivation) already as high-level, other refer to high-level features if they are estimated from score rather than from audio, supported by music theory, or if any sophisticated domain-specific algorithm was incorporated in feature estimation. In our work we consider a feature to be high-level if it is related to music theory and describes harmony, melody, instrumentation, rhythm, tempo or structural characteristics of a piece of music. To name maybe one of the most controversial example, a chroma vector estimated from wrapped frequency bin amplitudes is labeled as low-level, whereas the tone with the maximum chromagram amplitude is labeled as high-level since it relates closer to harmonical characteristics.

All features can be extracted by AMUSE (Vatolkin et al. 2010). The most of the low-level features are described in Theimer et al. (2008) and the manual of the MIR Toolbox (Lartillot and Toivainen 2007).

The LOW-LEVEL FEATURE SET consists of timbral, harmonical and temporal characteristics. The first group relates to 38 features from time, spectral, phase, cepstral and ERB domains: zero-crossing rate, spectral characteristics (centroid, skewness, slope, flux etc.), mel frequency cepstral coefficients (MFCCs), centroid from ERB bands etc. The low-level harmonical descriptors build a further group of 14 features, e.g. chroma vector, fundamental frequency, bass chroma, amplitudes and position of spectral peaks. The seven tempo-related features describe periodicity of peak amplitudes, correlated components and fluctuation patterns.

HIGH-LEVEL TEMPO & RHYTHM SET consists of estimated beat, tatum and onset numbers per minute, rhythmic clarity, tempo based on onset events and song duration.

HIGH-LEVEL HARMONIC FEATURES comprise several MIR Toolbox features such as estimated key and its clarity, tonal centroid and harmonic change detection function. Another set is based on Sonic Annotator NNLS chroma plugin (Mauch and Dixon 2010): local tuning, harmonic change and consonance. Further we implemented several own features based on NNLS chroma chord extractor and semitone spectrum: number of different chords and chord changes in a music interval, shares of the most frequent chords with regard to their duration and interval strengths estimated from the semitone peaks and chroma DCT-reduced log pitch representation (Müller and Ewert 2010).

INSTRUMENT-RELATED SET is based on our instrument recognition study (Vatolkin et al. 2012): there we ran a large-scale multi-objective feature selection by evolutionary algorithms for optimization of instrument classification in polyphonic audio mixtures. Four binary classification tasks were to identify guitar, piano, wind and strings. We saved up to 237 classification models classifying these 4 groups. This high number has two reasons: at first we used four different classifiers for model estimation (decision tree, random forest, naive bayes and support vector machine). Further, the goal of our previous survey was to optimize classification error and feature rate at the same time—small feature sets save disc space as well as training and classification times and may tend to provide more generalizable models. All the best models found (the first one with the smallest classification error and the largest feature number; the last one with the largest classification error and smallest feature amount) were used for high-level feature estimation. For all short audio frames around onset events the overall rate of models in a 10 s music interval was saved which detected an instrument, i.e. a value of 0.9 for piano rate meant, that 90 % of audio frames from a large 10 s window were classified as containing piano.

Another important issue to mention is feature processing. For short-framed features we apply two possibilities for feature aggregation as illustrated by Fig. 1.

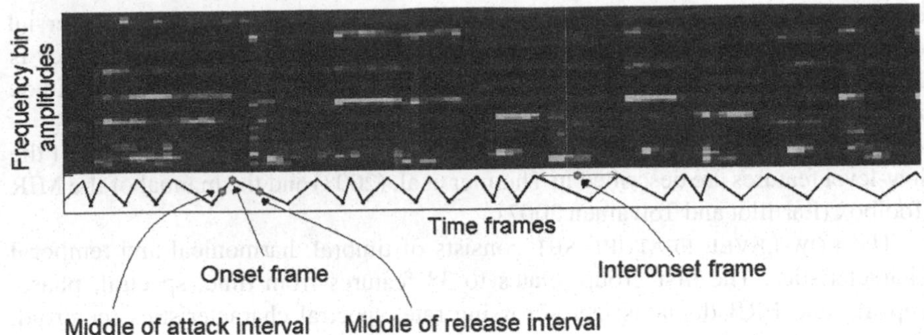

Fig. 1 Selection of short audio frames based on attack, onset and release events. Attack intervals are marked by *ascending lines* below the spectrum; release intervals by *descending lines*

This approach is based on the attack-onset-release (AOR) interval, where the *attack* phase corresponds to the increasing energy of a new note, *onset* event to the moment with the largest energy and *release* phase to the decreasing sound. The first method saves the features only from the frames between the previously extracted onset events (so that the noisy and inharmonical components around the beginning of a new note are disregarded) and estimates mean and deviation of each feature for large frames of 4 s with 2 s overlap. The second method creates three new different feature dimensions from each feature: from the frames in the middle of attack intervals, onset event frames and the middle of release intervals. Mean and deviation are again estimated for 4 s windows, so that the final number of feature dimensions is increased by factor 6.

3 Experimental Study

Six genre and eight style classification tasks from the labelled song set described in (Vatolkin et al. 2011) were used to test the different low-level and high-level feature sets. Random forest (RF) and support vector machine (SVM) were used for training of classification models. The feature sets were:

- MFCCs-BetwOns: 92-dimensional vector with MFCCs and delta MFCCs (mean and deviation saved for 4 s frames, see previous section).
- MFCCs-AOR: 276-dim. vector with the same original features which were used to build three new groups (attack, onset and release features as described above).
- LL-BetwOns: 808-dim. vector estimated from all original low-level characteristics.
- LL-AOR: 2424-dim. vector estimated from all original low-level characteristics.
- HLRhythmTempo: 6-dim. vector with rhythm and tempo features (since the original feature series were calculated for larger time frames or even complete song no further aggregation method was applied).

Table 1 The best three feature sets for all classification tasks and random forest models

Task	1st best set	2nd best set	3rd best set
CLASSIC	LL-BetwOns 0.0581	LL-AOR 0.0589	Complete-AOR 0.0600
ELECTRONIC	MFCCs-AOR 0.1866	LL-AOR 0.2128	MFCCs-BetwOns 0.2194
JAZZ	HLInstrumentsAll 0.0877	HL-AOR 0.1076	Complete-AOR 0.1119
POP & ROCK	LL-BetwOns 0.1957	MFCCs-BetwOns 0.1985	LL-AOR 0.2018
RAP	MFCCs-BetwOns 0.0642	MFCCs-AOR 0.0691	HLHarmony-AOR 0.0904
R&B	HLInstrumentsBest 0.2342	MFCCs-AOR 0.2457	MFCCs-BetwOns 0.2715
ADULT CONTEMPORARY	HLHarmony-AOR 0.2546	HLHarmony-BetwOns 0.2600	MFCCs-AOR 0.2988
ALBUM ROCK	LL-AOR 0.1876	LL-BetwOns 0.1892	Complete-AOR 0.1955
ALTERNATIVE POP ROCK	LL-BetwOns 0.1397	HLHarmony-BetwOns 0.1408	HLHarmony-AOR 0.1512
CLUB DANCE	HLHarmony-AOR 0.3473	HLHarmony-BetwOns 0.3474	LL-BetwOns 0.3875
HEAVY METAL	HLInstrumentsBest 0.1049	MFCCs-BetwOns 0.1141	MFCCs-AOR 0.1241
PROG ROCK	HLInstrumentsAll 0.4228	HLHarmony-AOR 0.4261	MFCCs-AOR 0.4361
SOFT ROCK	LL-AOR 0.2347	Complete-AOR 0.2511	LL-BetwOns 0.2572
URBAN	HLInstrumentsAll 0.1253	MFCCs-AOR 0.1299	LL-AOR 0.1310

The numbers below feature set descriptions correspond to classification error on a validation song set

- HLHarmony-BetwOns: 268-dim. vector with high-level harmony characteristics.
- HLHarmony-AOR: 804-dim. vector with high-level harmony characteristics.
- HLInstrumentsBest: 16-dim. vector with instrument recognition features: four models with the smallest classification errors from four classifiers.
- HLInstrumentsAll: 237-dim. vector with instrument recognition features: all models with the best tradeoff solutions resp. to classification error and selected feature rate from four classifiers.
- HL: 377-dim. vector with all high-level features (only mean values saved for 4 s frames).
- HL-AOR: 2262-dim. vector with all high-level features (mean and deviation saved for larger frames after generation of attack, onset and release features).
- Complete-BetwOns: 1562-dim. vector from all original features.
- Complete-AOR: 4686-dim. vector from all original features.

Table 2 The best three feature sets for all classification tasks and support vector machine models

Task	1st best set	2nd best set	3rd best set
CLASSIC	HL-AOR	HL	Complete-AOR
	0.0472	0.0502	0.0509
ELECTRONIC	LL-AOR	LL-BetwOns	Complete-BetwOns
	0.1815	0.1836	0.1955
JAZZ	HLHarmony-AOR	Complete-AOR	Complete-BetwOns
	0.1042	0.1128	0.1143
POP & ROCK	LL-AOR	LL-BetwOns	MFCCs-AOR
	0.1591	0.1901	0.1991
RAP	MFCCs-AOR	MFCCs-BetwOns	LL-AOR
	0.0614	0.0746	0.1147
R&B	MFCCs-AOR	MFCCs-BetwOns	HLInstrumentsAll
	0.2528	0.2840	0.2933
ADULT CONTEMPORARY	Complete-BetwOns	LL-BetwOns	Complete-AOR
	0.2117	0.2134	0.2450
ALBUM ROCK	LL-AOR	Complete-AOR	HLHarmony-BetwOns
	0.1754	0.2172	0.2213
ALTERNATIVE POP ROCK	LL-BetwOns	LL-AOR	Complete-AOR
	0.1641	0.1656	0.1752
CLUB DANCE	LL-BetwOns	HLHarmony-BetwOns	LL-AOR
	0.2671	0.2841	0.2853
HEAVY METAL	MFCCs-BetwOns	HLInstrumentsBest	HL-AOR
	0.1356	0.1667	0.1769
PROG ROCK	HLRhythmTempo	LL-AOR	LL-BetwOns
	0.3051	0.3213	0.3308
SOFT ROCK	LL-AOR	HLHarmony-BetwOns	Complete-AOR
	0.2368	0.2557	0.2569
URBAN	MFCCs-BetwOns	MFCCs-AOR	LL-BetwOns
	0.1504	0.1584	0.1764

The numbers below feature set descriptions correspond to classification error on a validation song set

Tables 1 and 2 list the best feature sets for each category. It can be clearly observed that for different tasks different feature sets perform best. No clear statement can be claimed that high-level features are more preferable with respect to classification error. This holds even for the same category—e.g. R&B identification with RF works well with high-level instrument features, but the second best MFCCs feature set is rather close. Feature sets may perform well or completely fail—as an example the high-level harmony sets are among the best three for Rap, Adult Contemporary, Alternative Pop Rock, Club Dance and Prog Rock, but they are among the worst for Heavy Metal (HLHarmony-AOR has classification error of 0.4933 and HLHarmony-BetwOns of 0.4730) or Pop & Rock (errors of 0.3416 resp. 0.3321).

It can be also observed, that the original features have a tendency to be more important than a processing method: for 11 of 14 tasks classified by RF as well as for 11 tasks classified by SVM the same original feature set is selected twice among the best three sets using two different feature aggregation methods (LL-BetwOns and LL-AOR for RF and Classic, MFCCs-AOR and MFCCs-BetwOns for Electronic etc.)—even if sets with "AOR"-method consist of three times more final features than from interonset frames.

Comparing RF and SVM models, SVM seem to perform slightly better on average (for 8 of 14 categories and the very best feature sets)—however the differences are sometimes rather small, and the advantage of RF is that it is faster and may provide more generalizable models as observed in Vatolkin et al. (2012). The two classifiers require often different feature sets for the same task, only for three tasks the same feature set is the best for both classifiers.

4 Conclusions and Outlook

Concluding our feature set comparison, we can clearly confirm, that different features are required for different problems. But also the choice of a classifier and a feature aggregation method may lead to varying results. At the current stage the reliable recognition of music categories by interpretable high-level features is still work in progress. In future, we plan to apply category-adapted feature selection and refer to this work as a baseline method with pre-defined feature sets. On the other side, it can be reasonable to decrease the computing time by identification of relevant segments in songs, so that it would not be necessary to extract and process features from complete audio recordings.

Another serious drawback of high-level features at least at the current research state is that many of them perform well only for some limited set of music pieces, and also are hard to identify in general for polyphonic recordings: tempo recognition may estimate the tempo as twice the correct value if it is based on autocorrelation, instrument recognition models may not work well if different playing styles are performed, and harmonic analysis becomes complicated if the number of simultaneously playing instruments is too large. However the robustness of high-level audio features will doubtlessly increase in future, and only these features can provide the best interpretability for human listeners. Also the combination with features from other domains (community tags, metada) could be promising for further improvements of classification performance.

Acknowledgements We thank the Klaus Tschira Foundation for the financial support.

References

Ahrendt, P. (2006). *Music Genre Classification Systems—A Computational Approach* (Ph.D. thesis). Informatics and Mathematical Modelling, Technical University of Denmark, Kongens Lyngby.

Berenzweig, A., Ellis, D. P. W., & Lawrence, S. (2003). Anchor space for classification and similarity measurement of music. In *Proceedings of the 2003 International Conference on Multimedia and Expo (ICME)* (pp. 29–32). New York: IEEE.

Bertin-Mahieux, T., Eck, D., Maillet, F., & Lamere, P. (2008). Autotagger: A model for predicting social tags from acoustic features on large music databases. *Journal of New Music Research, 37*(2), 115–135.

Lartillot, O., & Toivainen, P. (2007). MIR in Matlab (ii): A toolbox for musical feature extraction from audio. In *Proceedings of the 8th International Conference on Music Information Retrieval (ISMIR)* (pp. 127–130). Austrian Computer Society (OCG).

Mauch, M., & Dixon, S. (2010). Approximate note transcription for the improved identification of difficult chords. In *Proceedings of the 11th International Society for Music Information Retrieval Conference (ISMIR)* (pp. 135–140). International Society for Music Information Retrieval.

Mckay, C., & Fujinaga, I. (2008). Combining features extracted from audio, symbolic and cultural sources. In *Proceedings of the 9th International Conference on Music Information Retrieval (ISMIR)* (pp. 597–602). Philadelphia: Drexel University.

Müller, M., & Ewert, S. (2010). Towards timbre-invariant audio features for harmony-based music. *IEEE Transactions on Audio, Speech, and Language Processing, 18*(3), 649–662.

Rötter, G., Vatolkin, I., & Weihs, C. (2011). Computational prediction of high-level descriptors of music personal categories. In *Proceedings of the 2011 GfKl 35th Annual Conference of the German Classification Society (GfKl)*. Berlin: Springer.

Theimer, W., Vatolkin, I., & Eronen, A. (2008). *Definitions of audio features for music content description*. Algorithm Engineering Report TR08-2-001, Technische Universität Dortmund.

Vatolkin, I., Preuß, M., & Rudolph, G. (2011). Multi-objective feature selection in music genre and style recognition tasks. In *Proceedings of the 2011 Genetic and Evolutionary Computation Conference (GECCO)* (pp. 411–418). New York: ACM.

Vatolkin, I., Preuß, M., Rudolph, G., Eichhoff, M., & Weihs, C. (2012). Multi-objective evolutionary feature selection for instrument recognition in polyphonic audio mixtures. *Soft Computing, 16*(12), 2027–2047.

Vatolkin, I., Theimer, M., & Botteck, M. (2010). AMUSE (Advanced MUSic Explorer): A multitool framework for music data analysis. In *Proceedings of the 11th International Society on Music Information Retrieval Conference (ISMIR)* (pp. 33–38). International Society for Music Information Retrieval.

Weihs, C., Ligges, F., Mörchen, F., & Müllensiefen, D. (2007). Classification in music research. *Advances in Data Analysis and Classification, 1*(3), 255–291.

Part VII
LIS WORKSHOP: Workshop on Classification and Subject Indexing in Library and Information Science

Using Clustering Across Union Catalogues to Enrich Entries with Indexing Information

Magnus Pfeffer

Abstract The federal system in Germany has created a segmented library landscape. Instead of a central entity responsible for cataloguing and indexing, regional library unions share the workload cooperatively among their members. One result of this approach is limited sharing of cataloguing and indexing information across union catalogues as well as heterogeneous indexing of items with almost equivalent content: different editions of the same work. In this paper, a method for clustering entries in library catalogues is proposed that can be used to reduce this heterogeneity as well as share indexing information across catalogue boundaries. In two experiments, the method is applied to several union catalogues and the results show that a surprisingly large number of previously not indexed entries can be enriched with indexing information. The quality of the indexing has been positively evaluated by human professionals and the results have already been imported into the production catalogues of two library unions.

1 Introduction

Unlike the US, where the Library of Congress acts as a central authority for cataloguing and indexing of all kinds of materials, the German library system is based on a cooperative approach. The national library is only responsible for German works or works on Germany, while regional library unions share the workload of cataloguing all other materials. At the present time there are five such unions, each with its own catalogue. While sharing records is possible in theory as all unions use the same cataloguing rules and central authority files, it is in practice often limited (Lux 2003).

M. Pfeffer (✉)
Stuttgart Media University, Nobelstraße 10, 70569 Stuttgart, Germany
e-mail: pfeffer@hdm-stuttgart.de

M. Spiliopoulou et al. (eds.), *Data Analysis, Machine Learning and Knowledge Discovery*, Studies in Classification, Data Analysis, and Knowledge Organization, DOI 10.1007/978-3-319-01595-8_47,

Table 1 Koecher, M.: Lineare Algebra und analytische Geometrie (BVB catalogue)

Year	Subject headings (SWD)	Classification (RVK)
2003	Analytische Geometrie; Lehrbuch; Lineare Algebra	QH 140; SK 220; SK 380
1997	Analytische Geometrie; Lehrbuch; Lineare Algebra	QH 140; SK 220; SK 380
1992	Analytische Geometrie; Lehrbuch; Lineare Algebra	SK 220
1992	Analytische Geometrie; Lineare Algebra	SK 220
1985	–	–
1985	–	–
1985	Analytische Geometrie; Lehrbuch; Lineare Algebra	SK 110; SK 220
1983	–	–
1983	Analytische Geometrie; Lehrbuch; Lineare Algebra	SK 110; SK 220
1983	–	–
1981	Analytische Geometrie; Lehrbuch; Lineare Algebra	SK 220

Table 2 Koecher, M.: Lineare Algebra und analytische Geometrie (SWB catalogue)

Year	Subject headings (SWD)	Classification (RVK)
2003	–	SK 220
1997	Analytische Geometrie; Lehrbuch; Lineare Algebra	QH 140; SK 220
1992	Analytische Geometrie; Lineare Algebra	SK 220
1985	Analytische Geometrie; Einfhrung; Lehrbuch; Lineare Algebra	SK 220; SK 380
1983	Analytische Geometrie; Lineare Algebra	SK 110; SK 220

The needless redundant handling of the same items combined with slight variations in the application of the cataloguing rules result in databases that are hard to aggregate on the national level as the detection of duplicates will often fail. Within a single union catalogue, another consequence of the cooperative approach is visible: heterogeneous indexing of titles with almost equivalent content, as in different editions of the same work.

The following example illustrates this particular problem: The work *Lineare Algebra und analytische Geometrie* by Max Koecher has been republished in new editions over several decades. Tables 1 and 2 show the editions in two union catalogues. Both have entries without subject headings or classification numbers and there are variations in the terms and classes. Indexing information for identical editions across the two catalogues also varies.

In this paper, we will show that a simple clustering method that is applied to a combined data set of several union catalogues can be utilized to reduce this heterogeneity and also enrich a significant amount of previously not indexed entries with subject headings or classification information. The paper is structured as follows: After a short discussion of the related work by other authors, the prevalent subject headings and classification system currently used in German libraries are introduced. Next, a matching algorithm is proposed that can be used to reliably cluster different editions of the same work. Both the encouraging numerical results of the application of the proposed algorithm to several union catalogues as well as the results of an evaluation by indexing professionals will be presented next. The paper closes with an outlook on possible other applications for similar clustering approaches as well as a discussion of the further work needed to apply this kind of data homogenisation to a large number of catalogues on a regular basis.

2 Related Work

In previous experiments, the author has tried to apply case-based reasoning methods to automatically index entries in library catalogues that had not been indexed by a human specialist. Pfeffer (2010) gives an overview of the method and the results. While the first results were intriguing, later evaluation by human professionals showed significant flaws in the generated indexing information and motivated the experiments described in this paper.

In general, matching entries in catalogues or bibliographic databases has mostly been done to find duplicates, i.e. different entries that describe the identical resource. Sitas and Kapidakis (2008) give an exhaustive overview of the used algorithms and their properties. Even though the algorithms are geared towards *not* matching different editions of the same work they are still useful to gain insight into what metadate elements can differ between very similar editions. But there are also projects concerned with matching editions: Hickey et al. (2002) propose an algorithm to automatically create abstract "work" entities from different manifestations in a catalogue according to the FRBR entity-relationship model. Their algorithm is very similar to one used this paper, but based on the data fields and cataloguing rules in the US. Accordingly there is an optional step included that allows to look up the authors and editors in the Library of Congress authority file and use the established form of the name to improve the matching results. Taniguchi (2009) is working on the same problem and uses a similar approach and likewise adapts it to the peculiarities of the Japanese MARC variant and the local cataloguing rules. In both cases, only the data fields for author's and editor's names, main title of the work and uniform title of the work are used in the matching process. Both report success in grouping the expressions into homogenised clusters, but show no further applications for the resulting clusters.

Some next generation opacs or resource discovery systems allow the grouping of different editions of the same when presenting search result lists. Dickey (2008) describes the process and its foundation in the FRBR model. The paper lists several implementations in commercial products, but apparently the matching algorithms used are proprietary.

3 Preliminaries

The most commonly used indexing system in Germany is a controlled vocabulary, the *Schlagwortnormdatei (SWD)*[1] and its accompanying set of rules, the *Regeln für den Schlagwortkatalog (RSWK)*.[2] It contains more than 600.000 concepts and records preferred terms, alternative terms as well as broader and narrower terms. SWD and RSWK are in use since the late 1980s and have been adopted by almost all scientific libraries in Germany and Austria and some libraries in the German-speaking part of Switzerland. In most catalogues, search by subject is based solely on this vocabulary, so a higher proportion of indexed entries would improve retrieval results in subject based queries.

There is more variation in the classification systems used in Germany to create call numbers for printed books and other media. Many libraries still use a self-developed one that is tailored to their local situation. On the other hand, the Dewey Decimal Classification (DDC) (See Dewey (2005)) and the *Regensburger Verbundklassifikation (RVK)*[3] are systems used by many libraries. The first has been used by the German National Library for items published since 2007 and this decision by the national library has rekindled interest in DDC. The latter has grown from a local classification system developed in 1964 to become the classification system used by almost all university libraries in the state of Bavaria in the 1990s. It is currently being adopted by a growing number of libraries in the German-speaking countries to replace the aforementioned locally developed classification systems. Replacement usually includes reclassification and new call numbers for *all* items located in publicly accessible stacks of the adopting library. A higher proportion of entries with RVK classes in the union catalogue would alleviate this task considerably.

4 Matching Algorithm

The basic idea is to find all other editions of a given work to aggregate their indexing and classification information and attach it to those editions that lack subject headings or classification. With this goal in mind it should be noted that any proposed algorithm should at all cost avoid false positive matches: overly

[1]Subject Headings Authority File. In a recent development, the SWD is combined with authority files for persons and corporate bodies to form the *Gemeinsame Normdatei (GND)* (engl.: Universal Authority File). This file will be suitable to be used with the proposed RDA cataloguing rules. The catalogue data used for the experiments described in this paper still used the SWD, but the results can be applied to catalogues using the GND.

[2]Rules the for subject catalogue. See Scheven et al. (2012) for a complete reference.

[3]Regensburg Union Classification System. See Lorenz (2008) for an introduction.

eager matching would lead to clusters containing different works and thus in our application to misleading aggregated indexing information. Hickey et al. (2002) showed several examples for this case when a non-exact matching variant was used and caution against using loose matching. False negatives, on the other hand, would only reduce the potential benefits of the basic idea: in the worst case, editions of a given work form several clusters and no indexing information can be shared.

The proposed algorithm utilizes only the data fields for title and subtitle and all persons or corporate bodies involved with creating the work. All other information, like publisher, date and place of publication, format or standard numbers vary over different editions and cannot be used. As explained before, title and subtitle must match exactly. But for the creators and contributors, it is only required that one of them matches exactly when two entries are compared. This allows matching of editions that have been reworked by a new editor or author but still mention the original ones. Unlike the algorithm described in Hickey et al. (2002) an optional look-up in an authority file is not needed, as the German rules for cataloguing require that all creators and contributors to a work are cross-referenced with a central authority file when the item is catalogued. These rules also help in another case: when a work has been translated from another language, the rules require that the title of the original work is recorded as the uniform title. This field can accordingly be used to match editions across different languages.

The consolidated algorithm contains two phases:

- Phase 1: Key generation
  - For each entry of the catalogues, the following data fields are prepared:
    - If a uniform title exists, it is used, else title and subtitle are concatenated.
    - Authors, contributors and corporate bodies involved as creator or contributor are identified and their authorized name is retrieved.
  - Comparison keys are created from all combinations of the normalized form of the used title and one of the creators/contributors.
- Phase 2: Matching
  - All keys are compared and entries with matching keys are grouped in a cluster.

From a catalogue entry for the book by Scheven et al. (2012), the resulting keys are:

- `scheven esther|regeln fuer den schlagwortkatalog rswk`
- `kunz martin|regeln fuer den schlagwortkatalog rswk`
- `bellgard sigrid|regeln fuer den schlagwortkatalog rswk`

Table 3 Results from first experiment

Union	No. of monographs	With RVK	With SWD	Added RVK	Added SWD
SWB	12,777,191	3,235,958	3,979,796	959,419	636,462
HeBIS	8,844,188	1,933,081	2,237,659	992,046	1,179,133

5 Experiments

In a first experiment in 2010, the complete data from two library union catalogues was retrieved and filtered for monographs. The catalogues were from the *Sdwestdeutscher Bibliotheksverbund (SWB)*[4] and the *Hessisches Bibliotheks- und Informationssystem (HeBIS)*.[5]

Prior to the experiment, the proportion of indexed entries is 25–30 %. After applying the algorithm described in Sect. 4, indexing information is shared among the members of the individual clusters. Entries that had no indexing information before and get indexed by this transfer are counted. The results are shown in Table 3. There is a surprisingly large number of newly indexed entries and the proportion of indexed entries is increased by up to 50 %.

5.1 Evaluation

The results were offered as CSV files for download and imported as RDF triples into a prototype linked open data server, which allowed for browsing the entries and viewing the generated clusters on the web. Eckert (2010) has additional information on this service and the benefits of a presentation of experimental results as linked data.

The SWB also added a large part of the new subject headings and classification results into a test database to allow librarians to use them in retrieval and cataloguing. Several experts in subject indexing used both resources to evaluate the quality of the experimental results. Both random checks as well as systematic queries were performed and not a single false positive match was identified. The experts recommended importing the results to the production catalogues, which was completed by both SWB and HeBIS library unions in late 2011.

5.2 Extended Experiment

In a second experimental run in 2012, the algorithm was applied to a larger set of union catalogue databases. In addition to SWB and HeBIS, catalogues from the

---

[4]Library Union of South-West Germany. Catalogue: http://swb.bsz-bw.de/.

[5]Hessian Library Information System. Catalogue: http://www.portal.hebis.de/.

Table 4 Results from the extended experiment

Union	No. of monographs	With RVK	With SWD	Added RVK	Added SWD
SWB	13,330,743	4,217,226	4,083,113	581,780	957,275
Hebis	8,844,188	1,933,081	2,237,659	1,097,992	1,308,581
HBZ	13,271,840	1,018,298	3,322,100	2,272,558	1,080,162
BVB	22.685.738	5.750.295	6.055.164	2,969,381	2,765,967

Hochschulbibliothekszentrum NRW (HBZ)[6] and the *Bibliotheksverbund Bayern (BVB)*[7] were used. There was no data available from the fifth German library union, the *Gemeinsamer Biblitheksverbund.*[8] The SWB catalogue data set was new and already contained most of the results from the first run, while the HeBIS catalogue data was the same as in the first run.

Table 4 shows the final result for this run. Again, a huge increase can be seen. From the HeBIS data it can be seen that adding more catalogues to the data pool results in diminishing, but still significant returns. The HBZ can benefit extraordinarily from the experiment, as its member libraries have been more reluctant to switch to RVK than those of other unions and accordingly a smaller proportion of the entries in the catalogue have RVK information. The data will be made available to the unions in the coming months. Both HBZ and BVB have already expressed their intent to integrate the results into their production systems as soon as possible.

6 Conclusion and Outlook

In this paper, a simple matching algorithm for entries in catalogues or bibliographic databases is proposed that is based solely on comparison of uniform title, title, subtitle and authorized name of creators and contributors. It is used to cluster entries from several library union catalogues. In the resulting clusters, indexing and classification information is shared among all members in order to enrich previously not indexed entries. The resulting numbers from two experimental runs show that the proportion of indexed entries can be increased significantly. Independent experts have validated the results and the quality of the indexing information. The results have already been imported into production catalogue databases of two library unions.

[6]University Library Center of North-Rhine Westphalia. Catalogue: http://okeanos-www.hbz-nrw.de/F/.

[7]Bavarian Library Union. Catalogue: http://www.gateway-bayern.de/.

[8]Common Library Network. Catalogue: http://gso.gbv.de/.

Further work is needed in fine-tuning the matching algorithm to also include editions that have slight alterations in the title. Publisher and other information could be used in this case to prevent false positive matches. Also, there is no reason to limit the approach to German library union catalogues. Swiss and Austrian libraries have already expressed interest in using it and as an increasing number of library catalogues is becoming available under open licences, the data pool could be enlarged even further. As indicated in the related work section, the problem of different rules for authorized names and used authority files needs to be addressed, if international catalogues should be included. Furthermore, the approach can easily be extended to other indexing and classification systems, like the Library of Congress Subject Headings and the Dewey Decimal Classification.

But with an ever growing database and more applications, one must also consider the software side of the matching algorithm. For the experiments, a set of Perl scripts was written to transform catalogue records into an internal representation and do the two-phase matching on this internal representation. This kind of software works in an academic context, but is neither maintainable nor scalable to process very large data sets. Luckily, there are several projects working on adopting "Big Data" methods and development frameworks to metadata management. One of these projects is *culturegraph*,[9] which is the basis for a prototype linked open data service of the German National Library. It is written in Java and utilizes Hadoop, a framework for scalable, distributed applications. It is open source and comes with comprehensive documentation. Integrating the matching algorithm and possible variants into this project would enable interested parties to apply them to their own data and build on top of a common infrastructure. It is the next logical step to further develop this approach.

References

Dewey, M. (2005). *Dewey-Dezimalklassifikation und Register*, Mitchell J. S. (Ed.). Munich: Saur.

Dickey, T. J. (2008). FRBRization of a library catalog: better collocation of records, leading to enhanced search, retrieval, and display. *Information Technology and Libraries*, *27*(1), 23–32.

Eckert, K. (2010). Linked open projects: Nachnutzung von Projektergebnissen als Linked Data. In: M. Ockenfeld et al. (Eds.), *Semantic Web & Linked Data: Elemente zukünftiger Informationsstrukturen; 1. DGI-Konferenz, proceedings* (pp. 231–236). Frankfurt: DGI.

Hickey, T. B., O'Neill, E. T., & Toves, J. (2002). Experiments with the IFLA functional requirements for bibliographic records (FRBR). *D-Lib Magazine*, *8*(9).

Lorenz, B. (2008). *Handbuch zur Regensburger Verbundklassifikation : Materialien zur Einführung* (2nd edn). Wiesbaden: Harrassowitz.

Lux, C. (2003) The German library system: structure and new developments. *IFLA Journal*, *29*(2), 113–128.

Pfeffer, M. (2010). Automatische Vergabe von RVK-Notationen mittels fallbasiertem Schließen. In: U. Hohoff et al. (Eds.), *97. Deutscher Bibliothekartag in Mannheim 2008 - Wissen*

[9]Reprository: http://culturegraph.sf.net.

bewegen. Bibliotheken in der Informationsgesellschaft, proceedings (pp. 245–254). Frankfurt: Klostermann.

Scheven, E., Kunz, M., & Bellgardt, S. (Eds.) (2012). *Regeln für den Schlagwortkatalog : RSWK*. Frankfurt: Dt. Nationalbibliothek.

Sitas, A., & Kapidakis, S. (2008). Duplicate detection algorithms of bibliographic descriptions. *Library Hi Tech, 26*(2), 287–301.

Taniguchi, S. (2009). Automatic identification of "Works" toward construction of FRBRized OPACs: An experiment on JAPAN/MARC bibliographic records. *Library and Information Science, 61*, 119–151.

Text Mining for Ontology Construction

Silke Rehme and Michael Schwantner

Abstract In the research project *NanOn: Semi-Automatic Ontology Construction—a Contribution to Knowledge Sharing in Nanotechnology* an ontology for chemical nanotechnology has been constructed. Parts of existing ontologies like CMO and ChEBI have been incorporated into the final ontology. The main focus of the project was to investigate the applicability of text mining methods for ontology construction and for automatic annotation of scientific texts. For this purpose, prototypical tools were developed, based on open source tools like GATE and OpenNLP. It could be shown that text mining methods which extract significant terms from relevant articles support conceptualisation done manually and ensure a better coverage of the domain. The quality of the annotation depends mostly on the completeness of the ontology with respect to synonymous and specific linguistic expressions.

1 Introduction

The main objective of the NanOn project was to explore text mining methods for the semi-automatic construction of an expressive ontology for an important field of science. Another aim was to explore the benefits of an ontology for scientists. Nanotechnology is well suited for both purposes because it is interdisciplinary and heterogeneous and the relevant knowledge is expressed in various terminologies. To construct the ontology, experts for ontology design worked together with experts from the nanotechnology domain. To cooperate, these two groups had to find a common understanding of the possibilities and limitations of an ontology.

S. Rehme · M. Schwantner (✉)
FIZ Karlsruhe, 76344 Eggenstein-Leopoldshafen, Germany
e-mail: silke.rehme@fiz-karlsruhe.de; michael.schwantner@fiz-karlsruhe.de

M. Spiliopoulou et al. (eds.), *Data Analysis, Machine Learning and Knowledge Discovery*, Studies in Classification, Data Analysis, and Knowledge Organization,
DOI 10.1007/978-3-319-01595-8_48,

2 Approach

As a guideline for the ontology construction the NeOn methodology (Suárez-Figueroa et al. 2012) was applied, which included specification, conceptualisation, re-using existing knowledge sources and the final implementation. For the conceptualisation, existing text mining methods were modified and new ones were developed. Overall, this was an iterative process where the individual steps were partly conducted in parallel.

2.1 Specification

As a result of the requirements analysis done mainly by the nanotechnology experts the purpose of the ontology was to provide a widely accepted knowledge representation of nanotechnology. Initially, the whole area of nanotechnology was envisaged, but it soon became clear that this domain is so complex and extensive, that more experts would have been necessary than available for this project. As a consequence, the scope was limited to chemical nanotechnology, a scientific subject area of highly interdisciplinary character as well. It was further decided that the ontology should be designed for users from science and industry, helping them to improve their search for materials, properties, or processes. These users were represented also in the project by the participating experts. Finally, OWL 2 DL was selected as the descriptive language for authoring the ontology.

The specification included also a glossary of the most relevant terms of the defined subject area, which served as a first basis for the conceptualisation. Another basis was the compilation of 45 competency questions, exemplary questions from all areas to be covered selected by the domain experts, which the ontology should be capable of answering. These questions were also used to evaluate the completeness of the ontology at the end of the project. Examples are *"Which metals show surface plasmon resonance?"* or *"Which process is used for achieving property P in material M?"*.

2.2 Conceptualisation

As a preparatory step, the description of the domain was sub-divided into four modules which at the same time were intended as modules for the ontology:

- Materials: relevant inorganic materials in nanotechnology,
- Properties: physical and chemical properties and process properties,
- Processes: chemical processes like analytical and production processes,
- Equipment: e.g. equipment used for the implementation of processes.

This modularisation ensured high flexibility for further development and future updates of the ontology and also for the incorporation of available sources. After modularisation, the process of conceptualisation consisted of the following four steps:

1. Selection and integration of existing ontologies and thesauruses,
2. Identification of relevant terms by domain experts,
3. Manual annotation,
4. Text mining.

For the first step, structured knowledge in the form of existing ontologies or thesauruses of related knowledge fields were identified, as well as full texts of scientific publications and patents, from which knowledge can be extracted and used for the ontology. Mainly three ontologies were found, parts of each were incorporated into the NanOn ontology. The Chemical Methods Ontology (CMO) is an ontology comprising analytical processes and production processes used in chemistry and materials science. The Chemical Entities of Biological Interest Ontology (ChEBI) was also partly integrated because it contains chemical elements and substances and provides the background knowledge of chemistry for chemical nanotechnology. Finally, sections of the ChemAxiomMetrology, which comprises methods of measurement and measuring instruments, were adopted.

As a second step relevant terms known from the day-to-day work of the nanotechnology experts were identified. In a third step, the ontology was checked for completeness: documents were automatically annotated with concept expressions already in the ontology. These annotations were examined by the experts for completeness, and relevant terms not assigned automatically were manually annotated and incorporated into the ontology as classes or as synonyms of existing classes. Steps 2 and 3 were iterated several times in order to ensure that the ontology would be as comprehensive as possible. During this process it became clear that not all expectations of the experts could be implemented. The chemical and physical properties of nanomaterials are often highly interdependent—cf. the dependency of the boiling point from the material's purity and the ambient pressure. Thus they neither can be modelled satisfactorily in OWL 2 DL nor is it possible to populate the ontology automatically with their respective instances.

3 Text Mining

Text mining methods played a crucial role in various stages of the ontology building process. Initially, text mining was used to assist the ontology designers in finding concepts. With the necessary concepts determined, synonymous and quasi-synonymous expressions for the concepts and the semantic relations between them had to be found. And finally, text mining was used to populate the ontology with instances.

3.1 Identification of Concept Candidates

The results of the manual annotations proved that the modelling by the experts leaved significant gaps. So the idea was to identify automatically, on the basis of a corpus of scientific articles, relevant terms which can give indications for ontology concepts. The text corpus used was composed of 16,676 full-texts of scientific articles, 26,892 scientific abstracts, and 4,142 patents, comprising together about 107 million word tokens. The extraction of significant terms was a multi-stage process which involved tools of the OpenNLP library. After applying the sentence detector and the tokeniser, the tokens were tagged with the part-of-speech (PoS) Tagger. On this basis the NP Chunker was used, generating a list of noun phrases (NP) corresponding to patterns like

```
NP = (DET) (ADJP) Noun+      with ADJP --> (ADV) ADJ+ (ADJP)
where DET = Determiner, ADV = Adverb, and ADJ = Adjective.
```

The resulting noun phrases were skimmed, deleting leading or trailing words specified in a stop word list. Then the phrases were normalised morphologically to their singular form. After deleting duplicates, terms with a very low frequency were discarded, as most of them tend to be misspelled words. To eliminate terms of general language which are supposedly not specific for nanotechnology, the British National Corpus (BNC) was used as a contrast corpus. Terms with a frequency above a cut-off value of 2,000 in the BNC were discarded. The remaining phrases were weighted, using the tf/idf measure. The weight for terms found in the titles or abstracts of the documents was multiplied by 2. The cut-off values were chosen heuristically with the aim to limit the list of terms to a manageable number.

By this procedure, a list of 6,653 terms was extracted from the above mentioned text corpus. This list was reviewed by the domain experts who classified each term into one of the three categories “highly relevant as ontology concept”, “relevant as concept”, and “not relevant at all”. 1,789 terms were judged as highly relevant, 2,628 as relevant, and 2,236 terms as not relevant at all. Compared to the 6,653 terms of the initial list, this results in a precision of 66,4 %. 3,923 of the relevant or highly relevant terms had not been contained in the ontology before and were incorporated as new concepts.

3.2 Finding Synonymous Expressions

To populate the ontology with instances, documents had to be annotated with concepts and relations between them. To do so, it is important to include into the ontology as many synonymous or quasi-synonymous expressions for the concepts and relations as possible. Three algorithmic approaches were employed:

- Identifying acronyms and their definitions. The corpus was parsed for token sequences like, e.g., *flux transfer event (FTE)* or *gradient refractive index*

(GRIN). If the acronym was present in the ontology, the definition was incorporated and vice versa.
- Using OpenNLP text stemmers, spelling variants for concept labels were identified, i.e. singular and plural forms and other derived forms. Spelling variants like English and American diction were included as well.
- Hearst Patterns (Hearst 1992) were used to identify noun phrases connected by narrower and broader terms. These patterns are sequences of PoS-tags and functional words like *NPb such as NPn* or *NPn and other NPb*, where *NPb* would be a broader term to *NPn*, like in "*their molecular properties such as molecular mass could also*" the phrase "*molecular properties*" is a broader term to "*molecular mass*". Relevant (NPb, NPn) pairs were checked against the ontology and added where appropriate.

Essential for an ontology are semantic relations between the concepts beyond those which are found in thesauri. These relations were determined by domain experts. As corpus exploration tools, AntConc and Sketch Engine were applied to help the experts to amend these relations with synonymous expressions. AntConc (Anthony 2008) is capable of finding concordances and offers among others flexible statistical text analyses. Sketch Engine (Kilgarriff et al. 2004) provides information about the grammatical and collocational behaviour of words in a text corpus.

3.3 Automatic Annotation

For optimum use of the ontology, preferably all concepts of the ontology should be connected with as many instances as possible, where an instance would be a concrete reference of the concept in a document. This requirement would involve a very high effort for domain experts, so that the identification of the instances should preferably be done automatically. For this task, a plugin for GATE (Cunningham et al. 2011) was developed, which enables automatic annotation of documents based on a user-defined set of linguistic patterns (Nikitina 2012). For concepts, string matching algorithms are used. Whenever a label, synonym, acronym, or other expression for a concept recorded in the ontology was found in the documents, the location was accepted as an instance for the concept. In a next step, relations were identified on the basis of linguistic patterns and a distance dimension. Such a linguistic pattern would be, e.g., *<material>* "acting as" *<chemicalRole>*, where *<material>* and *<chemicalRole>* would be expressions for the concepts material or chemical role or respective sub-classes. If all parts of the pattern were found within a given distance, the relation *hasChemicalRole* was identified in this location.

Table 1 Contribution of the ontology construction steps

Modules	Adopted concepts		Modelling by experts		Text mining		Total
Properties	114		597		1,801		2,512
Processes	1,676		220		954		2,850
Materials	29		605		538		1,172
Equipment	84		140		174		398
Suppl. concepts	0		0		456		456
Total	1,903	(26 %)	1,562	(21 %)	3,923	(53 %)	7,388

Table 2 Examples for semantic relations and their linguistic expressions

Relation	Description	Linguistic expressions
isMeasuredByMethod	Connects a measuring method with a property	Detected using, evaluated by, quantified by
hasAnalyticalProcess	Connects an analytical process with a material	Analysed by, characterised using
hasChemicalRole	Connects a chemical role with a material	Acting as, serves as

4 The NanOn Ontology

After re-using existing ontologies, modelling by domain experts and, finally, discovering concepts with the help of text mining, the ontology comprised 7,388 concepts. For the interpretation of the numbers given in Table 1 the sequence of the processing is important: initially, 1,903 concepts were adopted from other ontologies; in hindsight these amounted to 26 % of all classes. The domain experts added 1,562 concepts, and text mining as the last step produced 3,923 more concepts which had not been in the ontology before. This again shows the importance of the text mining pass. Moreover, text mining revealed the necessity of supplementary concepts, which include mainly concepts for theories and models relevant for nanotechnology.

Each of the 7,388 concepts was labelled with a natural language term. To enhance automatic annotation, 11,149 additional natural language expressions were added, which included 9,405 synonyms (3,783 by text mining and 5,622 adopted from other ontologies), 196 automatically collected acronyms, 342 simple chemical formulas (like TiO, $BaMgF_4$), and 1,206 other linguistic expressions like verbal phrases.

As to the semantic relations, the experts identified 47 relations, of which 28 were designed for automatic annotation. To enable this, a total of 140 linguistic expressions for these relations were determined (Table 2 gives some examples). The other 19 relations are partly superordinate relations, like *hasProcess* is superordinate to *hasAnalyticalProcess*. Partly they are relations between concepts only, which are not intended for describing relations between instances, e.g., *hasReciprocalValue* ties concepts describing a physical property with their reciprocal property.

5 Evaluation

Within the project, first analyses of the ontology's quality and the automatic annotation of concepts and relations were carried out. Because of the high manual effort required for these investigations, comprehensiveness for the latter was not possible within the limits of the project. Still, they are useful to reveal basic tendencies.

For the evaluation of the final ontology its completeness is a vital criterion. Here, the 45 competency questions, defined at project start, came into play. The domain experts verified that each question could be answered on the basis of the ontology, so the ontology was judged to cover the application domain completely.

As to the precision of the concept annotation, 29 scientific articles were automatically annotated, resulting in 26,040 concept instances. The review by the experts revealed 294 wrong annotations, meaning a precision of 99 %.

This high precision could not be expected for relation annotation as sophisticated linguistic knowledge is necessary to detect relations between concepts in a text. 400 scientific articles were automatically annotated with an outcome of 50,463 annotations for relation instances. Of these, a sample with 16,095 annotations was inspected by the experts. 8,112 were correct annotations, so the precision is 50.4 %. However, the precision varies quite strongly among the several relations, from e.g. 92 % for *isMetaPropertyOf*, 54 % for *isProcessPropertyOf* to 26 % for *isAnalyticalProcessOf* and 0 % for *hasStartingMaterial*. The reason for these huge differences is that some relations connect very specific concepts or are expressed by very specific linguistic phrases. The identification of corresponding text passages succeeds well. Other relations cannot be separated easily from each other, because they either connect the same concept classes or they are phrased with the same words.

The added value, an ontology can contribute to information retrieval could be shown by the experts too. For example, the question *"Which are the applications for apolar materials?"* is hardly solved with standard retrieval. First, all apolar materials have to be found, then all passages, where an application of such a material is described. With the NanOn ontology this can be answered with a formal query like

```
<material> hasMaterialProperty ''apolar'' AND
                    <material> hasApplication <application>
```

The first part of the query identifies all instances of the material property *apolar* with relations to a material. The second part then looks for all instances of any such material, which are related to an application. The applications found are the result of the query. It should be emphasised, that the information found for the first and the second part of the query do not need to originate from the same document. Hence, the ontology can improve both recall and precision of information retrieval.

6 Conclusion

During this project it could be shown that the use of an ontology with its classes and relations turns out to be useful for information retrieval. It was demonstrated that some of the competency questions which had been defined could not be searched with traditional methods with an adequate result. Here, the NanOn ontology really adds value as it can give answers to such questions.

On the other hand, it was also shown that the construction of an ontology which fulfils all requirements in a specific scientific area, requires a lot of effort by domain experts. The text mining methods definitely support the construction of the ontology, but in all cases the extracted terms have to be evaluated by the experts. The automatic annotation relies significantly on the comprehensiveness of the ontology concerning synonyms. Automatic annotation of relations is more challenging, especially as the achieved results in our project were completely heterogeneous in terms of quality, so there is place for improvement. A further problem which was not dealt with in the project but became very clear in our cooperation with scientists was that the queries which make use of the ontology have to be defined in a formal language like SPARQL. This is too complicated for the average user, so it will be necessary to work on a sophisticated user interface if an ontology is to be used in scientific databases.

Overall, the domain experts were convinced by the possibilities of the ontology. It is planned to expand the ontology and to apply it in future projects for research data management.

Acknowledgements The NanOn project was funded by the German Joint Initiative for Research and Innovation. The authors thank their colleagues in the partnering institutes for their contribution: Nadejda Nikitina, Achim Rettinger (Institute AIFB of Karlsruhe Institute of Technology), Elke Bubel, Peter König, Mario Quilitz (Institute for New Materials, Saarbrücken), Matthias Mallmann (NanoBioNet, Saarbrücken), and Nils Elsner, Helmut Müller, and Andrea Zielinski (FIZ Karlsruhe).

References

Anthony, L. (2008). *From language analysis to language simplification with AntConc and AntWordProfiler*. JAECS Newsletter (issue 63, p. 2).

Cunningham, H., Maynard, D., & Bontcheva, K. (2011). *Text processing with GATE (Version 6)*. Univ. of Sheffield, Dept. of Computer Science.

Hearst, M. A. (1992). Automatic acquisition of hyponyms from large text corpora. In *Proceedings of the 14th International Conference on Computational Linguistics* (pp. 539–545).

Kilgarriff, A., Rychly, P., Smrz, P., & Tugwell, D. (2004). The sketch engine. In *Proceedings of the EURALEX 2004* (pp. 105–116).

Nikitina, N. (2012). *Oba : Supporting ontology-based annotation of natural language resources*. Technical report, KIT, Institute AIFB, Karlsruhe.

Suárez-Figueroa, M. C., Gómez-Pérez, A., & Fernández-López, M. (2012). *The NeOn methodology for ontology engineering. Ontology engineering in a networked World, Part 1* (pp. 9–34). Heidelberg: Springer.

Data Enrichment in Discovery Systems Using Linked Data

Dominique Ritze and Kai Eckert

Abstract The Linked Data Web is an abundant source for information that can be used to enrich information retrieval results. This can be helpful in many different scenarios, for example to enable extensive multilingual semantic search or to provide additional information to the users. In general, there are two different ways to enrich data: client-side and server-side. With client-side data enrichment, for instance by means of JavaScript in the browser, users can get additional information related to the results they are provided with. This additional information is not stored within the retrieval system and thus not available to improve the actual search. An example is the provision of links to external sources like Wikipedia, merely for convenience. By contrast, an enrichment on the server-side can be exploited to improve the retrieval directly, at the cost of data duplication and additional efforts to keep the data up-to-date. In this paper, we describe the basic concepts of data enrichment in discovery systems and compare advantages and disadvantages of both variants. Additionally, we introduce a JavaScript Plugin API that abstracts from the underlying system and facilitates platform independent client-side enrichments.

1 Introduction

Today, library catalog systems are commonly replaced by resource discovery systems (RDS). They use search engine technology to provide access to all kinds of resources; not only printed books, but also e-books, articles and online resources.

D. Ritze (✉)
Mannheim University Library, Mannheim, Germany
e-mail: dominique.ritze@bib.uni-mannheim.de

K. Eckert
University of Mannheim, Mannheim, Germany
e-mail: kai@informatik.uni-mannheim.de

M. Spiliopoulou et al. (eds.), *Data Analysis, Machine Learning and Knowledge Discovery*, Studies in Classification, Data Analysis, and Knowledge Organization,
DOI 10.1007/978-3-319-01595-8_49,

Ideally, they are a one-stop shop for library patrons. To enhance the user experience and the retrieval quality in this large amount of data, the data can be enriched by additional, contextual information. Therefore, the Linked Data Web with huge amounts of mostly freely available data, e.g. DBpedia, GeoNames, FOAF, is a promising source due to its standardized structure and the available links between the resources. The enrichment can be driven by two strategies: first, additional information can be displayed to the user. Typical examples are Wikipedia articles or links to other relevant sources. Second, the search itself can be improved by including external data into the discovery system, e.g. further descriptions, subject headings or classifications that can be used to receive the most relevant resources.

In this paper, we start with a brief introduction of the two enrichment strategies, namely client-side and server-side enrichment. We discuss their advantages and disadvantages as well as appropriate use cases. For client-side enrichment, we describe a JavaScript Plugin-API for Primo,[1] an RDS provided by Ex Libris. The concept of this API is easily transferable to other RDSs. We demonstrate the application of the API by means of several plugins.

2 Data Enrichment in Discovery Systems

In the previous years, several ideas came up to improve library catalogs by enriching the data. Mostly, the goal is to increase the search efficiency or provide the user some additional value. By taking additional sources like Wikipedia into account, a semantic and multilingual search can be provided, e.g. SLUB Semantics (Bonte 2011). Other examples are recommender services, like by Geyer-Schulz et al. (Geyer-Schulz 2003) or BibTip (Mönnich 2008), where the user gets recommendations based on statistics what others used. Additionally, the enrichment with photos or news articles has already been performed, see Rumpf (2012) for an overview. Mostly techniques like CGI oder Ajax are used to display such additional information.

There are several ways to perform a data enrichment. Nowadays, most RDS, like Primo, base on a 3-tier architecture as illustrated in Fig. 1. It consists of the client, the business logic, and the database tier. The client is responsible for the communication with the user, i.e., the content presentation and the user input. The client in web-based systems consist of the browser with possible JavaScript logic and the presentation layer on the server that generates the HTML views. In contrast, the database tier provides access to the data. Between these two tiers, the business logic is located, i.e., the implementation of application-specific functionalities like performing a search, ordering a book, or accessing a resource. We refer to all enrichments in the client tier as client-side enrichments, and enrichments in the database tier as server-side enrichments.

[1] http://www.exlibrisgroup.com/category/PrimoOverview/.

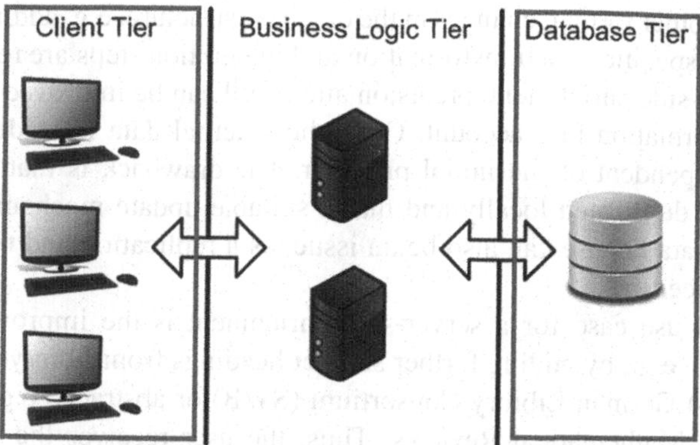

Fig. 1 Three-tier architecture of a system, including client, business logic and database tier

Client-Side Enrichment

Client-side enrichment typically takes place when some kind of additional information about a resource is presented to the user, be it in the list of search results or on a details page. External sources are queried for these further information. They are loaded on the fly and directly presented to the user without loading it into the database. Thus, there is no need for an adaption of the database or a transformation of the data to fit into the existing scheme.

A way to perform this enrichment, is the use of JavaScript, i.e., to actually enhance the response page in the browser. This is minimal invasive and works with every RDS, as almost no changes have to be made on the server-side.

This way, the network load is not only distributed to the single clients, but the additional data is also always up-to-date as it is requested on the fly. However, it has the drawback that this data cannot be indexed and is not searchable. Another drawback is the reliance on the availability of the data providers.

The typical use case for a client-side enrichment is the provision of additional information that is not needed for searching, but convenient for the user, e.g. links to Wikipedia articles or maps to geographical locations. They might be interesting and helpful, since the users get better informed about a resource and do not have to find these information manually. In turn, such data is mostly not useful for the document search because the huge amount of further information can result in the retrieval of non-relevant resources.

Server-Side Enrichment

On the other hand, there is a lot of data, e.g. additional descriptions or associated keywords, that actually can be used to enhance the retrieval itself. Therefore, it has to be loaded into the database and be part of the index. Since the data is mostly available in another format, changes in the database structure, e.g. adding new fields, and/or system-specific data transformation and integration steps are required.

With server-side enrichment, precision and recall can be improved by taking the additional information into account. Once the external data is loaded, the whole system is independent of the initial provider. The drawback is that all necessary data has to be duplicated locally and that a suitable update mechanism has to be devised. The data license can also be an issue, as a replication and transformation of the data is needed.

The typical use case for a server-side enrichment is the improvement of the actual retrieval, e.g., by adding further subject headings from library consortia like the South West German Library Consortium (SWB) or abstracts, e.g. provided by databases like Mathematical Reviews. Thus, the user receives the most relevant resources even if he or she does not use a proper search term. A comparable enrichment on the client-side would make this additional information visible, but for example the search results cannot be limited to a specific subject even if the subject headings are known.

Enrichments Using Linked Data

Linked Data is a theorem to publish structured and interlinked data in a common format. It bases on four principles: use URIs to identify things, use HTTP URIs so that the things can be referred to, provide useful information about the things in a standard format like RDF and include links to other data (Berners-Lee 2006). Everything that works with Linked Data works also without Linked Data but it makes everything easier. The more the data access, the format and the underlying data structure is unified and standardized, the more parts of the enrichment infrastructure are reusable. Otherwise, every time the wheel has to be reinvented and the infrastructure consists of a wild mixture of different parsers, access mechanisms, update scripts and transformation programs. With Linked Data, the integration of a new data source is very easy. Moreover, already existing links between linked data resources can be reused. By having so-called *same as links*, we can even overcome the integration problem to a certain degree. In a perfect world, where all data is Linked Data, you actually do not enrich your data *by Linked Data*, you enrich the Linked Data *by adding links*.

Today, no RDS provides a general support for Linked Data. This means, the internal database is not accessible as a Linked Data source and external (linked) data sources cannot get related to the internal data by means of URIs and simple links. From a Linked Data perspective, every data enrichment in current RDSs feels like a workaround. The possibilities to enrich the data are limited to some predefined ways and—most problematic—some of the data that is accessible via the RDS is not accessible for enrichment at all. This is especially the case for third-party metadata that is only merged into the search results, as provided by products like Primo Central[2] by Ex Libris or Summon[3] by Serials Solutions. At Mannheim University Library, we implemented a *JavaScript Plugin-API* that abstracts from the underlying RDS (Primo) and allows the enrichment even of data where we do not have access to, namely Primo Central.

3 JavaScript Plugin-API

The API is designed with two goals in mind: abstraction from the underlying RDS and easy development of new (enrichment) plugins. By replacing the RDS-specific part, all plugins should be transferable to a different RDS.

Figure 2 illustrates the enrichment process for a single plugin. There are three levels (Plugin, API, Primo) and all accesses of the plugin to Primo are handled via the API and vice versa. At first, the plugin registers itself for predefined events and specifies the required data to perform an enrichment, for instance *ResultListIsShown*. The API monitors the RDS and waits for the predefined events to take place. If the event happens for which the plugin registered, the API calls the plugin and provides the desired data. The plugin can now request data from external sources and perform the actual enrichment. To present the enriched data to the user, the plugin again uses predefined functions of the API, like *AddLinkToAuthor*. The API finally uses Primo-specific JavaScript code to actually insert the desired link into the HTML result page.

The API is developed alongside the plugins that we actually wanted to have. Currently, there are three plugins at the Mannheim University Library: PermaLink, Database Recommender, and Wikipedia.

PermaLink. The first plugin is a very simple one, but it provides a fundamental feature for Linked Data: a stable, persistent URI for each resource that is available in the RDS. This is not only important for Linked Data, it is also a feature often asked for by the patrons, as such a URI allows the bookmarking of resources, more concrete their metadata, and the sharing of resource information via email. To make the URI an actually working permanent link, we additionally need a link resolver, which is a script on a server that redirects from links like http://link.bib.uni-mannheim.de/primo/RECID to the corresponding details page in Primo.[4] The plugin itself is simple, it registers for the details page of a resource and asks for the internal identifier. Then, it adds the permalink to the details page (Fig. 3).

Database Recommender. The database recommender plugin proposes databases for a given query. It is our goal to provide Primo as a one-stop shop for our patrons, but unfortunately not all metadata of resources that we pay for is

---

[2]http://www.exlibrisgroup.com/category/PrimoCentral.

[3]http://www.serialssolutions.com/en/services/summon/.

[4]As RECID, the record ID of Primo is used. For example: MAN_ALEPH00143741.

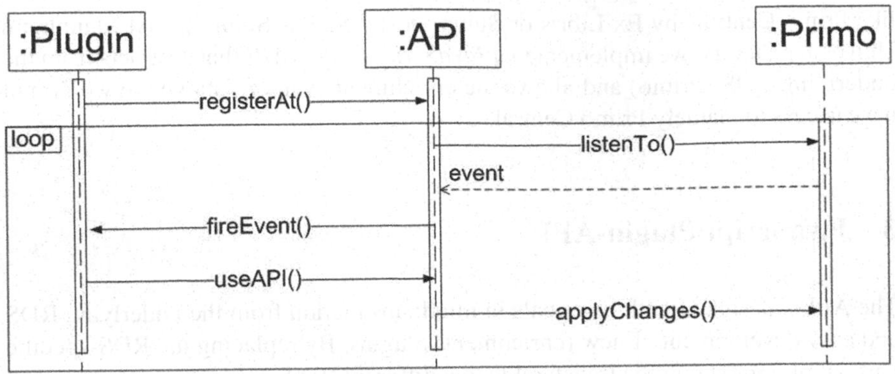

Fig. 2 Functionality of the Primo Plugin-API (UML sequence diagram)

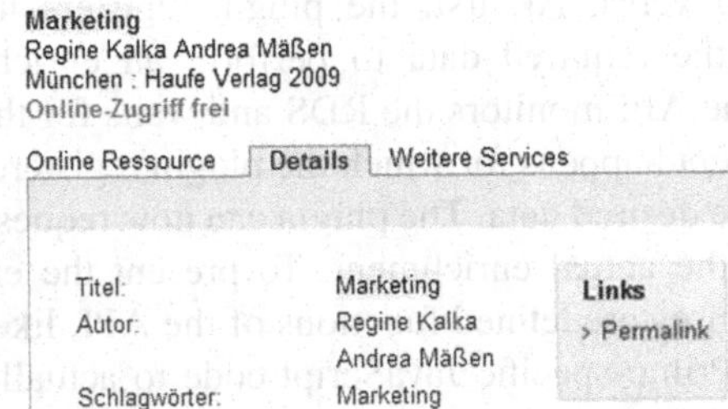

Fig. 3 PermaLink plugin in Primo

available to be included into Primo. We still have databases and other electronic resources that have to be searched separately, if all relevant literature has to be found. To determine suitable databases for a given query, we use a web service of the SuUB Bremen. The user query is sent to this service and a list of associated subject fields is returned (Blenkle 2009). Our subject specialists created a list of databases associated to each subject field. This information is used by the plugin to provide database recommendations which are inserted via the API below the result list (Fig. 4).

Wikipedia. Our third plugin links to the Wikipedia article of an author of a resource, if available. Therefore, we use an existing list with the assignments of Wikipedia articles to the GND[5] number of the described author (Danowski 2007). We use a caching server that resolves URIs containing the GND number and returns the first paragraph of the corresponding Wikipedia article, if available.

With the API, we provide a general function that allows to add an icon to elements of the details page and a text that is shown in a popup, when the user

[5]Gemeinsame Normdatei. Authority file provided by the German National Library.

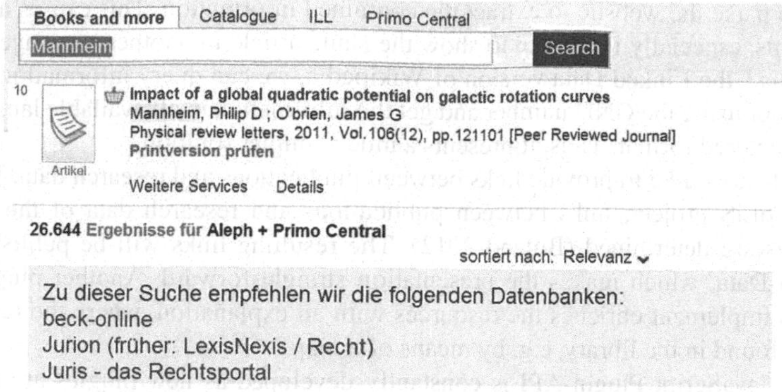

Fig. 4 Database recommender plugin in Primo

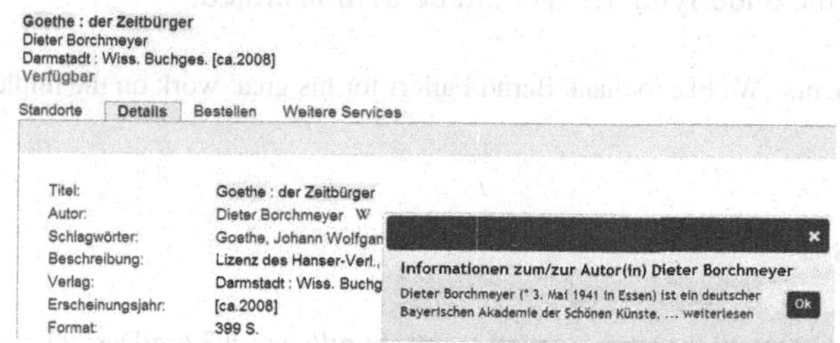

Fig. 5 Wikipedia articles of authors plugin in Primo

clicks on the icon. In this case, the plugin adds the Wikipedia icon right beside the author name. The text that is shown in the popup contains the first paragraph and the link to the full Wikipedia article (Fig. 5).

4 Conclusion and Future Work

In this paper, we explained the basic concepts of data enrichment in RDS. We focused on client-side enrichment, i.e., the presentation of additional information about a resource on-the-fly. We identified the minimal invasiveness of JavaScript-based client-side enrichment as a major advantage and provided an overview, how a JavaScript Plugin-API can be designed that abstracts from the underlying RDS. The goal of the API is to be able to present any kind of additional information to the user. Thus, the user does not need to search for this data manually.

The three implemented plugins gave us first insights of concrete enrichments that we implemented in Primo. We plan to develop more plugins, especially based on Linked Data. Currently, the plugins do not use Linked Data. For example, the Wikipedia Plugin "simply" fetches the Wikipedia article as website. However, we need to parse the website to extract the contained information. This causes a lot of problems, especially if we like to show the same article in another language. With DBPedia,[6] the Linked Data version of Wikipedia, we can query information about the author using the GND number and get the information in all available languages in a structured format. Thus, it presents a much simpler solution.

Further, we like to provide links between publications and research data. Within the InFoLiS project, links between publications and research data of the social sciences are determined (Boland 2012). The resulting links will be published as Linked Data, which makes the presentation straight-forward. Another plugin we plan to implement enriches the resources with an explanation, where the resource can be found in the library, e.g. by means of a map.

The JavaScript Plugin-API is constantly developed as new plugins are added. An official release under an open source license is planned. We are very interested in co-developers that want to use the API with a different RDS, so that its ability to abstract from the underlying RDS could be demonstrated.

Acknowledgements We like to thank Bernd Fallert for his great work on the implementation.

References

Berners-Lee, T. (2006). Linked data - design issues. http://www.w3.org/DesignIssues/LinkedData.html

Blenkle, M., Ellis, R. & Haake, E. (2009). E-LIB Bremen – Automatische Empfehlungsdienste für Fachdatenbanken im Bibliothekskatalog/Metadatenpools als Wissensbasis für bestandsunabhängige Services. *Bibliotheksdienst*, *43*(6), 618–627.

Boland, K., Ritze, D., Eckert, K., & Mathiak, B. (2012). Identifying references to datasets in publications. In P. Zaphiris, G. Buchanan, E. Rasmussen, & F. Loizides (Hrsg.) *Proceedings of the Second International Conference on Theory and Practice of Digital Libraries (TDPL 2012)*, Paphos, Cyprus, September 23–27, 2012. *Lecture notes in computer science* (Vol. 7489, pp. 150–161). Berlin: Springer.

Bonte, A., et al. (2011). Brillante Erweiterung des Horizonts: Eine multilinguale semantische Suche für den SLUB-Katalog. *BIS*, *4*(4), 210–213.

Danowski, P., & Pfeifer, B. (2007). Wikipedia und Normdateien: Wege der Vernetzung am Beispiel der Kooperation mit der Personennamendatei. In *Bibliothek Forschung und Praxis* (Vol. 31, pp. 149–156), Nr. 2 [ISSN 0341-4183].

Geyer-Schulz, A., Hahsler, M., Neumann, A., & Thede, A. (2003). An integration strategy for distributed recommender services in legacy library systems. In *Between data science and applied data analysis studies in classification, data analysis, and knowledge organization* (pp. 412–420).

Mönnich, M., & Spiering, M. (2008). Adding value to the library catalog by implementing a recommendation system. *D-Lib Magazine*, 14(5/6), May/June 2008.

Rumpf, L. (2012). Open Catalog: Eine neue Präsentationsmöglichkeit von Bibliotheksdaten im Semantic Web? *Perspektive Bibliothek*, *1*(1), 56–80.

[6] http://dbpedia.org.

Index

M. Spiliopoulou et al. (eds.), *Data Analysis, Machine Learning and Knowledge Discovery*, Studies in Classification, Data Analysis, and Knowledge Organization,
DOI 10.1007/978-3-319-01595-8,

Author Index

M. Spiliopoulou et al. (eds.), *Data Analysis, Machine Learning and Knowledge Discovery*, Studies in Classification, Data Analysis, and Knowledge Organization, DOI 10.1007/978-3-319-01595-8,